Lecture Notes in Earth Sciences   63

Editors:
S. Bhattacharji, Brooklyn
G. M. Friedman, Brooklyn and Troy
H. J. Neugebauer, Bonn
A. Seilacher, Tuebingen and Yale

Springer-Verlag Berlin Heidelberg GmbH

Bo Holm Jacobsen
Klaus Moosegard
Paolo Sibani (Eds.)

Inverse Methods

Interdisciplinary Elements of Methodology,
Computation, and Applications

Springer

Editors

Prof. Dr. Bo Holm Jacobsen
Dept. of Earth Sciences, University of Aarhus
Finlandsgade 8, DK-8200 Aarhus N, Denmark

Prof. Dr. Klaus Mosegaard
Dept. of Geophysics, NBI, University of Copenhagen
Juliane Maries Vej 30, DK-2100 Copenhagen East, Denmark

Prof. Dr. Paolo Sibani
Dept. of Physics, University of Odense
Campusvej 55, DK-5230 Odense M, Denmark

"For all Lecture Notes in Earth Sciences published till now please see final pages of the book"

Cataloging-in-Publication data applied for

Die Deutsche Bibliothek - CIP-Einheitsaufnahme

Inverse methods : interdisciplinary elements of methodology, computation, and applications / Bo Holm Jacobsen ... (ed.).
(Lecture notes in earth sciences ; 63)
ISBN 978-3-540-61693-1   ISBN 978-3-540-70687-8 (eBook)
DOI 10.1007/978-3-540-70687-8

NE: Jacobsen, Bo Holm [Hrsg.]; GT

ISSN 0930-0317
ISBN 978-3-540-61693-1

Typesetting: Camera ready by editors
SPIN: 10540638   32/3142-543210 - Printed on acid-free paper

ERRATA

Lecture Notes in Earth Sciences, Vol. 63
Bo Holm Jacobsen, Klaus Mosegaard,
Paolo Sibani (Eds.)
Inverse Methods

(ISBN 978-3-540-61693-1)

We regret that we have spelt the name of the co-editor, Prof. Dr. Klaus
Mosegaard, incorrectly on the cover, the title page, and in the back of the book.
The correct spelling is: MOSEGAARD.
We sincerely apologize for this mistake.

Preface

Through the last few decades inversion concepts have become an integral part of experimental data interpretation in several branches of science. In numerous cases similar inversion-like techniques were developed independently in separate disciplines, sometimes based on different lines of reasoning, and sometimes not to the same level of sophistication. This fact was realized early in inversion history. In the seventies and eighties "generalized inversion" and "total inversion" became buzz words in Earth Science, and some even saw inversion as the panacea that would eventually raise all experimental science into a common optimal frame.

It is true that a broad awareness of the generality of inversion methods is established by now. On the other hand, the volume of experimental data varies greatly among disciplines, as does the degree of nonlinearity and numerical load of forward calculations, the amount and accuracy of a priori information, and the criticality of correct error propagation analysis. Thus, some clear differences in terminology, philosophy and numerical implementation remain, some of them for good reasons, but some of them simply due to tradition and lack of interdisciplinary communication.

In a sense the development of inversion methods could be viewed as an evolution process where it is important that "species" can arise and adapt through isolation, but where it is equally important that they compete and mate afterwards through interdisciplinary exchange of ideas.

This book was actually initiated as a proceedings volume of the "Interdisciplinary Inversion Conference 1995", held at the University of Aarhus, Denmark. The aim of this conference was to further the competition and mating part of above-mentioned evolution process, and we decided to extend the effect through this publication of 35 selected contributions. The point of departure is a story about geophysics and astronomy, in which the classical methods of Backus and Gilbert from around 1970 have been picked up by helioseismology. Professor Douglas Gough, who is a pioneer in this field, is the right person to tell this success story of interdisciplinary exchange of research experience and techniques [1-31] (numbers refer to pages in this book). Practitioners of helioseismology like to stress the fact that the seismological coverage on the Sun in a sense is much more complete and accurate than it is on Earth. Indeed we witness vigorous developments in the Backus & Gilbert methods (termed MOLA/SOLA in the helioseismology literature) [32-59] driven by this fortunate data situation. Time may have come for geophysicists to look into helioseismology for new ideas.

Seismic methods play a key role in the study of the Earth's lithosphere. The contributions in [79 – 130, 139 – 150] relate to reflection seismic oil exploration, while methods for exploration of the whole crust and the underlying mantle are presented in [131 – 138, 151 – 166].

Two contributions [167 – 185] present the application of inversion for the understanding of the origin of petroleum and the prediction of its migration in sedimentary basins.

Inversion is applied to hydrogeophysical and environmental problems [186 – 222], where again developments are driven by the advent of new, mainly electromagnetic, experimental techniques.

The role of inversion in electromagnetic investigations of the lithosphere/astenosphere system as well as the ionosphere are exemplified in [223 – 238].

Geodesy has a fine tradition of sophisticated linear inversion of large, accurate sets of potential field data. This leads naturally to the fundamental study of continuous versus discrete inverse formulations found in [262 – 275]. Applications of inversion to geodetic satellite data are found in [239 – 261].

General mathematical and computational aspects are mainly found in [262 – 336].

Nonlinearity in weakly nonlinear problems may be coped with by careful modification of linearized methods [295 – 302]. Strongly nonlinear problems call for Monte Carlo methods, where the cooling scedule in simulated annealing [303 – 311, 139 – 150] is critical for convergence to a useful (local) minimum, and the set of consistent models is explored through importance sampling [89 – 90].

The use of prior information, directly or indirectly, is a key issue in most contributions, ranging from Bayesian formulations based a priori covariances e.g. [98 – 112, 122 – 130, 254 – 261], over more general but also less tractable prior probability densities [79 – 97], to inclusion of specific prior knowledge of shape [284 – 294, 312 – 319].

Given the differences and similarities in approach, can we benefit from exchange of ideas and experience? In practice ideas and experience seldom jump across discipline boundaries by themselves. Normally one must go and get them the hard way, for instance by reading and understanding papers from disciplines far from the home ground.

Look at the journey into the interdisciplinary cross-field of inversion techniques as a demanding safari into an enormous hunting ground. This book is meant to provide a convenient starting point.

Acknowledgements

The Interdisciplinary Inversion Conference 1995 was arranged by The Danish Interdisciplinary Inversion Group, and it was supported by the Danish Natural Science Research Council.

The Danish Interdisciplinary Inversion Group is a cooperation between the editors of this book and Per Christian Hansen, UNI•C Scientific Computing Group, Danish Technical University, Carl Christian Tscherning, Niels Bohr Institute, University of Copenhagen, and Jørgen Christensen-Dalsgaard, Theoretical Astrophysics Center, University of Aarhus.

A board of 17 specialists served as anonymous reviewers of the contributing papers. Their help is acknowledged with gratitude.

July 1996, Aarhus, Copenhagen and Odense

Bo Holm Jacobsen	Klaus Mosegaard	Paolo Sibani
Dept. of Earth Science	Niels Bohr Institute	Dept. of Physics
University of Aarhus	University of Copenhagen	University of Odense

Contents

List of Authors

Kamil L. Aminov,
Kazan Phys.-Tech. Institute RAS,
420020, Kazan, Russia.

Ole B. Andersen,
Kort & Matrikelstyrelsen,
Rentemestervej 8,
DK-2400 Copenhagen NV, Denmark.
(E-mail: oa@kms.min.dk)

Bjarne Andresen,
Ørsted Laboratory, Niels Bohr Institute,
University of Copenhagen
Universitetsparken 5,
DK-2100 Copenhagen Ø, Denmark.
(E-mail: andresen@fys.ku.dk)

Christophe Barnes,
Institut de Physique du Globe de Paris,
4 Place Jussieu,
75252 Paris Cedex 05, France

Tanja Barth,
Dept. of Chemistry, Univ. of Bergen,
Allegt. 41,
N-5007, Bergen, Norway.

Sarbani Basu,
Teoretisk Astrofysik Center,
Aarhus University,
Ny Munkegade,
DK-8000 Aarhus C, Denmark.

Oliver Bäumer,
Institut für Geophysik und Meteorologie,
Mendelssohnstrasse 3,
D-38106 Braunschweig, Germany.

Marwan Charara,
Institut de Physique du Globe de Paris,
4 Place Jussieu,
75252 Paris Cedex 05, France

Niels B. Christensen,
Dept. of Earth Sciences, Aarhus Univ.,
Finlandsgade 8,
DK-8200 Aarhus N, Denmark.

J. Christensen-Dalsgaard,
Teoretisk Astrofysik Center,
Aarhus University,
Ny Munkegade,
DK-8000 Aarhus C, Denmark.

Louis C. Dahl,
Dept. of Chemistry, Univ. of Bergen,
Allegt. 41,
N-5007 Bergen, Norway.

Peter Dalgaard,
Dept. of Biostatistics,
University of Copenhagen,
Blegdamsvej 3,
DK-2200 Copenhagen N, Denmark.

Dorte Dam,
Dept. of Earth Sciences, Aarhus Univ.,
Finlandsgade 8,
DK-8200 Aarhus N, Denmark.

Wojciech Debski,
Inst. of Geophysics, Polish Ac. of Sci.,
ul. Ksiecia Janusza 64,
01-452 Warsaw, Poland.

H. Lydia Deng,
Center for Wave Phenomena,
Colorado School of Mines,
Golden, CO 80401, USA.
(E-mail: hdeng@dix.mines.edu)

Lisbeth Engell-Sørensen,
Inst. Solid Earth Physics, Univ. Bergen,
Allegt. 41,
N-5007, Bergen, Norway.

Jerry Eriksson,
Dept. of Computing Science,
Umeå University,
901 87 Umeå, Sweden.
(E-mail:Jerry.Eriksson@cs.umu.se)

J.M. Gordon,
Center for Energy and Environmental
Physics, Jacob Blaustein Inst. for
Desert Research, Ben Gurion Univ. of
the Negev, Sede Boqer Campus 84993,
Israel
and The Pearlstone Center for Aeronaut-
ical Engineering Studies, Dept. of Mech.
Eng. , Ben Gurion Univ. of the Negev,
Beersheva, Israel.
(E-mail: jeff@menix.bgu.ac.il)

Douglas Gough,
Inst. of Astronomy,
Madingley Road,
Cambridge, UK
and Dept. of Applied Mathematics and
Theoretical Physics,
Silver Street, Cambridge, UK.
(E-mail: douglas@ast.cam.ac.uk)

Wences Gouveia,
Center for Wave Phenomena,
Colorado School of Mines,
Golden CO 80401, USA.
(E-mail: wgouveia@dix.mines.edu)

Lars Kai Hansen,
CONNECT, Electronics Institute,
Building 349, Tech. Univ. of Denmark,
DK-2800 Lyngby, Denmark.

Per Christian Hansen,
UNI-C, Building 304,
Technical Univ. of Denmark,
DK-2800 Lyngby, Denmark.
(E-mail:
Per.Christian.Hansen@uni-c.dk)

F. Pérez Hernández,
Instituto de Astrofísica de Canarias,
Tenerife, Spain.

Jens Hjort,
Inst. of Astronomy,
Madingley Road,
Cambridge CB3 OHA, UK.
(E-mail: jens@ast.cam.ac.uk)

Rachel Howe,
Astronomy Unit,
Queen Mary and Westfield College,
Mile End Road
London E14NS, UK.

Jens Martin Hvid,
Dept. of Geology and Geotech. Eng.,
Technical Univ. of Denmark, 204,
DK-2800 Lyngby, Denmark.

Bo Holm Jacobsen,
Dept. Earth Sciences, Aarhus Univ.
Finlandsgade 8,
DK-8200 Aarhus N, Denmark.

Jørgen S. Jørgensen,
Fysisk Institut, Odense Universitet,
DK-5230 Odense M, Denmark.

Wolfgang Keller,
Geodetic Inst., Stuttgart University,
Keplerstr. 11
D-70174 Stuttgart, Germany.
(E-mail: wolke@gi5.bauingenieure.uni-
stuttgart.de)

Manabu Kunitake,
Communications Research Laboratory,
4-2-1 Nukuikita-Machi Koganei,
Tokyo 184, Japan.
(E-mail: kunitake@crl.go.jp)

Pierre Maxted,
Astronomy Centre, Univ. of Sussex
Falmer, Brighton, BN1 9QH, UK
(E-mail: pflm@star.maps.susx.ac.uk)

Klaus Mosegaard,
Niels Bohr Inst. for Astronomy, Physics
and Geophysics, Copenhagen Univ.,
Juliane Maries Vej 30,
2100 Copenhagen Ø, Denmark.
(E-mail:klaus@gfy.ku.dk)

Tijmen-Jan Moser,
Inst. of Solid Earth Physics,
Univ. of Bergen,
Allegt. 41,
N-5007, Bergen, Norway.

M.S. Munkholm,
Dept. of Earth Sciences, Aarhus Univ.
Finlandsgade 8
DK-8200 Aarhus N, Denmark.

Ingelise Møller,
Dept. of Earth Sciences, Aarhus Univ.
Finlandsgade 8
DK-8200 Aarhus N, Denmark.
(E-mail: geofim@aau.dk)

Lars Nielsen,
Dept. of Earth Sciences,Aarhus Univ.
Finlandsgade 8
DK-8200 Aarhus N, Denmark.
(E-mail: geofln@aau.dk)

Søren B. Nielsen,
Dept. of Earth Sciences, Aarhus Univ.
Finlandsgade 8
DK-8200 Aarhus N, Denmark.
(E-mail: geofsbn@aau.dk)

Konstantin S. Osypov,
Dept. of Geophysics, Uppsala Univ.,
Sweden. On leave to Center for Wave
Phenomena, Colorado School of Mines,
Golden CO 80401, USA.
(E-mail: ko@geofys.uu.se)

Jan Pajchel,
Norsk Hydro, Bergen, Norway.

J. Boiden Pedersen,
Fysisk Institut, Odense Univ.,
DK-5230 Odense M, Denmark.

Peter Alshede Philipsen,
CONNECT, Electronics Institute
Technical University of Denmark, 349
DK-2800 Lyngby, Denmark.

Roland G. Roberts,
Dept. of Geophysics, Uppsala Univ.
Box 556, S75122 Uppsala 1, Sweden

F. Sacerdote,
Politecnico di Milano, Dipartimento di
Ingegneria Idraulica, Ambientale e del
Rilevamento - Sezione del Rilevamento
Piazza Leonardo da Vinci 32
20133 Milano, Italy.

F. Sansò,
Politecnico di Milano, Dipartimento di
Ingegneria Idraulica, Ambientale e del
Rilevamento - Sezione del Rilevamento
Piazza Leonardo da Vinci 32,
20133 Milano, Italy.

John A. Scales,
Dept. of Geophysics
Center for Wave Phenomena
Colorado School of Mines
Golden, CO 80401, USA.
(E-mail: jscales@dix.mines.edu)

Jesper Schou,
Stanford University
HEPL Annex A201
Stanford, CA 94305-4085, USA.
(E-mail: jschou@solar.stanford.edu)

Wolf-Dieter Schuh,
Technical University Graz
Math. Geodesy and Geoinformatics
Steyrergasse 30
A-8010 Graz, Austria.

A.A. Stepanov,
Teoretisk Astrofysik Center
Aarhus Univ., Denmark ,
and Inst. of Math. and Computer Sci.
University of Latvia, Rainis Berd. 29
LV-1459, Riga, Latvia.

Gabriel Strykowski,
Kort- og Matrikelstyrelsen
(National Survey and Cadastre)
Rentemestervej 8
2400 Copenhagen NV, Denmark.

Kurt I. Sørensen,
Dept. of Earth Sciences, Aarhus Univ.
Finlandsgade 8,
DK-8200 Århus N, Denmark.
(E-mail: geofkis@aau.dk)

Albert Tarantola,
Inst. de Physique du Globe de Paris,
4 Place Jussieu
75252 Paris Cedex 05, France.

Michael J. Thompson,
Astronomy Unit
Queen Mary and Westfield College
Mile End Road
London E14NS UK.

Peter Toft,
CONNECT, Electronics Institute
Technical University of Denmark, 349
DK-2800 Lyngby, Denmark.
(E-mail: ptoft@ei.dtu.dk)

Per-Åke Wedin,
Dept. of Computing Science
Umeå Univ.
901 87 Umeå, Sweden.
(E-mail:Per-Ake.Wedin@cs.umu.se)

Zhengsheng Yao,
Dept. of Geophysics, Uppsala Univ.
Box 556, S75122 Uppsala 1, Sweden
(E-mail: zy@geofys.uu.se)

The Success Story of the Transfer and Development of Methods from Geophysics to Helioseismology

Douglas Gough

Institute of Astronomy and Department of Applied Mathematics and Theoretical Physics, Cambridge, UK

Foreword

It is a privilege indeed to be the only speaker at this meeting charged with addressing a title that has been determined by the organizing committee, and not by himself, not so much because it has spared me the task of choosing a title, but more because along with the charge came the request to be anecdotal, and thereby to set the scene for an informal conference. This is evidently much simpler than telling an unbiased story, although my hope is that the results I shall present are as unbiased as can reasonably be expected under the circumstances. But I must point out right away that although I shall tell a story from my own perspective, the advances that have been made in helioseismology have been brought about by an understanding that has grown from many detailed investigations by a small – indeed minute, by geophysical standards – army of enthusiastic scientists. Explicit discussion of all that work is hardly appropriate here. However, I first emphasize that credit is due to those scientists who have carried it out. It is likely that most of what I have to say about inverse methods *per se* is known already to all at this meeting; however it is potentially useful at least to helioseismologists for me to present our view of the methods, because then the experts here will know where to advise us most fruitfully.

Helioseismology, in the form in which we know it today, started on 15 June 1975, as a consequence of a lecture by Henry Hill on his possible detection of global oscillations of the sun, which he delivered to a conference on Astrophysical Fluid Dynamics being held in Cambridge. Our conference here marks the twentieth anniversary of that event. Jørgen Christensen-Dalsgaard and I had been working on the theory of global solar oscillations, and, since he was a young and bright energetic student, it was easy for Jørgen to calculate immediately those eigenfrequencies of a theoretical model of the sun that might correspond to Henry's pioneering observations, and then to make a presentation at the very same conference comparing theory with observation. (Resources were such in those days that real computing had to be carried out at night, so Jørgen did not need to miss any of the meeting. I explained to him that had I not been organizing the conference I could have helped, but I'm not sure that he believed me.)

We subsequently published the results in an optimistically written article in *Nature*, by which time yet other observations of possible detections had been brought to our attention. The title of our paper was: 'Towards a heliological inverse problem', which I mention simply to demonstrate that we had already the goal of inversion in mind.

I should also point out that an appreciation of that goal was not entirely absent from the rest of the astronomical community at that time. Indeed, an eminent astronomer from my institute, Professor R.A. Lyttleton, who sadly died only last week, once remarked: 'If a modern astronomer were to meet a nineteenth-century chimney sweep, he would deduce that the sweep were made of pure carbon.' It is in the face of such false global deductions from superficial observation that we ply our trade.

Although my tale is supposed to be woven around the learning we have gained from geophysicists, I do have in the forefront of my mind the title of this conference, and the intention of the organizers that I should concentrate on inversion. My problem is that I don't know what inversion really is. Moreover, as will become apparent later, if I take a strict view of what I think it might be, the only true inversion that we might have carried out was not even an outcome of our contact with geophysicists. Immediately before this meeting I remarked to Philip Stark, somewhat impetuously, that had my talk been scheduled to be after that of Albert Tarantola, I might then know about what I should speak. But Philip, evidently not yet being in a mood to accept unjustified prior information, replied that there was no evidence for that to be so, because according to the programme Albert was merely going to ask the question. Therefore, I shall simply relate a story which involves the use of techniques some of which are used by some of those who claim to carry out inversions.

1 Introduction

To acquaint nonastrophysicists with the basic structure of our task, and to establish a notation, I very briefly summarize the helioseismological problem. The sun, like the earth, is a rotating self-gravitating body, approximately in isostatic balance (except in a near-surface region), which supports seismic waves. As in the earth, appropriate superpositions of those waves essentially constitute normal modes of oscillation. Only acoustic modes (p modes), whose restoring force is principally the reaction to compression, and gravity modes (g modes), whose restoring force is buoyancy, are currently of relevance. The sun is a fluid, and cannot support what geophysicists call shear waves (I refrain from discussing even the possibility of waves supported by shear in the background state); moreover, I shall discuss neither magnetohydrodynamic modes (for which Lorentz forces provide the restoring) nor inertial oscillations (restored by vorticity stretching), because they have not yet been utilized (and, in the latter case, perhaps not even observed) for diagnosing solar structure. However, I should hasten to add that Lorentz forces and vorticity stretching are taken into account when calculating degeneracy splitting of p modes and g modes, and that buoyancy forces are included in the computation of p modes and compression in the computation of g modes, as too is the self-gravity of the perturbations.

To a first approximation the background state of the sun, by which I mean the putative state in the absence of seismic modes, can be considered to be static, and the seismic motion can be considered to be adiabatic. Moreover, when rotation and magnetic fields are ignored, the background state is essentially spherically symmetrical. Therefore, because the amplitudes of the seismic waves are small, the equations of motion can be linearized, and one can describe each seismic mode in terms of a scalar wave function:

$$\Psi_{nlm}(\boldsymbol{r}, t) = \psi_{nl}(r) P_l^m(\cos\theta)\cos(m\phi - \omega t + \delta_{nlm}) \tag{1}$$

with respect to spherical polar coordinates (r, θ, ϕ), where t is time and δ_{nlm} is some initial phase. The degree l of the spherical harmonic is called the degree of the mode, and the order m is called the azimuthal order. The amplitude function ψ_{nl} satisfies a fourth-order differential equation, subject to regularity conditions at the coordinate singularity $r = 0$ and true boundary conditions at the surface $r = R$ of the star. This representation is good throughout all except the uncertain near-surface layers, to which I shall return later. Suffice it to say in the meanwhile that because these layers are thin they act essentially only to modify the outer boundary conditions in an uncertain way, so it is hardly necessary for me to get involved here in the details of what those conditions ought to be, nor what conditions are actually applied in practice. The derivation of the regularity conditions is straightforward. The formal outcome is a two-point boundary-value problem, admitting a sequence of eigenfrequencies $\omega = \omega_{nlm}$ for each (l, m), which we label with an integer n such that ω increases with n. For any (l, m) the g-mode frequencies all lie below all the p-mode frequencies (there is no g modes with $l = 0$, of course, because buoyancy arises from deviations from spherical symmetry) so we may conveniently label g modes with $n \leq 0$ and p modes with $n > 0$. To date, only p modes (and fundamental g modes, which have $n = 0$ and which we call f modes) have been unambiguously observed, so almost all of my discussion will be with p modes principally, though not exclusively, in mind. Figure 1 depicts the result of some early observations of p modes together with the theoretical predictions of two models of the sun.

The frequencies of a spherically symmetrical configuration are independent of m, which must be the case because m is related to the component of the horizontal variation along the equator of the coordinate system, and different orientations of the coordinate system are physically indistinguishable. When the symmetry is broken, as it is by rotation or a magnetic field, then the degeneracy in the eigenfrequencies is lifted: evidently the frequency splitting is a diagnostic of the symmetry-breaking agents. In that case there is also a deviation of the wave function from the form (1), but because there is an underlying variational property in the system one can obtain a first estimate of the relation between, say, angular velocity Ω and the degeneracy splitting without recourse to that deviation. Indeed, if Ω were the only symmetry-breaking agent, the variational principle can be written in the form:

$$\mathcal{I}\omega^2 - 2\mathcal{R}\omega - \mathcal{K} = 0, \tag{2}$$

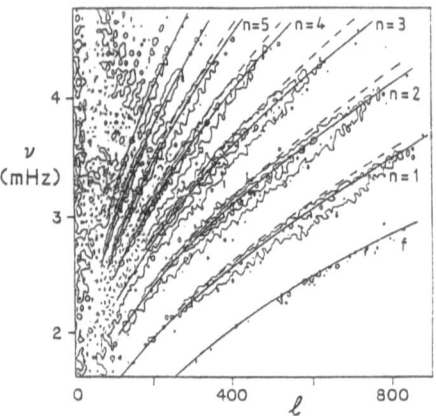

Fig. 1. Contours of constant power in the spectrum of high-degree solar oscillations obtained by Deubner, Ulrich and Rhodes (1979). The lowest ridge, labelled f, is the fundamental g mode; the f-mode multiplet frequencies contain very little information about the structure of the sun and are not considered in this address. Each of the other ridges is produced by p modes of like order n, the lower values of which are indicated on the diagram. The continuous curves represent the cyclic eigenfrequencies (in mHz) of a theoretical solar model with initial helium abundance $Y = 0.25$, the dashed curves are frequencies of a model with $Y = 0.19$ (from Gough, 1982).

where $\mathcal{I}$, $\mathcal{K}$ and $\mathcal{R}$ are functionals of the background state of the sun, which I denote by $\boldsymbol{X}(r)$, and of the seismic displacement eigenfunction $\boldsymbol{\xi}_{nlm}(r, \theta)$, which can be obtained directly in terms of $\psi_{nl} P_l^m(\cos\theta)$ using the equations of motion. The functionals $\mathcal{I}$, $\mathcal{K}$ and $\mathcal{R}$ do not depend explicitly on ω. (I am assuming that rotation is about a unique axis, so that the background state is axisymmetrical and the motion is sinusoidal in ϕ with wave number m.) $\mathcal{R}$ is also a functional of $\Omega(r, \theta)$, and vanishes when $\Omega = 0$. Since $\Omega/\omega \ll 1$ (the observed periods of seismic oscillation are typically about 5 minutes, the rotation period is about a month), $\mathcal{R}$ may safely be linearized with respect to Ω. In that case $\mathcal{I}$ and $\mathcal{K}$ can be evaluated for the corresponding nonrotating state, because the only steady rotational perturbation to the background structure $\boldsymbol{X}$ arises from the centrifugal terms, which are quadratic in Ω. Thus $\boldsymbol{X} = \boldsymbol{X}(r)$. The eigenfrequency ω can therefore be separated into a contribution ω_0 from the nonrotating state and a perturbation ω_1: $\omega = \omega_0 + \omega_1$, where

$$\omega_0^2 = \frac{\mathcal{K}(\boldsymbol{X})}{\mathcal{I}(\boldsymbol{X})} \tag{3}$$

and

$$\omega_1 \simeq \frac{\mathcal{R}(\boldsymbol{X}, \Omega)}{\mathcal{I}(\boldsymbol{X})}. \tag{4}$$

The zero-order frequency ω_0 depends on n and l; the correction, ω_1, depends on n and l and also on m, and has the property that it is an odd function of m.

The simplest helioseismic inversion problem can be stated as follows: from constraints of type (3) provided by observed mean multiplet frequencies ω_0 (i.e. uniformly weighted averages over m at fixed n and l), and other prior information, 'determine' the (spherically averaged) structure $\boldsymbol{X}(r)$ of the sun, and then, having determined $\boldsymbol{X}$, use the constraints (4) to 'determine' the angular velocity Ω from the degeneracy splitting frequencies ω_1. I shall postpone until the next section what I mean by the word 'determine'.

To make matters concrete permit me to specify what I mean by the structure function $\boldsymbol{X}(r)$. It is a vector of minimal state variables that determine the seismic structure of the sun, namely that aspect of the structure to which seismic waves are sensitive. Since the dynamics of both adiabatic p modes and adiabatic g modes concern only Newton's second law – material with mass density ρ being accelerated by gradients in pressure p and gravitational forces (which depend solely on ρ) – they are sensitive only to p, ρ and the adiabatic exponent $\gamma = (\partial \ln p / \partial \ln \rho)_s$, (the thermodynamic derivative being taken at constant specific entropy s), which relates linearized pressure perturbations to density perturbations. Hydrostatic equilibrium (isostacy) of the background state relates p to ρ (given $\rho(r)$, the pressure $p(r)$ would be determined uniquely were it not for the uncertainty of the physics of the outer layers, which in some senses is small but discussion of which I have in any case decided to postpone), and since that constraint is always used as part of the prior information (indeed, the wave equation and consequently the variational constraint (2) depend on it), the vector $\boldsymbol{X}$ has only two components, say ρ and γ, or any two independent functions of them. It is not uncommon to use $u := p/\rho$ as one of the variables, because that relates directly to sound speed c, which has the dominant influence on the propagation speed of p waves. (The variable $c^2 = \gamma p/\rho$ is not usually used directly, because of a technical difficulty that that would introduce).

Other prior information used in the diagnostic analyses are the total mass M of the sun and its radius R, which is the radius of some well defined level in the atmosphere. Sometimes the seismic structure variables are related (in the adiabatically stratified regions of the convection zone) to chemical composition, but that requires knowledge of the equation of state.

Broadly speaking, the sun can be divided into three regions: an innermost so-called radiative region, occupying $r \lesssim 0.7R$, an essentially adiabatically stratified convective region, occupying $0.7R \lesssim r \lesssim R$, and a geometrically thin superficial region. The innermost region is usually regarded as being quiescent and in thermal balance, with energy being transported through the optically dense material by radiation. There the stratification might be specified by, say, density, pressure and temperature, which are related through an equation of state which depends on chemical composition. The adiabatically stratified region of the convection zone, like the radiative interior, is essentially in hydrostatic equilibrium, except in its upper boundary layer where it abuts the superficial region. The outermost superficial region contains the solar atmosphere above the upper superadiabatic boundary layer of the convection zone, which is spatially inhomogeneous and locally time dependent, and is therefore not in hydrostatic equilibrium. Moreover, in this region the seismic motion is not adiabatic. As I have already pointed out, this region is ill understood. Its thickness is much less than the radius of the sun (of order $10^{-3}R$). However, because most of it is relatively cool (the temperature T is approximately 6×10^3 K, whereas near $r = 0$, where the nuclear reactions are converting hydrogen to helium and supplying the radiant energy, $T \simeq 1.5 \times 10^7 K$) and therefore the sound speed is low, its influence on p-mode eigenfrequencies is not negligible. That is why special precautions must be taken to avoid spurious inferences from seismic data.

Finally, I announce that the whole of my discussion will concern seismic analysis in terms of global free-oscillation frequencies. Some work has been carried out on local wave analyses to make inferences about structural inhomogeneities or horizontal and vertical flows. In addition, there has been some work on the forcing of the modes, which is believed to take place in the turbulent upper superadiabatic convective boundary layer. But all these studies are in their infancy, and discussion of them is best left to later conferences in this series.

2 A Brief Summary of the Methods Used for Analysing the Frequencies of Free Global Solar Oscillations

There are various ways of classifying the methods of analysis. One way of dividing them is to separate them into (i) what one might call calibration, namely comparing with observation the frequencies of a set of theoretical models that depend on a set of parameters λ and choosing λ such as to provide the best fit, and (ii) an attempt to 'determine' some function, such as $X(r)$ or $\Omega(r)$, that specifies that part of a solar model that does not lie in the seismic null space. This is the separation that I was originally led to believe distinguishes the usual methods of scientific inference from inversion. One can see immediately that Figure 1 provides an illustration of method (i): the model with $Y = 0.25$ fits the data better than the other, and thus we have the basis for a calibration of the helium abundance Y.

Indeed, the data illustrated in Figure 1 were the first to be used for estimating a property of the sun, though not by a calibration explicitly of the kind I have just mentioned. The pair of theoretical models plotted in the figure were chosen primarily for public presentation, because visual impact is often more widely persuasive than intellectual argument. I postpone until the next section a description of the first helioseismic inference.

At the risk of boring the experts, I make the commonly stated and almost banal point that because any function contains an infinite amount of information and there are never more than a finite number of real data available, those data are never sufficient to determine a function. It follows that inversion of data alone, as represented by (ii), is impossible. To render it possible requires supplementing the finite data with an infinite amount of additional information, and this, I believe, is the heart of the method. It is the choice of both what that information is and how it is used that makes inversion an art. I usually refer to that information as *prejudice*, and I shall continue to call it that throughout this talk. I gather that the technical term in the geophysical community is *prior information*, a term which I like less because I suspect it leaves open the possibility of concealing somewhat from the layman its true nature. Whether that prejudice is more or less constraining than the data depends on the measure of importance one chooses to adopt: that is always a subjective issue, which is why I consider it to be more appropriately called prejudice.

This discussion highlights what is undoubtedly an advantage of (ii) over (i): it is hardly possible to carry out (ii) without explicit application of the prejudice, and therefore one is forced at least to be aware of the importance of that prejudice. All too often one is made aware of the lack of awareness of some practitioners of (i) of the assumptions upon which the calibrated theoretical models depend, and therefore of the prejudices embodied in their acceptance.

When I first started to learn how geophysicists approached method (ii) I quickly became aware of the pioneering work by Backus and Gilbert. It certainly had the greatest influence on the progress of my work. I gratefully acknowledge the kindly help I received from Kathy Whaler at the time, who not only advised me what papers to read but also warned me of what seemed to me to be a stultifying divergence of opinion that had grown up between geophysicists on the two coasts of the United States. Indeed, the acceptance of helioseismic inverse methods by many American solar physicists was delayed by that dispute. But Kathy was immune to that, and was always prepared to consider and discuss what I had in mind to do, even when the received geophysical wisdom of the time forbade it.

The work of Backus and Gilbert concerned the accommodation of only linear constraints. Therefore to render it applicable to helioseismology the structure constraints (3) must be linearized about some reference model $X_0(r)$ of the sun. Then, if $\delta\omega_0^2$ is the difference between the squares of the mean multiplet frequencies of a specified mode of oscillation of the sun (assuming that the mode be identified) and the corresponding mode of the reference model,

$$\delta\omega_0^2 \simeq \mathcal{I}^{-1}(\delta\mathcal{K} - \omega_0^2\delta\mathcal{I}),\tag{5}$$

where ω_0 is the eigenfrequency of the reference model, and where

$$\delta\mathcal{K}(\delta X; X_0) = \mathcal{K}(X) - \mathcal{K}(X_0),\tag{6}$$

$$\delta\mathcal{I}(\delta X; X_0) = \mathcal{I}(X) - \mathcal{I}(X_0)\tag{7}$$

and $\delta X = X - X_0$ is the difference between the structure X of the sun and that of the reference model. After linearizing the right-hand sides of equations (6) and (7) with respect to δX, equation (5) can be written in terms of integrals of δX weighted by data kernels (Fréchet derivatives) which depend on X_0, ω_0 and the oscillation eigenfunction ξ_{nl} of the appropriate mode of the reference model, yielding

$$d_{\boldsymbol{i}} = \int_0^R K_{u,\gamma}(r, X_0; \xi_{\boldsymbol{i}}, \omega_{0\boldsymbol{i}})\delta\ln u\, dr + \int_0^R K_{\gamma,u}(r, X_0; \xi_{\boldsymbol{i}}, \omega_{0\boldsymbol{i}})\delta\ln\gamma\, dr\tag{8}$$

where the datum $d_{\boldsymbol{i}}$ is the scaled frequency difference $\frac{1}{2}\omega_{0\boldsymbol{i}}^{-1}\delta\omega_{0\boldsymbol{i}}^2$ between mode $\boldsymbol{i}$ of the sun and that of the reference, and I have now explicitly labelled the mode with $i = (n, l)$ to emphasize that there are constraints associated with many modes. In view of the variational property of equation (3) with respect to the eigenfunctions, it is not necessary to know (or even explicitly estimate) the corresponding eigenfunction for the sun. For this reason I did not adorn ξ_i unnecessarily with the subscript zero. If δX could be determined, an estimate of X could be obtained by incrementing X_0, and in principle the whole process could be repeated. Once X is known, equation (4) provides a linear constraint on $\Omega(r, \theta)$, so no further iteration is necessary. It too can therefore be written in a form similar to (8):

$$\omega_{1i} = \int_0^\pi \int_0^R K_\Omega(r, \theta, X_0; \xi_i, \omega_{0i}) \Omega(r, \theta) r^2 \sin \theta \, dr \, d\theta, \qquad (9)$$

except that now $i = (n, l, m)$ and, of course, the integral over the star is in two dimensions. Equations (8) and (9) can be written in the generic form:

$$d_i = < K^i f >, \qquad (10)$$

where the kernel K^i and the function f to be determined are either scalars or vectors and the angular brackets denote integration over the reference model.
It was evident that there had been two apparently separate routes that Backus and Gilbert had taken. The first was to represent in some way the projection of f into the function subspace $\mathcal{O}$ that is accessible to the data, and then to choose the projection that most closely reproduces the data, taking due account of the errors in the observations. There were various ways of taking errors into account, but one of the most mathematically natural seemed to be to take a linear transformation of data $d_i \rightarrow \tilde{d}_k$ such that the kernels $\tilde{K}_k$ associated with $\tilde{d}_k$ are orthogonal, and then to reduce $\mathcal{O}$ by excluding the space spanned by those $\tilde{K}_k$ associated with data combinations $\tilde{d}_k$ that were dominated by errors, leaving, say, $\tilde{K}_{k'}$, $d_{k'}$. The outcome was represented as a sum of projections of f onto $\tilde{K}_{k'}$, which has been called a spectral expansion. In practice it was carried out by expanding f in terms of the original kernels K^i, and solving the linear algebraic equations for the coefficients that are obtained from the constraints (10) by what is now called truncated singular-value decomposition, though I learnt the principles from the book on linear differential equations by Lanczos (1961), and for a while I named the procedure after him.

Although this procedure has some mathematical elegance, I, together with all other helioseismologists, I believe, have never really been endeared to it, although I do use it occasionally. What makes me uneasy is my failure to appreciate in a physical sense projections into a function subspace whose properties I cannot comprehend well enough to permit me to explain it to my grandmother. My inclination has been instead to project onto some other more convenient basis, which in some sense I think I do understand in some naive way, such as piecewise constant functions, piecewise linear functions, splines or even polynomials. Indeed, the first helioseismic inversions of real data, if inversions is what they can be called, were carried out in terms of expansions in both piecewise constant functions and polynomials. The former is illustrated in Figure 7.

Two criticisms of those analyses were levelled by geophysicists. The first was that by choosing an arbitrary basis for the expansion of f (which in this case was of an equatorially weighted average $\overline{\Omega}$ of the angular velocity, determined from only sectoral modes, for which $m = \pm l$), part of the solution was in the annihilator $\mathcal{A}$ (null space), namely the complement of $\mathcal{O}$, and since the data provided no information about $\mathcal{A}$ one has no right to trespass there. The second, which perhaps is hardly distinct from the first and might even have been an implied justification for it, was that because part of the 'solution' is in $\mathcal{A}$ one cannot assess the errors properly. Several critics had maintained that assessing the errors is central to inversion (to such an extent that one sometimes wonders whether they assign greater importance to error assessment than they do to considering the nature and implications of the solution), and is what distinguishes inversion from other methods of analysis.

My opinion as a physicist on penetrating $\mathcal{A}$ is that it has several obvious merits, and no real disadvantage that I can see. Since I do not really understand function subspace, I cannot comprehend more easily the meaning of a projection of f into $\mathcal{O}$ alone than I can of the projection into both $\mathcal{O}$ and some part of $\mathcal{A}$. To me the most natural subspace into which to project is one that is spanned by likely functions (likeliness is of course dependent on preconceived ideas, namely prejudice, of which one must, of course, be wary).

We are currently in an early discovery mode. When our subject is more mature, I am sure it will be necessary to take stock of the situation, and subject our inferences to more rigorous appraisal. Of course we accept that estimating uncertainty is extremely important, but most of us cannot really appreciate the projection of an error bound *per se* into a subspace that is in principle defined in rather abstract mathematical terms that relate only to data than one can of a projection into a subspace that is defined by other criteria. When it comes to estimating functional properties of the inversions, however, the situation could be quite different, particularly if the functional property just happened to be independent of the projection of f into $\mathcal{A}$.

On the whole, helioseismologists have little interest in confining themselves to $\mathcal{O}$, and indeed prefer not to. Aside from the reasons I have already outlined, the most influential argument was presented in a very nice comparison of methods by Christensen-Dalsgaard and Thompson (1993), in which it was demonstrated with artificial data that inversions using expansions in simple basis functions can yield results which, to the physicists' eyes, look more nearly like the true function that they are meant to represent than do expansions in the basis functions of $\mathcal{O}$. Indeed, Hansen, Sekii and Shibahashi (1992) have deliberated penetrated $\mathcal{A}$ in an explicit way in order to seek a representation that has a certain (aesthetically) desired property (smoothness).

So difficult, it seems, is it to interpret functional representations, that Backus and Gilbert also took a different approach, originally to obtain a measure of the power of seismic data to resolve localized properties. I need not explain precisely what they did, except to point out that the idea was to consider linear combinations of data whose associated kernels are unimodular and localized in space and, preferably, have no negative sidelobes, if possible. The data combinations are then the values of the averages weighted by the kernels, which one can plot. Indeed, so too are the original data, provided they are rescaled to render K^i unimodular, except that those kernels are usually not well localized. The beauty of representing f in this way, however, is that both I and my grandmother can understand it. If one succeeds in obtaining kernels with no sidelobes, the outcome is a blurred image of the true function, the degree of blurring being indicated by the characteristic widths of the kernels. As I see it, this technique, at least in principle, is perhaps the best one to try first, for then one obtains a simple visual image, albeit out of focus, of the function one is trying to determine. (In practice, however, one more often first tries techniques that are simpler to implement.) After that, one can tailor procedures that are particularly sensitive to some interesting aspect of the system that one might wish to investigate. For example, Pijpers and Thompson (1994) constructed kernels that exhibit discontinuities in the gradient of a function.

There is a fundamental difference between the explicit aims of the two approaches that Backus and Gilbert pioneered. The second approach is, from a scientific point of view, the more direct: one attempts to design a set of kernels that have some desired property, most commonly one of providing localized information about a function. Because one sets out to design a kernel, it is natural then to look at what one has achieved, and one is thereby almost forced into assessing in a relatively straightforward way the meaning of the averages of the function one is trying to determine. Least-squares data fitting, or any other data fitting procedure, on the other hand, aims merely at seeking a so-called solution that happens to fit the data. Not being contradicted by data is therefore the principle explicit objective, rather than determining some more useful property of the system under investigation.

For me, one of the biggest drawbacks of the data-fitting method is that it always works, in the sense that it always yields a representation of f. Optimally localized averaging often does not. If the data contain essentially no information about a function in some interval of the domain, then one quite explicitly fails to obtain localized kernels in that interval. This immediately tells us something very important. The data-fitting methods always give a representation of the function, and the danger is that one might be tempted to believe it. Believe it or not, I have met people who actually claim that such methods are necessarily superior because they are guaranteed to produce a 'solution'. Of course, for linearized problems, one can always construct the corresponding averaging kernels from the least-squares fits, a step which should be mandatory. However, not everyone is so fastidious.

Another disadvantage of least-squares data fitting concerns the criterion that is often used to gauge success. Because the explicit aim is to find some way to fit the data, rather than to address some scientifically important property of the system, the success is judged by how well that fitting has been achieved. Adding more data, and successfully fitting them, therefore appears to improve the inversion. But if one then asks of the solution a question about some property in an interval which has no influence on those new data, that improvement is dangerously illusory to the unwary. Indeed, indiscriminate fitting of all the data can bias against fitting those data that are directly pertinent to a question one might be trying to answer. Augmenting the data set with such irrelevant information can actually worsen the inference, even when care is taken in interpreting the results.

Notwithstanding my reservations about the data-fitting methods, I, together with most other helioseismologists, commend their use in moderation: they are computationally faster than superior methods that tailor kernels to have some more useful property. Indeed, experience has taught us that operationally it is a good idea to use a variety of methods, to give some idea of the robustness of the inferences. Contradictory inferences from different methods evidently reveal that at least all but one of those inferences are incorrect, and unless one can understand why, they should all be doubted: moreover, it must always be realized that consistency does not imply correctness.

I now summarize briefly the methods that helioseismologists currently most commonly employ. I shall not give detailed explanations, since the procedures are no doubt well known to the audience.

2.1 Analysing Linearized Constraints

Given a set of frequencies $\{\omega_i\}$, one subtracts from each member the corresponding eigenfrequency ω_{0i} of a reference model and considers linearized constraints such as (8) or (9) for the structure difference δX or the angular velocity Ω. Two basic methods are used: (i) seeking a function with desired properties based on smoothness (regularized) that reproduces the data adequately, and (ii) tailoring averaging kernels to have some desired property, such as localization, moderating both processes so as not to permit excessive contamination of the results by the errors ϵ_i in the data.

For simplicity of presentation, I shall assume the structure, or angular velocity, to be a scalar function $f(r)$ of radius alone, with kernels $K^i(r)$ such that the constraints may be written:

$$d_i = \int_0^R K^i f dr + \epsilon_i \, .\tag{11}$$

Generalization to two and perhaps three dimensions or to a vector structure function is straightforward in principle, though in practice some multidimensional inversions with adequate resolution using very large data sets might be computationally very difficult. Later, I shall mention briefly some two-dimensional inversion procedures that have been carried out on rotational-splitting data, both real and artificial, and present some results.

All of the procedures I shall describe have either directly or in some slightly different guise been either motivated or influenced by similar earlier work in geophysics. A complementary and somewhat more extensive review is provided by Christensen-Dalsgaard (1992) in the proceedings of the first of these workshops.

Regularized Least-Squares Fitting (RLSF). Regularized least-squares fitting is the most common of the inversion procedures currently in use to make reliable inferences about the sun's structure. The idea is first to represent the function f as a linear combination of a chosen finite set of N basis functions $\phi_j(r)$, $j = 1, \ldots N$:

$$f = \sum_{j=1}^{N} \alpha_j \phi_j, \tag{12}$$

where α_j are constants to be determined. Typically, ϕ_j are localized, such as the piecewise constant functions

$$\phi_j = \begin{cases} 1 & \text{if } r_{j-1} < r < r_j, \\ 0, & \text{otherwise} \end{cases} \tag{13}$$

on the dissection $\{r_j\}$ of the interval $[0, R]$ of r, or piecewise linear functions or cubic splines. Choosing ϕ_j to be the kernels K^i, with a 1-1 mapping between i and j, yields the spectral expansion, which, as I have already pointed out, is unfashionable. Given I data d_i, one then determines α_j by minimizing

$$\mathcal{E} = \chi^2 + \lambda P \tag{14}$$

for some tradeoff parameter λ, where

$$\chi^2 = \hat{I}^{-1} \sum_i \sigma_i^{-2} \left(d_i - \sum_j \alpha_j \int_0^R K^i \phi_j \, dr \right)^2, \tag{15}$$

$\hat{I}$ is an estimate of the number of degrees of freedom in the data, whose errors ϵ_i are presumed to be random and independent with zero means and variances σ_i^2, and P is a penalty functional which most commonly penalizes curvature. P is also usually quadratic in f, such as

$$P(f) = \int_0^R W(r) \left(\frac{d^2 f}{dr^2} \right)^2 dr, \tag{16}$$

so that, after substitution of the expansion (12) into (16), the minimization of $\mathcal{E}$ yields a set of linear equations for α_j:

$$\mathbf{M}\alpha = \delta, \tag{17}$$

where δ is the vector of error-weighted data, whose components are d_i/σ_i. Solving those equations is often accomplished via singular-value decomposition; I refer you to the recent review by Thompson (1995) for a discussion of technical details and variants.

Optimally Localized Averaging (OLA). The Backus-Gilbert (1968, 1970) philosophy for determining localized averages is the most common of the kernel-tailoring inversions. Linear combinations of the data

$$\bar{f}(r_0) = \sum_i c_i(r_0) d_i \tag{18}$$

are chosen such that the associated averaging kernels

$$K(r; r_0) = \sum_i c_i(r_0) K^i(r) \tag{19}$$

are both well localized and free from sidelobes, yet with coefficients c_i that are not so large in magnitude as to cause the combination (18) to be dominated by data errors. Initially the Backus-Gilbert (1968) prescription of minimizing

$$12 \int_0^R (r - r_0)^2 K^2(r) dr + \lambda \sum_{ij} E_{ij} c_i c_j \tag{20}$$

for some tradeoff parameter λ was universally adopted, where E_{ij} is the covariance matrix of the errors. Now it is becoming more common to tailor the kernels by an error-moderated measure of the difference between K and a function $\mathcal{D}$ with the desired properties:

$$\int_0^R [K(r; r_0) - \mathcal{D}(r; r_0)]^2 dr + \lambda \sum_{ij} E_{ij} c_i c_j. \tag{21}$$

Oldenberg (1976) and Pijpers and Thompson (1992) used this procedure for computational convenience. Pijpers and Thompson named it subtractive optimally localized averaging (SOLA). It has subsequently been found that a greater advantage of the method is that if $\mathcal{D}$ is chosen to be a realistic paradigm of what might actually be achieved (Pijpers and Thompson chose a Gaussian function with adjustable width), a kernel K might be constructed with lesser undesirable sidelobes than is achieved when $\mathcal{D}$ is, say, the delta function $\delta(r - r_0)$, which is what Oldenberg had used.

Linearized Asymptotic Inversions (LAI). If JWKB asymptotic representations of the eigenfunctions ξ_i are substituted into the formulae for the kernels, and only the terms that contribute in leading order to the integral in the constraint (11) (when f is smooth) are retained, the outcome can often be transformed into an Abel integral, which is invertible. For example, if f is the relative perturbation to the sound speed,

$$\frac{d_i}{\omega_i} \sim \frac{w}{S} \int_{r_1}^R (w^2 - a^2)^{-\frac{1}{2}} f(r) \frac{dr}{c} \tag{22}$$

(e.g. Gough, 1993), in which $a = c/r$, $w = \omega/(l + \frac{1}{2})$, and I have omitted the subscript i from w; also

$$S(w) = \int_{r_1}^R (w^2 - a^2)^{-\frac{1}{2}} \frac{dr}{c} \tag{23}$$

and in this case $f(r) = \delta c/c$. The function $a(r)$ is evaluated for the reference model. Normalized rotational splitting frequencies ω_{1i}/m resulting from the spherically symmetrical angular velocity $\Omega(r)$ about a unique axis also satisfy an equation like (22), with $\delta c/c$ replaced by Ω.

Equation (22) indicates that each of the data d_i/ω_i (or ω_{1i}/m in the case of rotation) is a function of w alone as $i = (n, l)$ varies. Therefore, with adequate data, a smooth function F of w can be fitted to the data, and equation (22), which with the substitution $\eta = a^2$ becomes an Abel equation, can be inverted, yielding

$$f(r) = \frac{2a^3}{\pi} \left(-\frac{d\ln a}{d\ln r} \right) \int_{a_s}^{a} \frac{(dF/dw)dw}{w(a^2 - w^2)^{1/2}} , \qquad (24)$$

where $a_s = a(R)$. In practice, data are not available for w as small as a_s; that would require modes with zero penetration depth. Therefore, some assumption must be made about the contribution to the integral in the range of w from a_s to the smallest value for which there is a datum. I shall not describe what is done, except to say that at great depths (relative to the depth of penetration of the shallowest mode for which there is a frequency measurement) the function $f(r)$ is insensitive to that assumption.

2.2 Model-Independent Inversion

If one uses the averages of f obtained by one of the methods described in section 2.1 to determine an improved structure function $\mathbf{X}$, one can use that structure to define a new reference model and iterate. Although the outcome may not be unique, in the sense that it might depend on the starting point, and perhaps on the manner in which the iterated structure $\mathbf{X}$ is determined, it is likely to be more weakly dependent on the initial reference model than a single iteration.

An alternative procedure is to dispense with linearization about a reference model, and work with the basic constraint (2), which, in the absence of rotation, is given by equation (3). To my knowledge, the only serious attempt to do that is for the sound speed $c(r)$ using JWKB asymptotics. Then, equation (3) becomes

$$\frac{(n + \alpha)\pi}{\omega} \sim \mathcal{F}(w) = \int_0^R (1 - a^2/w^2)^{1/2}\frac{dr}{c} , \qquad (25)$$

which, after interpolating the data $(n + \alpha)\pi/\omega$ to yield the continuously defined function $\mathcal{F}(w)$, can be differentiated into an Abel equation and inverted to yield

$$\frac{r}{R} = \exp \left[-\frac{2a}{\pi} \int_{a_s}^{a} \frac{w(d\mathcal{F}/dw)dw}{(a^2 - w^2)^{1/2}} \right] \qquad (26)$$

(e.g. Gough, 1993) which, as in the case of LAI, must be approximated by assuming some form for $a(r)$, or $\mathcal{F}(w)$, associated with locations above the penetration depth of the shallowest mode. Equation (26) expresses r as a function of $a = c/r$, which is immediately invertible to yield $a(r)$, and hence $c(r)$.

3 Principal Helioseismic Inferences Concerning the Stratification of the Sun

The most revealing inversions to date have been for the sound speed $c(r)$, and I shall restrict my discussion just to those.

3.1 The General Sound-Speed Variation and the Internal Helium Abundance

One of the first structure inversions to be carried out applied the asymptotic formula (26) to high-degree data obtained by Duvall (1982), which yielded $c(r)$ to a depth of about 70 Mm (Gough, 1985). Subsequently, the data set was augmented with observations by Duvall and Harvey (1984) of modes of degree down to $l = 1$, from which the inversion presented in Figure 2 was obtained by Christensen-Dalsgaard et $al.$ (1985). The sound speed was found to agree to within about $\frac{1}{2}$ per cent with that in a standard theoretical model of the sun computed with a helium abundance Y of about 0.25, except in the interval $0.3R \lesssim r \lesssim 0.5R$, where the solar sound speed exceeded the model sound speed by about 1 per cent.

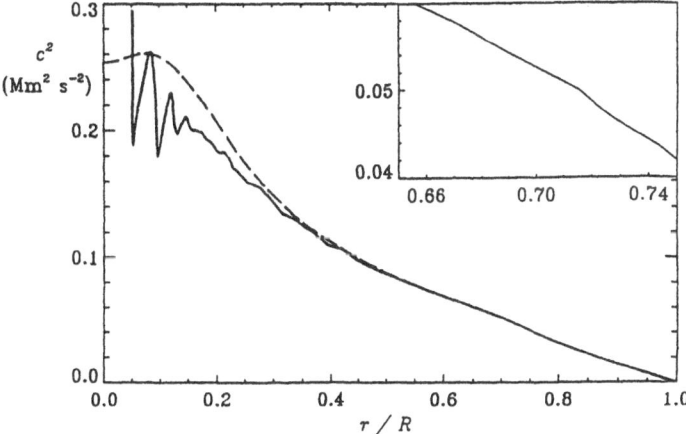

Fig. 2. Asymptotic inversion of solar p-mode frequencies by Christensen-Dalsgaard et $al.$ (1985). The continuous curve is the square of the inferred solar sound speed; the dashed curve is the square of the sound speed of a theoretical model with $Y \approx 0.25$ computed by Christensen-Dalsgaard (1982). The inset is an enlargement of the inversion in the vicinity of the base of the convection zone.

That inversion graphically resolved an issue of considerable importance: that the structure of the sun was closer to a standard solar model with $Y = 0.25$ than it was to a model with a much lower value of Y. Two somewhat more modern models of the present-day sun are illustrated in Figure 3, to give an

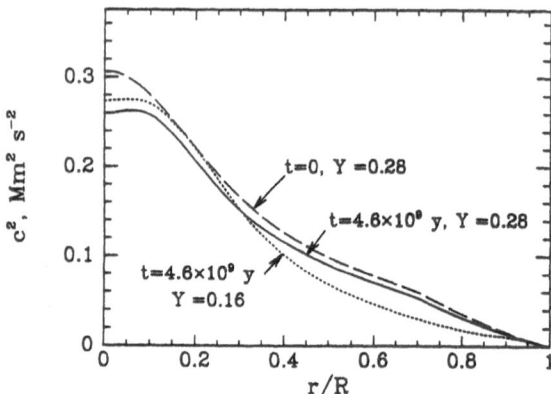

Fig. 3. Square of the sound speed in three theoretical models of the sun: the ages t and the initial helium abundances Y are indicated on the diagram. The depression in sound speed near the centre in the two models with $t = 4.6 \times 10^9 y$ is a consequence of the increased mean molecular mass of the fluid resulting from the extra helium that has been generated by the nuclear reactions. Since the central sound speed c_0 is a decreasing function of Y, one might have expected the model with $Y = 0.25$ in Figure 2 to have c_0 lying between the values for the two evolved models in this diagram. The reason it does not is that the model in Figure 2 is older, having been computed with opacity and nuclear-reaction cross sections that had been updated by the time the models in this figure were constructed.

idea of how c varies with Y. The importance of the result lies in the fact that Y is safely greater than the generally believed primordial universal value at the time nuclear reactions ceased after the Big Bang, and is inconsistent with the lower value illustrated in Figure 3, which yields a neutrino flux roughly in agreement with observation. Therefore, the solar neutrino problem really does, at first sight, appear to be a problem in understanding either neutrino production or neutrino propagation, and not in explaining a low solar helium abundance. As I pointed out in section 2, this conclusion had already been reached by calibrating theoretical models of the sun with the shallow high-degree data illustrated in Figure 1; but the strength of the new result is that the inference concerns the structure deep in the radiative interior directly.

A second inference from the sound speed determination illustrated in Figure 2 was the location of the base of the convection zone. This can be seen as a change in curvature of $c^2(r)$, and is more evident in the inset. Subsequent more careful analysis by Christensen-Dalsgaard, Gough and Thompson (1991) yielded for the radius r_c of the transition, $r_c/R = 0.713 \pm 0.003$.

A third important inference that came out of the analysis was that the sound speed in the sun significantly exceeded that in the model, by up to about 1 per cent, in the outer regions of the radiative envelope. This is illustrated by Figure 4, which is a more recent inversion for $\delta u/u$ (where $u = p/\rho$), in the form of

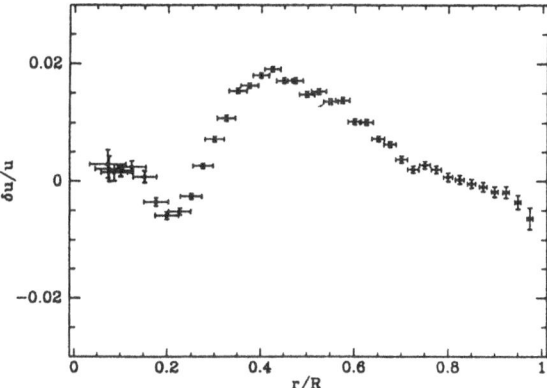

Fig. 4. Optimally localized averages of the relative deviation $\delta u/u$ of the function u in the sun from that in a standard solar model of Christensen-Dalsgaard. 650 solar oscillation frequencies between 1.5 and 3.0 mHz of modes with $0 \leq l \leq 140$ were used for the inversion. They were obtained from the compilation by Libbrecht, Woodard and Kaufman (1990).

optimally localized averages. Christensen-Dalsgaard *et al.* (1985) concluded that that could be explained only if the values of opacity that had been used in the theoretical model were too low by up to 20 per cent at temperatures between about 10^6 K and 4×10^6 K; they could say nothing about temperatures lower than 10^6 K (actually below about 2×10^6 K) because material that cool is in the convection zone, which is adiabatically stratified essentially irrespective of opacity. The alternative possibility could have been that the sun was out of thermal balance, but the implied thermal relaxation (in the absence of macroscopic material motion) would involve luminosity excursions on timescales of 10^7 years of a magnitude that would not only be hard (though perhaps not impossible) to explain, but would cause severe difficulties to climatologists. (However, climatologists did, and still do have unsurmounted difficulty in explaining the absence of continual terrestrial glaciation prior to 1 Gy or less ago, when the solar constant must have been less than 95 per cent of the present-day value if quantum and classical Newtonian physics is correct, so perhaps the climatological issue should not be weighted very highly.) Yet another possibility is that perhaps there is a slow meridional circulation in the radiative interior which advects heat downwards against the augmented thermal gradient that would be required to carry the extra heat flux necessary to maintain the observed luminosity. However, these alternatives seem not to be the case. Because the prediction about the opacity was in such a limited range of conditions, it was possible to persuade Iglesias and Rogers to make a spot check of the astrophysical opacity tables that had been provided from elsewhere. They found that the helioseismic prediction was more-or-less correct (Iglesias and Rogers, 1991), though not quite sufficient to remove the discrepancy entirely. Moreover, after careful comparison with the calculations that produced the astrophysical tables, it was discovered that those earlier calculations had misrepresented spin-orbit coupling in the radiative atomic transitions. Thus, helioseismology had successfully been able to use the sun as a laboratory for atomic physics. An interesting corollary to this story is that Iglesias and Rogers then went on to compute the opacity at lower temperatures, and found that the spin-orbit coupling had an even greater influence near 10^5 K. The astrophysical consequence of that was to resolve several outstanding problems in the understanding of pulsations in several other categories of stars.

3.2 The Equation of State in the Convection Zone

Whilst I am on the subject of microphysics and the stratification of the solar envelope, I shall describe another adventure in inference. Once one penetrates (in one's mind's eye, of course) beneath the superadiabatic boundary layer of the solar convection zone, the specific thermal capacity of the fluid becomes so great that convection becomes gentle: the motion required to transport all the heat is extremely subsonic, and consequently Reynolds stresses and fluctuations in pressure and density are small. The stratification is essentially isostatic and adiabatic. It can be shown that under such circumstances a seismically accessible quantity W is equal to a purely thermodynamic quantity Θ:

$$W := \frac{r^2}{G\tilde{m}}\frac{dc^2}{dr} = \Theta := \frac{1 - \gamma_\rho - \gamma}{1 - \gamma_{c^2}}, \tag{27}$$

where $\tilde{m}$ is the mass of the equilibrium background solar model within the radius r, and where γ_ρ and γ_{c^2} are the partial logarithmic derivatives of γ with respect to ρ and c^2 at constant c^2 and ρ respectively. Since the mass in the convection zone is only about $0.02\,M$, where M is the mass of the sun, one does not require a very sophisticated procedure for estimating the deviation of $\tilde{m}$ from M; indeed, it is common practice even to set $\tilde{m} = M$ when comparing solar inversions with inversions of eigenfrequencies of theoretical model solar envelopes.

Equation (27) is valid in the region of the second ionization of helium, where γ drops below its perfect-gas value of 5/3 (at which $\Theta = -2/3$), and Θ rises. Figure 5 shows Θ and W in a variety of model envelopes. To a first approximation, increasing the helium abundance Y simply increases the height of the ionization hump, as one would expect because it increases the amount of material undergoing ionization. There is another parameter, α (the so-called mixing-length parameter which influences the properties of the superadiabatic boundary layer), which specifies the envelope model, the most pronounced effect of which is to change the location of the hump. Thus, by comparing inversions of solar data with inversions of corresponding eigenfrequencies of a grid of theoretical models, it should be possible to calibrate both γ and Y. The first calibration was carried out asymptotically (Däppen *et al.*, 1988), using again equation (26) and differentiating $c(r)$ afterwards to substitute into equation (27). The result of that calibration, $Y = 0.233 \pm 0.003$, was not reported at the time, because it was feared that systematic errors might have contaminated the result. It was

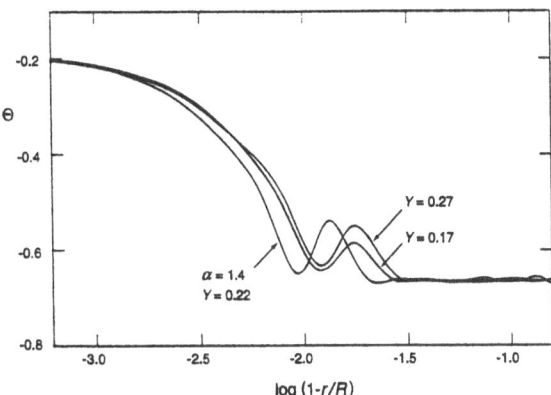

Fig. 5. The quantity Θ, defined by equation (27), plotted as a function of depth below the photosphere on a logarithmic scale for three model solar envelopes defined by the helium abundance Y and a parameter α, the ratio of mixing-length to pressure scale height, which occurs in the prescription for determining the temperature gradient in terms of the heat flux. The models with $Y = 0.17, 0.27$ have $\alpha = 1.9$. All models have the same total heavy-element abundance $Z = 0.02$.

eventually announced in 1992, by Kosovichev *et al.* They performed analyses with the same intention of determining Y, and obtained similar results when the same equation of state was used. This rather low value of the convection-zone abundance is now thought to have been brought about by slow helium settling through the base of the convection zone over the lifetime of the sun. Indeed, Christensen-Dalsgaard, Proffitt and Thompson (1993) have demonstrated that such settling also substantially improves the agreement between the sound speed in the radiative interior of the best-fitting theoretical model with that determined helioseismically for the sun (see Figure 6.)

Another outcome of the investigation by Kosovichev *et al.* (1992) was that solar abundance determinations by OLA and RLSF did not agree. The discrepancy was unlikely to be a programming error, because when the two methods were applied to eigenfrequencies of a proxy model of the sun, constructed with the same physics as the reference and with the same data kernels, the determinations did agree; only when the proxy model and its oscillations were constructed with a different equation of state (thereby rendering the proxy inconsistent with the reference) did they not. It was concluded that there must therefore be a deficiency in the equation of state. Dziembowski, Pamyatnykh, and Sienkiewicz (1992) came to the same conclusion. This provides an example of the importance of using more than one inversion method on the same problem. Although information that there is something wrong with the equation of state might in principle have been extracted by only one of the methods, had enough effort been expended on varying tradeoff parameters or penalty functionals, that is

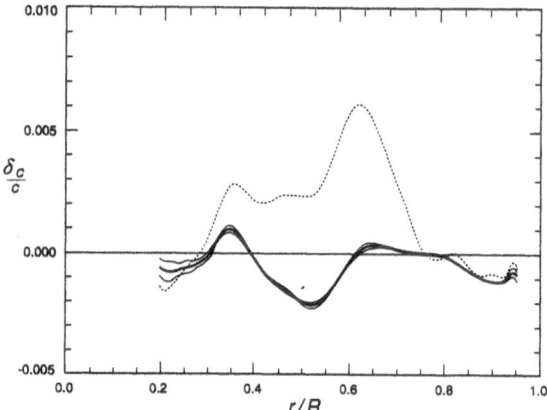

Fig. 6. The thick continuous curve represents the relative sound-speed difference $\delta c/c$ between the sun and a solar model in which a presciption for gravitational settling of helium has been included. The two thin continuous curves are $\pm 1 \times \sigma$ either side of it, where σ^2 is the variance of the pointwise probable errors. The dotted line is the difference $\delta c/c$ between the sun and a model without helium settling (after Christensen-Dalsgaard, Proffitt and Thompson, 1993).

not what happened: what actually happened is that attention was drawn to the issue by the discrepancy.

Subsequently, a calibration of an equation of state (in which bound-state partition functions are truncated by putative electrostatic interactions with neighbours when the mean interparticle distance becomes less than some value representing the effective sizes of the bound species) has been found to give almost perfect agreement between theoretical and solar values of $W(= \Theta)$, inferred by asymptotic inversion, if the bound-state radii are taken to be about 1.7 Bohr radii (Baturin *et al.*, 1994). My claim here is not that atomic and ionic dimensions have actually been measured seismologically, for there may be deficiencies in the equation of state used by Baturin *et al.* that have been masked by false bound-state sizes. What is important, however, is that effects of a magnitude that result from modifications to the states of atoms and ions in the plasma have been detected seismologically. This really does herald the use of helioseismology to do observational plasma physics at densities that cannot be attained in a controlled way on earth.

3.3 Sound Speed in the Core

In the absence of data from g modes, p-mode frequencies of the lowest degrees are required to penetrate into the energy-generating core. Knowing the structure of the core is of great importance to solar physics, partly because of the neutrino problem. Unfortunately, it is very difficult to measure frequencies with sufficient accuracy to make meaningful inversions in the core. One reason is that the sound speed is highest in the core; waves spend relatively little time there and consequently their amplitudes are low. Therefore the detailed structure of the core has only a minor effect on the frequencies. Another reason is that there are relatively few low-degree modes, so there is less almost redundant information and therefore less scope for averaging out measurement errors. Nevertheless, apparently quite precise measurements have recently been made. Those which have attracted the most attention are a six-month continuous data set obtained from the IPHIR instrument on the Phobos spacecraft (Toutain and Fröhlich, 1992) and ground-based measurements from the BiSON network of observatories (Elsworth et al., 1994). Interestingly, there are systematic differences, of order 1 part in 10^4, between the measurements, which impart differences of several parts in 10^3 on the inferences of core sound speeds (Gough, Kosovichev and Toutain, 1995). Uncertainties at this level are too great for us to make subtle inferences about interesting dynamical aspects of the core, such as whether there has been any redistribution of the products of the nuclear reactions, an issue of importance to understanding neutrino production. However, given the state of knowledge prior to helioseismology, one might marvel at the fact that we are even discussing measuring core conditions to better than 1 per cent.

4 The Internal Angular Velocity

The first serious attempt at inverting rotational splitting data (Duvall et al., 1984) was poorly resolved, because there were only 37 data, each of which was an average over many modes. Both RLSF, with a variety of basic functions, and OLA were used, the variations amongst the resulting representations providing a (no doubt optimistic) indication of the pointwise uncertainty in the angular velocity. Some of the results are illustrated in Figure 7. The piecewise functions were chosen on a dissection which roughly reflects the resolution of the data for $r > 0.5R$, but is somewhat relatively over-resolved at lower radii.

A comment should be made about the OLA. The averages plotted are genuine combinations of the rotational-splitting kernels only in the convection zone, $r \gtrsim 0.7R$. At smaller radii it was impossible to obtain well localized averaging kernels. What was done, therefore, was to draw a smooth curve, not displayed in Figure 7, through the localized averages in the convection zone, subtract the contribution that it made to the splitting from the data, and then to carry out OLA in the radiative interior, in which it was then possible to construct localized kernels centred at radii down to $0.4R$. Of course the full kernels, over the entire range of r, had strong sidelobes in the convection zone, but by this device the averages become in some sense simpler to interpret in physical terms; they are averages of the angular velocity, over kernels whose characteristic width are indicated by the horizontal bars, subject to the assumption that the angular velocity in the convection zone is given by a smooth curve through the data.

I must point out another difficulty that must be faced in the interpretation of Figure 7. The rotational-splitting data were of sectoral modes ($m = l$) only,

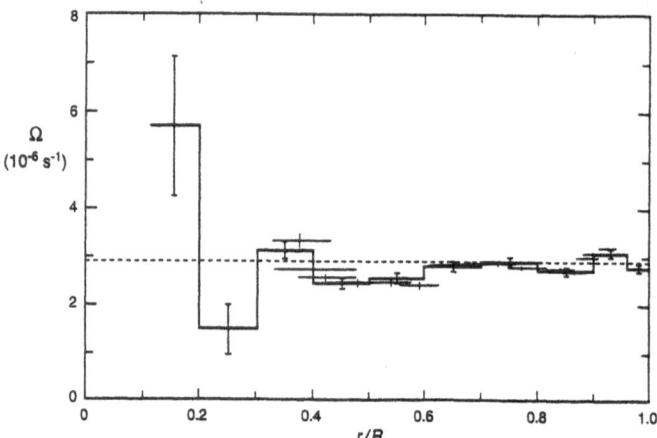

Fig. 7. Equatorially confined latitudinal averages of the solar angular velocity, inferred from the averaged rotational-splitting data of Duvall and Harvey (1984). The piecewise contant function was calibrated by least-squares fitting; the dissection was guided by the OLA resolution, except for the innermost two divisions, which were chosen arbitrarily. The innermost division extended to $r = 0$ in the computation, but because the kernels are very small near the centre the data contain almost no information about Ω in the inner half of that division, and to emphasize that only the outer half is plotted. The isolated crosses represent the OLA: the horizontal limbs represent the widths of the averaging kernels. The vertical bars represent pointwise standard errors (after Duvall *et al.*, 1984).

whose kernels are concentrated near the equatorial plane. However, the degree of equatorial concentration decreases as l decreases, and consequently as the characteristic penetration depth increases. Therefore the more deeply penetrating modes sample regions near the surface that are not reached by the high-degree modes whose splittings dominate the averages centred near the surface. The inversions illustrated in Figure 7 assume no latitudinal variation of Ω.

As an aid to judging how serious the unaccounted latitudinal variation of Ω is, I present in Figure 8 two inversions of the same data that were used for Figure 7, this time by LAI, for the equatorial angular velocity. The lower curve assumes Ω is independent of colatitude θ, and the upper curve assumes $\Omega = f(r)W(\theta)$, where $W(\theta)$ is the variation observed in the photosphere: $W(\theta) \approx 1 - 0.2 \cos^2 \theta$. What is plotted is $f(r)$. Only in the central regions, where in any case the inversions are quite uncertain, do the two curves diverge noticeably. Subsequently, more extensive splitting data have enabled us to gain a fair idea of the latitudinal variation beneath the photosphere, as is illustrated in Figures 9 and 10. It appears that reality is actually between the two assumptions adopted for Figure 8. But before I discuss those later inversions, I shall comment further on the inversions displayed in Figures 7 and 8.

One should never carry out an investigation, either theoretical or observational and especially both, without prejudice. The prejudice provides a goal

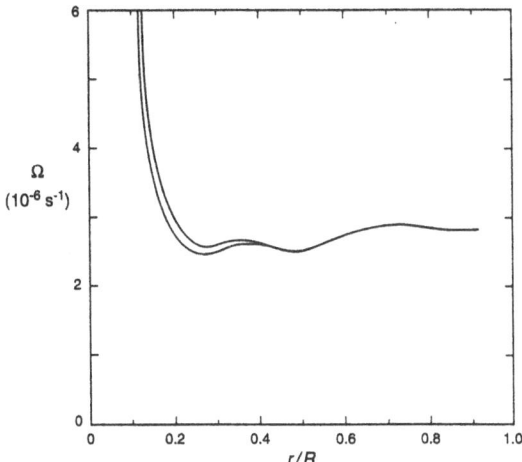

Fig. 8. Two estimates of the solar equatorial angular velocity Ω_e inferred by asymptotic inversions of the averaged rotational-splitting data of Duvall and Harvey (1984). The lower curve was obtained under the assumption that Ω is independent of latitude, the upper curve by assuming the latitudinal dependence to be the same at all depths and equal to that observed at the surface (from Christensen-Dalsgaard, Gough and Toomre, 1985).

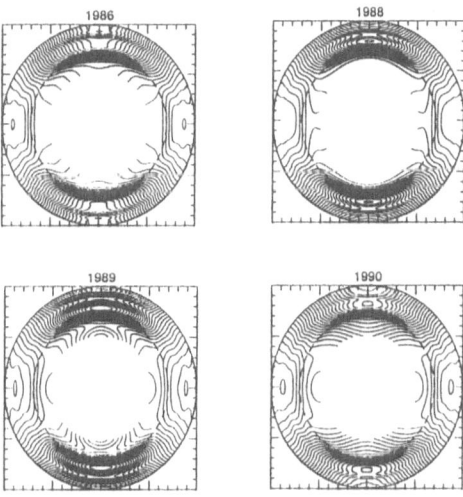

Fig. 9. Contours of constant angular velocity in a meridional plane through the rotation axis of the sun, inferred by Gough *et al.*, (1993) from rotational-splitting data obtained in the summers of 1986, 1988, 1989 and 1990, first using the spectral expansion in latitude and subsequently RLSF with respect to radius. The continuous circles indicate the location $r = R$ of the photosphere, the dashed circles the base of the convection zone. The axis of rotation is vertical. Only the north-south symmetric component of Ω is obtainable from the linearized constraints.

(a)   (b)

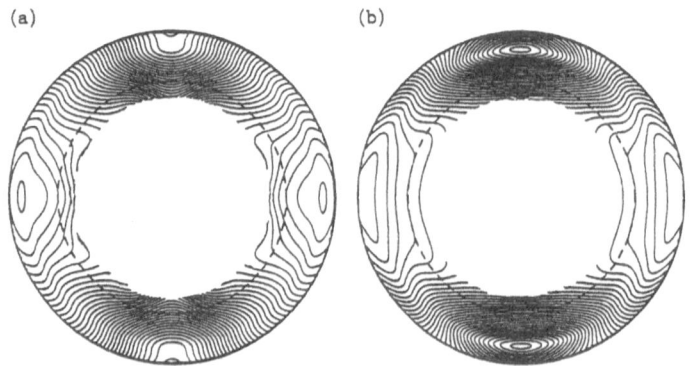

Fig. 10. Contours of constant solar angular velocity, computed from the average of the four data sets in the manner of Figure 9 (from Sekii, Gough and Kosovichev, 1995). Inversion (a) used RLSF with a regularization that penalizes the second derivative of the coefficients $\Omega_s(r)$ of the expansion (29). The regularization in inversion (b) penalized also any variation parallel to the rotation axis in the partial shell $0.7 < r/R < 0.9$ and $r \sin \theta > 0.7R$.

towards which to work, and the incentive to check one's procedures carefully when the results do not accord with expectation. Admittedly, there is the danger of complacency, which causes error to be missed if the results turn out to be as anticipated. Indeed if, finally, the results do confirm the prejudice, one finds oneself in a position merely to be content that one has apparently understood the system under investigation. But if they do not, one is eventually forced to admit that one was wrong, and to revise one's ideas. That is a memorable experience, and results in true learning.

Unlike the structure inversions that I discussed in the previous section, which are disappointingly more-or-less in accord with standard solar models, the inversions illustrated in Figures 7 and 8 are an example of a case in which a revision of ideas is being forced upon us. But in order to discuss that, I must first acquaint you with the ideas that were prevalent at the time when the inversions were carried out.

Stars like the sun are losing angular momentum. A magnetic field causes the ionized solar wind to corotate with the sun out to a distance of about $5R$, reacting back on the sun with a retarding torque. That explains why solar-type stars (indeed, essentially all types of main-sequence star) exhibit a surface rotation rate which is a decreasing function of age. But to what extent is the torque transmitted into the interior? That question had been addressed by many people. A plethora of hydrodynamical instabilities were invoked, leading to different conclusions about the rate at which Ω should decrease with increasing r. The most (rather, in some people is minds, least) extreme case was made by those who believed the solar interior to be pervaded by a large-scale magnetic field which has caused rotation in the radiative interior to be uniform: it takes a field of only about $3\mu G$ – not very different from the intensity of the interstellar galactic field – to spin down the core on a timescale comparable with the age of the sun, and I find it difficult to imagine how the field could be substantially lower, given that the characteristic decay time of the largest-scale components of the field is not much less than the sun's age. (It is, of course, not out of the question that some of the magnetohydrodynamical mechanisms invoked by dynamo theorists to operate in the putative dynamo in the convection zone to maintain the magnetic field could also accelerate its decay in the radiative interior, but there is little real evidence to support such an hypothesis). Predictions about the variation of Ω ranged at that time from a quite high angular velocity in the core to one that was hardly greater than the surface value. Yet all had Ω decreasing monotonically with r. It appears that nobody predicted the existence of a region in the radiative zone that rotates more slowly that the surface, as is indicated in Figure 7 and 8. How could that be?

Before continuing, I should point out that there have been several other reports of rotational splitting of low-degree p modes in the following decade, with some discrepancies; although the very uncertain rapid rotation of the core is not found universally, the region of relatively slow rotation does seem to be a robust result.

I have two suggestions for possibly accounting for the phenomenon. The first comes about from the fact that all appropriate stability calculations of the sun that have been carried out in the last 23 years have found the cores of solar models to be intrinsically unstable to g modes. Yet most solar modellers resolutely ignore that result. The nonlinear development of the instability has been discussed in a few instances, but none of the theoretical models of it is sufficiently realistic for one to be able to draw from it how the sun must behave. It is, however, not entirely out of the question that there is a local laminar overturning motion, possibly oscillatory, possibly direct, which advects angular momentum between the core and its surroundings. The outcome could be an augmentation of the angular velocity in the core at the expense of a compensating diminution in the surrounding environment, as is exhibited in the figures.

Perhaps for many people a more plausible possibility is that the interior is connected dynamically to the high-latitude regions of the convection zone by a predominantly dipole-like magnetic field; although the slowly rotating region has a lower angular velocity than the photosphere at low latitudes, it is rotating substantially more rapidly than does the photosphere near the poles. If such magnetic coupling does take place, then contours of constant Ω must connect the slowly rotating interior near the equator to appropriate locations at higher latitudes at the base of the convection zone. We must await more accurate data before we can ascertain whether or not that is the case. I should also point out that Eckman circulation induced by the rotational shear in the vicinity of the convection zone, as considered, for example, by Spiegel and Zahn (1992), also leads to a latitudinal redistribution of angular momentum of the kind required to explain the phenomenon, though the actual model presented by Spiegel and Zahn does not appear to accord with the observations.

I now turn to the determination of the dependence of Ω on both r and θ. That requires knowledge of the m dependence of the rotational splitting. At present that is presented by the observers as an expansion in terms of Legendre functions, such as

$$\omega_{nlm} - \omega_{nl0} = l \sum_{k=1}^{k_m} a_{lk} P_k(m/L), \qquad (28)$$

where L is perhaps l or perhaps $\sqrt{l(l+1)}$, depending upon the observer. Such an expansion suggests a similar expansion for $\Omega(r, \theta)$:

$$\Omega = \sum_{s=0}^{s_m} \Omega_s(r) P_{2s}(\cos \theta). \qquad (29)$$

Indeed, if expansion (29) is substituted into the rotational-splitting formula (9), there results an expansion of the type (28), with only odd terms, and with $k_m = 2s_m + 1$. The relationships for each l can easily be solved to yield integrals of $\Omega_s(r)$ as linear combinations of a_{lk}, each of which presents a one-dimensional linear inverse problem which can be 'solved' in one of the usual ways. Thus, with respect to θ this is simply a spectral expansion, which in a sense excludes any contribution to the θ dependence of Ω from the annihilator. Amongst the two-dimensional inversion procedures that are available, this is perhaps the most commonly adopted, which is rather surprising in view of the general reluctance by helioseismologists to use spectral expansions when inverting with respect to r.

An alternative procedure is to penetrate the annihilator $\mathcal{A}$ by including more than $\frac{1}{2}(k_m - 1)$ terms in the expansion (29) and incorporating a regularizing penalty functional in a least-squares fitting procedure. Full two-dimensional OLA is impracticable, because the number of data for well resolved inversions is too great to handle directly. The most promising technique devised so far is to try to obtain, within limited ranges of r (or potentially θ) almost separable optimally localized kernels as a product of two 1-d OLA (in practice SOLA seems to be best). Sekii (1993 a,b; 1995) has pioneered this technique. An alternative procedure is to work with a reduced set of data obtained by projecting the error-reduced data $\boldsymbol{\delta}$ onto just the singular vectors of the matrix $\mathbf{M}$ appearing in equation (17) that are associated with relatively large singular values (Christensen-Dalsgaard and Thompson, 1993; Christensen-Dalsgaard et al., 1995). The procedure should produce a reduced set of data that contains all the information that is accessible from the original data. The preprocessed data, and their associated kernels, can then be used in either 2-d OLA or 2-d RLSF.

Contour diagrams from inversions using the spectral expansion in θ are illustrated in Figure 9, from data obtained by Libbrecht and Woodard over the summers of 1986, 1988, 1989 and 1990. It is evident that the angular velocity inferred in and immediately beneath the convection zone changes with time. It is to be hoped that those changes are not merely a product of unrecognized measurement errors. Unfortunately, the data are too widely separated in time to ascertain whether or not the variation, if real, is a result of propagating waves, as some dynamo theorists predict. Imminent continuous data from the GONG project and from the SOI on the spacecraft SOHO are bound to answer that question.

Another question of interest to dynamo theorists is the functional form of the contours. From diagrams similar to that in Figure 9 corresponding to 1986, the angular velocity has been described as being approximately independent of radius through the convection zone and only very weakly dependent on latitude in the radiative interior (e.g. Christensen-Dalsgaard and Schou, 1988; Brown *et al.*, 1989). That description has caused dynamo theorists to rethink their theories, since their models tend to lead to an angular velocity in the convection zone which is roughly constant on complete cylinders (that do not penetrate the radiative interior) aligned with the rotation axis, except possibly in a boundary layer near the surface. It must be realized, however, first, that the inversions use spectral expansions in θ with only a very limited basis – s_m was only 2 in the original inversions, and is only 5 in the most recent – and second, that at low latitudes, where dynamo theorists wanted contours of constant Ω to be parallel to the axis, except near $r = R$ the gradient of Ω is relatively small, so quite small pointwise changes to the value of Ω can distort the contours substantially. Indeed, Gough *et al.* (1993) showed that for any Ω variation of the kind illustrated in Figure 10a (asymptotically) seismically equivalent variation exists for which Ω is precisely constant on complete cylinders in the convection zone. Earlier (and subsequent) statements to the contrary, based on spectral expansions in θ, have presumably arisen as a result of a refusal to countenance contributions to Ω from the annihilator (or perhaps from the failure to recognize that such contributions are seismically acceptable). I should point out that if one insists on Ω being strictly constant on complete cylinders, the resulting variation of Ω near the poles needs to vary quite dramatically in the polar regions. That is true also of the dynamo models, but the inversions vary much more, and with some data sets with $s_m = 5$ have even resulted in retrograde jets (Schou and Brown, 1994). However, precise rotation on cylinders all the way out to the surface is not what dynamo theorists predict. The inversion by Sekii *et al.* (1995) in Figure 10b is not very different from many of the dynamo models at low latitudes, though at high latitudes it is actually somewhat too smooth.

It is evident that the form of Ω is not yet determined well enough to determine whether or not modern ideas about the putative solar dynamo are correct. It is to be hoped that much more will be learnt from the imminent next generation of helioseismic observations. What is clear, however, is that the radiative interior is not rotating very rapidly, as had previously been suspected. A consequence of that is that the quadrupole moment J_2 of the sun's gravitational potential is not very different from the value associated with uniform rotation, about $2 \times 10^{-7} GMR$, where G is the gravitational constant. To within the accuracy of planetary ranging measurements, that is consistent with the planetary orbital precessions predicted by General Relativity. It is worth noting that J_2 is calculated from the centrifugal distortion resulting from the function Ω that was deduced from the rotational splitting, and, partly because the latter is linear in the 'small' quantity Ω (rather than quadratic, as is the centrifugal force), rotational-splitting inversion affords a more accurate method of determining the quadratic functional J_2 than does any existing (or even proposed) direct method.

A final remark about the mean speed of rotation of the radiative interior is in order. It is evident from Figures 9 and 10 that there is a region of intense shear at the base of the convection zone, particularly at high latitudes. The data have not been able to resolve its structure; it could be very much thinner than what is indicated – almost discontinuous. It is interesting to investigate a possible consequence. To maintain the shear requires a stress. What form might that stress take? If it acts locally, perhaps as a Reynolds stress provided by small-scale turbulence, then the stress is likely to be proportional to the shear. In that case one can easily evaluate the total torque exerted between the convection zone and the radiative interior, to within an unknown multiplicative factor in the stress-strain relation. If, in addition, one adopts the simplistic model with $\Omega = \Omega(\theta)$ in the convection zone and Ω independent of θ in the radiative interior, separated by a discontinuity, one can calibrate the discontinuity, as did Brown *et al.* (1989). The outcome is that the torque is zero, which indeed one would expect if the sun is more-or-less in a steady state. Moreover, that result might perhaps be taken as evidence for a local stress-strain relation. That would certainly limit stresses transmitted nonlocally by the meridional flow of the kind considered by Spiegel and Zahn (1992) which is driven by the tachocline. It would also constrain the possible Lorentz stresses. Just what limits the present splitting data actually do impose on such processes has not yet been investigated. I suspect that current data are inadequate to provide usefully stringent constraints, and that once again we must await the imminent next generation of helioseismic observations.

5 Afterword

I hope that I have provided a flavour of what helioseismologists are up to, and what some of their endeavours are. I have not covered the entire arena of the subject. There is, of course, an industry devoted to calibrating models, as is the case in almost all branches of quantitative science, but I have not discussed that, partly because this is a conference on inversion. Another reason is that uncritical model calibration, without an understanding of what information the eigenfrequencies carry, can be misleading. In practice, the requisite information frequently comes from studying inversion, though in skilled hands that need not necessarily be required, as Christensen-Dalsgaard (e.g. 1995) has (often) demonstrated.

I have restricted attention to aspects of the subject that have been relatively well developed. Most notable of my omissions is the study of structural asphericity. This is brought about not solely by rotation, but also by material motion which generates horizontal inhomogeneities in temperature and density, and by magnetic fields. Analysis of just the frequencies of global modes is likely not to be the most profitable avenue to take in this pursuit; indeed, one cannot formally distinguish in the values of eigenfrequencies alone between flow with east-west symmetry, a magnetic field, and variations in sound speed. To make the distinction requires at least a study of the shapes of the eigenfunctions, and to this end various local analyses are being pursued. In addition, time-delayed correlations, studied in so-called time-distance helioseismology, can provide information about some aspects of velocity fields, by concentrating attention on specific rays that are constituent parts of a complete mode: opposite Doppler shifts induced in waves travelling with and against the flow produce effects which are linear in the flow velocity, whereas in the superposition of two such propagating waves that constitutes the standing wave (mode), such effects cancel to leading order. These studies are in their infancy, but will no doubt develop into powerful tools to supplement the more traditional analyses of the kind I have discussed in this address.

References

Backus, G. and Gilbert, F.: Numerical applications of a formalism for geophysical inverse problems, Geophys. J. R. astr. Soc. **13** (1967) 247–276

Backus, G. and Gilbert, F.: The resolving power of gross earth data, Geophys, J. R. astr. Soc. **16** (1968) 169–205

Backus, G. and Gilbert, F.: Uniqueness in the inversion of inaccurate gross Earth data, Phil. Trans. Roy. Soc. A **266** (1970) 123–192

Baturin, V.A., Gough, D.O., Vorontsov, S.V. and Däppen, W.: Towards a helioseismic calibration of the equation of state of the plasma in the solar convective envelope, in *The equation of state in astrophysics* (ed. G. Chabrier and E. Schatzmann, CUP, Cambridge) *Proc. IAU Colloq.*147 (1994) 545–549

Brown, T.M., Christensen-Dalsgaard, J., Dziembowski, W.A., Goode, P.R., Gough, D.O. and Morrow, C.A.: Inferring the Sun's internal angular velocity from observed p-mode frequency splittings, Astrophys. J. **343** (1989) 526–546

Christensen-Dalsgaard, J.: On solar models and their periods of oscillation, Mon. Not. R. astr. Soc. **199** (1982) 735–761

Christensen-Dalsgaard, J.: Study of solar structure based on p-mode helioseismology, in *Seismology of the sun and sun-like stars* (ed. Domingo, V. and Rolfe, E.J., ESA SP-286, Noordwijk) (1988) 431–450

Christensen-Dalsgaard, J.: Solar oscillations and the physics of the solar interior, in *Challenges to theories of the structure of moderate-mass stars* (ed. D.O. Gough and J. Toomre, Springer, Heidelberg) *Lecture notes in Physics* **388** (1991) 11 – 36

Christensen-Dalsgaard, J.: Inverse analysis in helioseismology, in *Proc, Interdisc. inv. workshop* (ed. B.H. Jacobsen) **1** Geo Skrifter, 41 (1992a) 9 – 27

Christensen-Dalsgaard, J.: Solar models with enhanced energy transport in the core, Astrophys. J. **385** (1992b) 354 – 362

Christensen-Dalsgaard, J.: Testing a solar model: the forward problem, in *The structure of the sun* (ed. T. Roca Cortes, Cambridge Univ. Press) in press (1995)

Christensen-Dalsgaard, J. and Berthomieu, G.: Theory of solar oscillations, in *Solar interior and atmosphere* (ed. A.N. Cox, W.C. Livingstone, M.S. Matthews, Univ. Arizona Press, Tucson) (1991) 401 – 478

Christensen-Dalsgaard, J. and Däppen, W.: Solar oscillations and the equation of state, Astron. Astrophys. Rev. **34** (1992) 267 – 361

Christensen-Dalsgaard, J. and Gough, D.O.: Towards a heliological inverse problem, Nature **259** (1976) 89 – 92

Christensen-Dalsgaard, J. and Schou, J.: Differential rotation in the solar interior, in *Seismology of the sun and sun-like stars* (ed. E.J. Rolfe, ESA SP - 286, Noordwijk) (1988) 149 – 153

Christensen-Dalsgaard, J. and Thompson, M.J.: A preprocessing strategy for helioseismic inversions, Astron. Astrophys. **272** (1993) L1 – L4

Christensen-Dalsgaard, J. Gough, D.O. and Thompson, M.J.: The depth of the solar convection zone, Astrophys. J. **378** (1991) 413 – 437

Christensen-Dalsgaard, J., Gough, D.O. and Toomre, J.: Seismology of the Sun, Science **229** (1985) 923 – 931

Christensen-Dalsgaard, J., Hansen, P.C. and Thompson, M.J.: Generalized singular value decomposition analysis of helioseismic inversions Mon. Not. R. astr. Soc. **264** (1993) 541 – 564

Christensen-Dalsgaard, J., Proffitt, C.R. and Thompson, M.J.: Effects of diffusion on solar models and their oscillation frequencies, Astrophys. J. **403** (1993) L75 – L78

Christensen-Dalsgaard, J., Schou, J. and Thompson, M.J.: A comparison of methods for inverting helioseismic data, Mon. Not. R. astr. Soc. **242** (1990) 353 – 369

Christensen-Dalsgaard, J., Larsen, R.M., Schou, J. and Thompson, M.J.: Optimally localized kernels for 2D helioseismic inversion, in *GONG'94: Helio - and astero-seismology* (ed. R.K. Ulrich, E.J. Rhodes Jr and W. Däppen, ASP Conf. Ser., San Francisco) **76** (1995) 70 – 73

Christensen-Dalsgaard, J., Duvall, T.L. Jr, Gough, D.O., Harvey, J.W. and Rhodes, E.J. Jr: Speed of sound in the solar interior, Nature **315** (1985) 378 – 382

Däppen, W., Gough, D.O. and Thompson, M.J.: Further progress on the helium abundance determination, in *Seismology of the sun and sun-like stars* (ed. E.J. Rolfe, ESA SP - 286, Noordwijk) (1988) 505 – 510

Deubner, F. L., Ulrich, R.K. and Rhodes, E.J. Jr: Solar p-mode oscillations as a tracer of radial differential rotation, Astron, Astrophys. (1979) **72** 177 – 185

Duvall, T.L. Jr: A dispersion law for solar oscillations, Nature **300** (1982) 242 – 243

Duvall, T.L. Jr and Harvey, J.W.: Rotational frequency splitting of solar oscillations, Nature **310** (1984) 19 – 22

Duvall, T.L. Jr, Dziembowski, W.A., Goode, P.R., Gough, D.O., Harvey, J.W. and Leibacher, J.W.: The internal rotation of the Sun, Nature **310** (1984) 22 – 25

Dziembowski, W.A., Pamyatnykh, A.A. and Sienkiewicz, R.: Seismological tests of standard solar models calculated with new opacities, Acta Astr. **42** (1992) 5 – 15

Elsworth, Y., Howe, R., Isaak, G.R., McLeod, C.P., Miller, B.A., New, R., Speake, C.C. and Wheeler, S.J.: Solar p-mode frequencies and their dependence on solar activity: recent results from the BiSON network, Astrophys. J. **434** (1994) 801 – 806

Gough, D.O.: Diagnostics of the solar interior, Europhys. News **13** (1982) 3 – 5

Gough, D.O.: Recent advances in helioseismology, in *Proc. Kunming workshop solar phys. and interplanetary travelling phenomena* (ed. B. Chen and C. de Jager, Science Press, Beijing) **1** (1985) 137 – 164

Gough, D.O.: Course 7. Linear adiabatic stellar pulsation, in *Astrophysical fluid dynamics* (ed. J.-P. Zahn and J. Zinn-Justin, Elsevier, Amsterdam) (1993) 399 – 560

Gough, D.O., Kosovichev, A.G., Sekii, T., Libbrecht, K.G., and Woodard, M.F.: The form of the angular velocity in the solar convection zone, *GONG 1992: Seismic investigation of the sun and stars* (ed. T.M. Brown, ASP Conf. Ser., San Francisco) **42** (1993a) 213 – 216

Gough, D.O., Kosovichev, A.G., Sekii, T., Libbrecht, K.G., and Woodard, M.F.: Seismic evidence of modulation of the structure and angular velocity of the Sun associated with the solar cycle, in *Inside the stars* (ed. W.W. Weiss and A. Baglin, Ast. Soc. Pac. Conf. Ser., San Francisco) *Proc. IAU Colloq.* **40** (1993b) 93 – 96

Gough, D.O., Kosovichev, A.G. and Toutain, T.: Constrained estimates of low-degree mode frequencies and the determination of the interior structure of the Sun, Solar Phys. **157** (1995) 1 – 15

Hansen, P.C., Sekii, T. and Shibahashi, H.: The modified truncated svd method for regularization in general form, SIAM J Sci. Stat. Comp. **13** (1992) 1142 – 115

Hill, H.A., Stebbins, R.T. and Brown, T.M.: Recent oblateness observations: Data, interpretation and significance for earlier work, *Atomic masses and fundamental constants* (ed. J.H. Sanders and A.H. Wapstra, Plenum, New York) (1976) 622 – 628

Iglesias, C.A., Rogers, F.J. and Wilson, B.G.: Reexamination of the metal contribution to astrophysical opacities, Astrophys. J. Lett. **322** (1987) L45 – L48

Iglesias, C.A., Rogers, F.J. and Wilson, B.G.: Spin-orbit interaction effects on the Rosseland mean opacity, Astrophys. J. **397** (1992) 717 – 728

Kosovichev, A.G., Christensen-Dalsgaard, J., Däppen, W., Dziembowski, W.A., Gough, D.O., and Thompson, M.J.: Sources of uncertainty in direct seismological measurements of the solar helium abundance, Mon. Not. R. astr. Soc. **259** (1992) 536 – 558

Lanczos, C.: Linear differential equations, (Van Nostrand, London) (1961)

Libbrecht, K.G., Woodard, M.F. and Kaufman, J. M.: Frequencies of solar oscillation, Astrophys, J. Suppl. **74** (1990) 1129 – 1149

Oldenberg, D.W.: Calculation of Fourier transforms by the Backus-Gilbert method, Geophys. J.R. astr. Soc. **44** (1976) 413 – 431

Pijpers, F.P. and Thompson, M.J.: Faster formulations of the optimally localized averages method for helioseismic inversion, Astron. Astrophys. **262** (1992) L33 – L36

Pijpers, F.P. and Thompson, M.J.: The SOLA method for helioseismic inversion, Astron. Astrophys. **281** (1994) 231 – 240

Schou, J. and Brown, T.M.: On the rotation rate in the solar convection zone, Astrophys. J. **434** (1994) 378 – 383

Schou, J., Christensen-Dalsgaard, J. and Thompson, M.J.: On comparing helioseismic two-dimensional inversion methods, Astrophys. J. **433** (1994) 389 – 416

Sekii, T.: On an $IR \otimes IR$ inversion technique for solar rotation, *GONG 1992: Seismic investigation of the sun and stars* (ed. T.M. Brown, ASP Conf., Ser., San Francisco) (1993a) 237 – 240

Sekii, T.: A new strategy for 2d inversion for solar rotation, Mon. Not, R. astr. Soc. **264** (1993b) 1018 – 1024

Sekii, T.: $IR^1 \otimes IR^1$ inversion for solar rotation, Mon. Not, R. astr. Soc., submitted (1995)

Sekii, T., Gough, D.O. and Kosovichev, A.G.: Inversions of BBSO rotational splitting data, *GONG '94 : Helio - and asteroseismology from the earth and space* (ed. R.K. Ulrich, E.J. Rhodes, Jr and W. Däppen, ASP Conf. Ser., San Francisco) **76** (1995) 59 – 62

Spiegel, E.A. and Zahn, J-P.: The solar tachocline, Astron. Astrophys. **265** (1992) 106 – 114

Toutain, T. and Fröhlich, C.: Characteristics of solar p-modes: results from the IPHIR experiment, Astron. Astrophys **257** (1992) 287 – 297

Filtering in Inversion for Solar Internal Structure

Sarbani Basu[1], *J. Christensen-Dalsgaard*[1], *F. Pérez Hernández*[2] *and M.J. Thompson*[3]

[1] Teoretisk Astrofysik Center, Danmarks Grundforskningsfond, Institut for Fysik og Astronomi, Aarhus Universitet, Denmark
[2] Instituto de Astrofísica de Canarias, Tenerife, Spain
[3] Astronomy Unit, Queen Mary and Westfield College, London, UK

1 Introduction

The Sun oscillates in many thousands of different modes. The frequencies of these modes are related to the internal structure of the Sun, in particular to the sound-speed profile. Hence, the solar oscillation frequencies can be inverted to determine the internal structure of the Sun to a very high precision (e.g. Gough & Thompson 1991). Theoretically, both acoustic waves (p modes) and buoyancy waves (g modes) can be present in the Sun. However, only p modes have been definitely detected so far. Modes of stellar oscillation are characterized by the degree l and azimuthal order m of the spherical harmonic which describes the behaviour of the mode over spherical surfaces, as well as the radial order n. The observed solar modes can be described throughout most of the solar interior by the equations describing linear adiabatic oscillations (cf. Unno et al. 1989), as the superposition of acoustic waves, each travelling in a resonant cavity, with a lower turning point at a depth determined by the degree and frequency of the mode and an upper turning point at a depth just beneath the photosphere.

Unlike the Earth, the structure of the Sun is normally assumed to be sufficiently simple that realistic solar models can be computed from first principles. Matter in the Sun behaves like a gas with only modest interactions between the constituents; thus the ideal gas law, with relatively small modifications, provides an adequate description of the thermodynamic state. In much of the solar interior, energy is transported by radiative diffusion, with a diffusion coefficient which may be computed from atomic physics. By contrast, in the convection zone – which occupies the outer ~ 28 per cent by radius – energy is transported by convection and the temperature gradient is essentially adiabatic (except in a very thin region near the top of the convection zone). Finally, the rates of energy generation and nuclear transmutation are determined by nuclear parameters which are either measured or computed with reasonable accuracy. On the basis of this description of the physics of the solar interior, models of the present Sun can be computed by following solar evolution, resulting from the fusion of hydrogen into helium, from an assumed initially chemically homogeneous model to the present solar age. Given such a model, it is straightforward to compute its *adiabatic* oscillation frequencies. By linearising the equations of stellar oscillation around the model, the differences between the structure of the Sun and this model can be determined through inversion of the frequency differences between the Sun and the model. This provides a test of the physics and other assumptions used in the computation. Such procedures have been used extensively to invert for solar structure using the oscillation frequencies (e.g., Gough & Kosovichev 1988; Däppen et al. 1991; Antia & Basu 1994; Dziembowski et al. 1994).

Unfortunately, although the above procedure apparently represents the Sun's structure and the mode dynamics well throughout most of the solar interior, we run into problems near the surface: here uncertain details of convective energy transport and corresponding dynamical effects should be taken into account; also, the assumption of adiabaticity of the oscillations breaks down, with no definite way of treating the nonadiabatic effects being currently available. Thus we are faced with the challenge of solving an inverse problem, without knowing fully how to solve the forward problem.

Fortunately, although we cannot completely solve the realistic forward problem, we can characterize the effect of the outer layers on the mode frequencies. Our aim is then to invent an inversion procedure the results of which are insensitive to uncertainties of the form of those arising from the superficial layers. Comparison of theoretical p-mode frequencies of different models shows that, when appropriately scaled, the frequency differences for low- to moderate-degree modes due to changes in the surface structure are largely a function of frequency (e.g. Christensen-Dalsgaard & Berthomieu 1991). This comes about because such p modes travel almost vertically in the superficial layers of the Sun, regardless of the degree of the mode; hence the signature of these layers on the frequencies depends little on degree, except for an inverse proportionality to the inertia of the modes. (Shallowly penetrating modes have a small inertia and their frequencies are easily perturbed by near-surface changes; while deeply penetrating modes have a large inertia and are relatively insensitive to changes in the surface layers.) Changes in the equilibrium structure of a solar model introduce an oscillatory signal in the frequencies, which for modes of a given degree can be shown to have a 'frequency' which increases with the acoustic depth τ of the layers in which the signals arise (Gough 1990; Christensen-Dalsgaard & Pérez Hernández 1992). Hence, the effect of unknown physics in the surface layers of the Sun is expected to be the introduction of frequency shifts which are slowly varying functions of frequency.

These arguments indicate that the inverse problem of finding the structure of the solar interior can be written as

$$\frac{\delta\omega_i}{\omega_i} = \int K^i_{c^2,\rho}(r)\frac{\delta c^2(r)}{c^2(r)}\mathrm{d}r + \int K^i_{\rho,c^2}(r)\frac{\delta\rho(r)}{\rho(r)}\mathrm{d}r + \frac{F_{\mathrm{surf}}(\omega_i)}{S_i} \ , \qquad (1)$$

where $\delta\omega_i$ is the difference in the frequency ω_i of the ith mode between the solar data and a reference model, and c and ρ are, respectively, sound speed and density in the interior. The kernels $K^i_{c^2,\rho}$ and K^i_{ρ,c^2} are known functions of the reference model which relate the changes in frequencies of adiabatic oscillations to the changes in c and ρ respectively. The term in F_{surf} results from the near-surface errors in the physics; thus we assume that F_{surf} is a slowly varying function of frequency, with S_i being proportional to the inertia of the mode (cf. Christensen-Dalsgaard 1991).

Although techniques have been developed to remove the surface term during inversion, these depend on the technique of inversion used. In the regularised least-squares technique, for instance, the function $F_{\mathrm{surf}}(\omega)$ is obtained simultaneously with the functions $\delta c^2(r)/c^2(r)$ and $\delta\rho(r)/\rho(r)$ in the least-squares fit to the data (cf. Antia & Basu 1994). When using optimally localised averaging or its variants, where the inferred solution is obtained by forming explicitly linear combinations of the data, the surface term has been taken care of by constraining those linear combinations to be insensitive to the presence of slowly varying functions of frequency of the form given in equation (1) (e.g., Däppen et al. 1991;Christensen-Dalsgaard & Thompson 1995). Since the surface term is not treated in the same way, it often becomes very difficult to compare the inversion results. We aim to present a way of filtering out the effects of near-surface uncertainties from the frequency differences, prior to the inversion. Once the surface term is removed, any standard inversion procedure can be applied. Since the method unavoidably also suppresses those components of $\delta\omega_i/\omega_i$ which arise from the actual differences near the surface in structure (i.e., $\delta c^2/c^2$ and $\delta\rho/\rho$ in equation 1), the corresponding contributions from the kernels $K^i_{c^2,\rho}$ and K^i_{ρ,c^2} are also suppressed for consistency. We do this by constructing a filter which removes low-frequency components from the frequency differences while simultaneously suppressing the kernels at the surface.

2 Formulation

If we have a *low-pass* filter $F_{\alpha\beta}$ on a suitable frequency mesh $\{\omega_\alpha\}$, then we can construct a suitable filter $\mathcal{F}$ which can be applied directly to the frequency differences $\delta\omega/\omega$. The details of this procedure can be found in Basu et al. (1996).

By supposition, the low-pass filter has no effect on F_{surf}. Hence, applying the filtering to (1) and subtracting the result from (1) we obtain

$$\sum_j \mathcal{G}_{ij} \frac{\delta\omega_j}{\omega_j} = \int \left[\tilde{K}^i_{c^2,\rho}(r) \frac{\delta c^2(r)}{c^2} + \tilde{K}^i_{\rho,c^2}(r) \frac{\delta\rho(r)}{\rho(r)} \right] \mathrm{d}r \ , \qquad (2)$$

where

$$\mathcal{G}_{ij} = (\delta_{ij} - \mathcal{F}_{ij}) \frac{S_j}{S_i} \ , \quad \text{and, e.g.,} \quad \tilde{K}^i_{c^2,\rho} = \sum_j \mathcal{G}_{ij} K^j_{c^2,\rho} \ . \qquad (3)$$

Hence we achieve the desired consistent filtering.

The filter $F_{\alpha\beta}$ is specially formulated to have a sharp cut-off frequency. The resulting filter $\mathcal{G}$ suppresses all signals arising from the near surface region with $r \geq 0.995\ \mathrm{R}_\odot$. The details of the construction of the filter F was provided by Pérez Hernández & Christensen-Dalsgaard (1994).

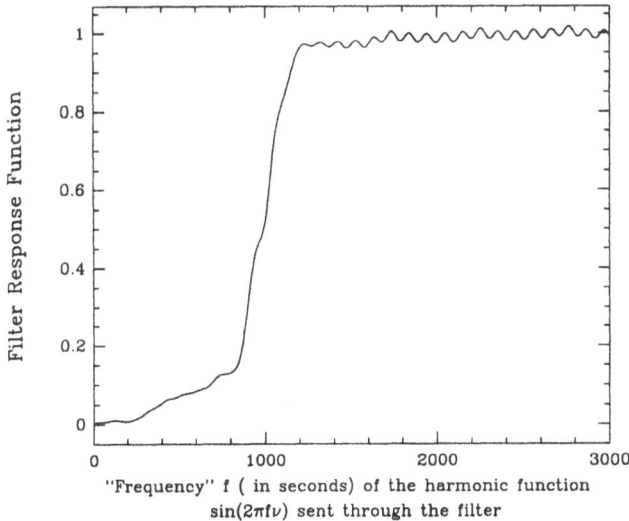

Fig. 1. The frequency response function of the filter $\mathcal{G}$ used in this work, defined as the normalized response of the filter to a harmonic function. Note that the filter has a fairly sharp cut-off frequency

The frequency response function of the high-pass filter $\mathcal{G}$ is shown in Fig. 1. We see that it indeed has a sharp cut-off frequency.

We have used two solar models to test the method. In the test model, taking the role of the unknown Sun, the surface contributions to the frequencies have been artificially increased by modifying the atmosphere, and adding an extra frequency dependent component to the scaled frequencies. We use about 1900 modes for which observed solar frequencies are available. This enables us to use observational errors, although we invert artificial data. Most of the solar frequencies in this range have been measured with an accuracy of a few parts in 10^5. By using the observers' error estimates, we can get a realistic estimate of the accuracy with which we can invert helioseismic data.

3 The Inversion Technique

We use the method of subtractive optimally localised averages (henceforth SOLA; cf. Pijpers & Thompson 1994; Christensen-Dalsgaard & Thompson 1995), to carry out the inversions. Thus we choose coefficients $c_i(r_0)$ in such a way that the linear combination

$$\mathcal{K}_{c^2,\rho}(r_0,r) = \sum_i c_i(r_0) K^i_{c^2,\rho}(r) \tag{4}$$

is a well-localised averaging kernel at target radius r_0. In the SOLA technique this is achieved by minimizing the difference between $\mathcal{K}_{c^2,\rho}(r_0,r)$ and a target kernel $\mathcal{T}(r_0,r)$. We have used target kernels of Gaussian shape. The minimization is carried out subject to the condition that the sum $\mathcal{C}_{c,\rho}(r_0,r) \equiv \sum_i c_i(r_0) K^i_{\rho,c^2}(r)$ and the error in the corresponding combination of the data are small (for details, see Christensen-Dalsgaard & Thompson 1995). In this case the sum $\left(\delta c^2/c^2\right)_{\mathrm{inv}} = \sum_i c_i \delta\omega_i/\omega_i$ provides a well-localised average of the first function $\delta c^2/c^2$. Normally in SOLA, the surface term is removed by putting an additional constraint that $\sum c_i \Phi_\lambda(\omega_i) = 0$ for $\lambda = 1, \ldots, \Lambda$, where Φ_λ is a polynomial of degree λ and Λ is the largest degree of polynomial used. However, if the surface term is removed prior to the inversion, then of course, this condition is not required.

In addition to the inversion results between two solar models, we also show the results of inverting real solar data. The data we used consist of low-degree ($l = 0$ to $l = 3$) frequencies from Elsworth et al. (1994) and data from Libbrecht et al. (1990) on intermediate- and high-degree modes in the range $l = 4$ to $l = 100$.

4 The Results

The effect of filtering the frequency differences is shown in Fig. 2. We can see that filtering dramatically reduces the spread in the scaled frequency differences by removing the smooth frequency-dependent components. The residual frequency differences are almost purely a function of $w = \omega/(\ell+1/2)$, which is a measure of the lower turning-point of the mode. Indeed it is largely this variation with w that the inversion uses to infer the variation with radius of the structural differences. The remaining frequency dependence are the higher frequency components that are not cut off by the filter: these correspond to differences situated close to the surface but beneath the very superficial layers whose effect we have sought to suppress.

The results of the inversion for the sound-speed difference between the two solar models are shown in Fig. 3. We show the results obtained both with and without filtering the data beforehand. As can be seen from the figure, the inversion results after filtering are better than those obtained without filtering.

The vertical error bars are an indication of the error in the inversion due to errors in the data. As can be seen, the errors are very small indeed. The horizontal error bars indicate the radial resolution of the inversion.

In Fig. 4, we show the relative sound-speed difference between the Sun and a reference solar model, in the sense (Sun minus model). The precision with which we are able to deduce the solar sound speed is evident. The reference model used for this inversion had been constructed with up-to-date physics including gravitational settling of helium. Note that the difference in sound speed between the Sun and the model is less than 0.5 % through most of the solar interior. The dip in the inferred difference around 0.7 $R_\odot$ indicates that the model has a slightly deeper convection zone than the Sun.

5 Conclusions

We have shown here that by applying suitable filters to the frequency differences and the mode kernels it is possible to preprocess the frequency differences prior

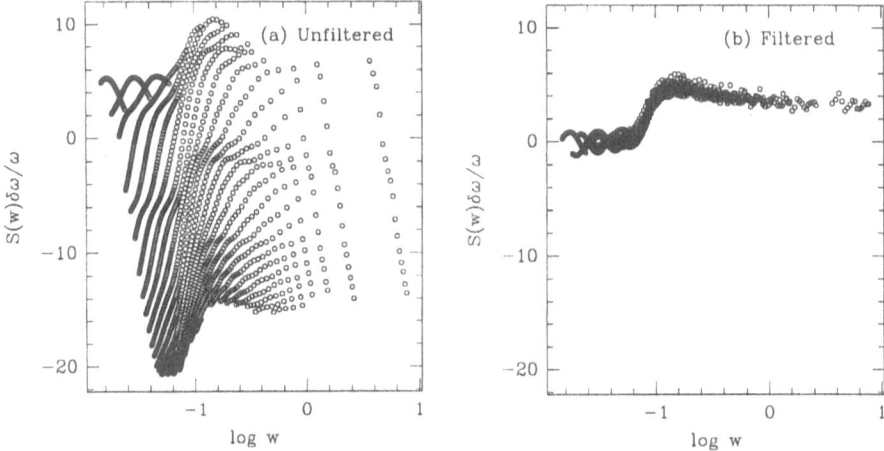

Fig. 2. The scaled relative frequency differences between the reference and test model in the sense test model minus reference model plotted as a function of $w(=\omega/(l+1/2))$. Panel (a) shows the original frequency differences, while panel (b) shows the filtered frequency difference. Note that the filtered scaled frequency differences are almost purely a function of w alone.

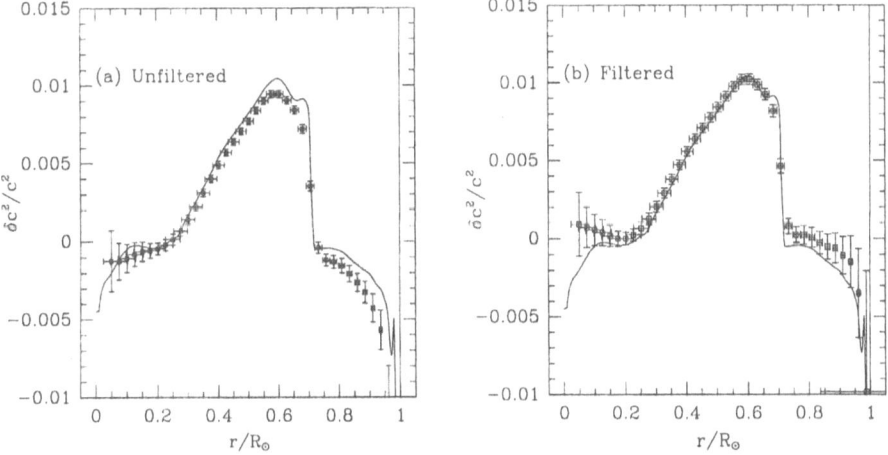

Fig. 3. The sound-speed inversion results for the test model. Panel (a) shows the results of inversion without removing the surface terms from the data, panel (b) shows the inversion results of the filtered data. In each panel the continuous line is the exact relative sound-speed difference between the two models. The points mark inversion results at each target radius. The vertical error-bars indicate the error in inversion due to the errors in the data. The horizontal error-bars are half the distance between the quartile points of the averaging kernel at that radius, and is an indication of the resolution of the inversion.

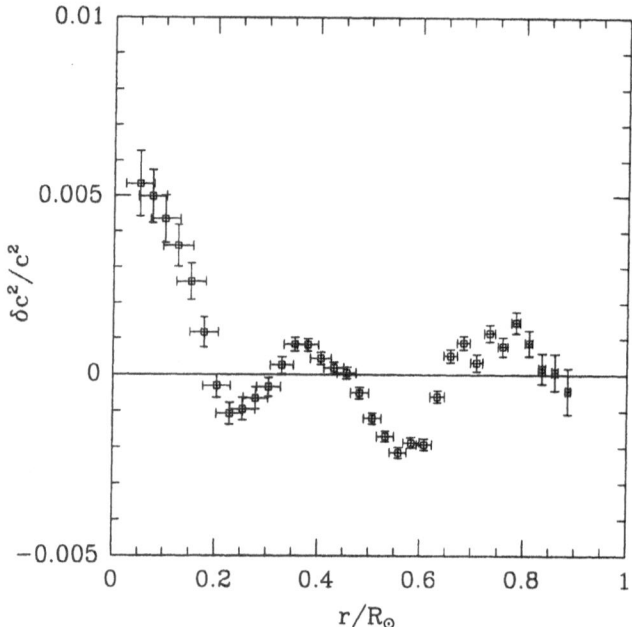

Fig. 4. The sound-speed inversion results for the Sun, relative to a solar model with up-to-date physics. The inversion used a combination of 1899 frequencies from Elsworth et al 1994 ($l = 0$ to $l = 3$) and from Libbrecht et al. 1990 ($l = 4$ to $l = 100$)

to applying any of the standard inversion techniques in order to eliminate the frequency-dependent component which arises from the near surface uncertainties. In the process we also suppress the corresponding contributions from the kernels relating frequency differences to differences in structure. The results of inverting such pre-processed data compare very favourably with those obtained without removing the surface contributions.

Although other ad hoc methods have been developed to handle the surface terms while inverting helioseismic data (e.g., Däppen et al. 1991), we believe that filtering out the surface contribution before doing the inversion provides us with the advantage of being able to compare directly the results obtained by different inversion techniques. The ease with which the filter response can be tuned to the problem at hand makes this approach very flexible.

We have also shown that we are able to invert real solar data to determine the sound speed within the Sun to high precision.

Acknowledgements

This work was supported by the Danish National Research Foundation through its establishment of the Theoretical Astrophysics Center, and by the UK Particle Physics and Astronomy Research Council through grant GR/K09526.

References

Antia H. M., Basu S., 1994, Nonasymptotic helioseismic inversion for solar structure, *Astron. Astrophys. Suppl.*, **107**, 421

Basu S., Christensen-Dalsgaard J, Pérez Hernández F., Thompson, M.J., 1996, Filtering out near-surface uncertainties from helioseismic inversions, *Mon. Not. R. Astron. Soc.*, in press

Christensen-Dalsgaard J., 1991, Solar oscillations and the physics of the solar interior, in Gough D. O. and Toomre J., eds, *Lecture Notes in Physics*, **388**, Springer, Heidelberg, p11

Christensen-Dalsgaard J., Berthomieu G., 1991, Theory of solar oscillations, in Cox A. N., Livingstone W. C. and Matthews M., eds, *Solar Interior and Atmosphere, Space Science Series*, University of Arizona Press, p401

Christensen-Dalsgaard J., Pérez Hernández F., 1992, Phase-function differences for stellar acoustic oscillations – I. Theory, *Mon. Not. R. Astron. Soc.*, 257, 62

Christensen-Dalsgaard J., Thompson M. J., 1995, Sola inversions for the radial structure of the Sun, in Ulrich R.K., Rhodes Jr. E.J., Däppen W., eds, *GONG'94: Helio- and Astero-seismology from Earth and Space*, PASPC, **76**, 144

Däppen W., Gough D. O., Kosovichev A. G., Thompson M. J., 1991, A new inversion for the hydrostatic stratification of the Sun, in Gough D. O., Toomre J., eds, *Lecture Notes in Physics*, **388**, Springer, Heidelberg, p111

Dziembowski W. A., Goode P. R., Pamyatnykh A. A., Sienkiewicz R., 1994, Seismic model of the sun's interior, *Astrophys. J.*, **432**, 417

Elsworth Y., Howe R., Isaak G.R., McLeod C.P., Miller B.A., New R., Speake, C. C., Wheeler S.J. 1994, Solar p-mode frequencies and their dependence on solar activity: recent results from the BISON network, *Astrophys. J.*, **434**, 801

Gough D. O., 1990, Comments on helioseismic inference, in Osaki Y., Shibahashi H., eds., *Lecture Notes in Physics*, **367**, Springer, Berlin, p.283

Gough D. O., Kosovichev A. G., 1988, An attempt to understand the Stanford p-mode data, in Domingo V., and Rolfe E. J., eds., *Seismology of the Sun and Sun-like Stars*, ESA SP-286, Noordwijk, p195

Gough D. O., Thompson M. J., 1991, The inversion problem, in Cox A. N., Livingston W. C., Matthews M., eds, *Solar interior and atmosphere*, Space Science Series, University of Arizona Press, p. 519

Libbrecht K. G., Woodard M. F. & Kaufman J. M. 1990, Frequencies of solar oscillation, *Astrophys. J. Suppl.*, **74**, 1129

Pérez Hernández F. & Christensen-Dalsgaard J., 1994, The phase function for stellar acoustic oscillations. II. Effects of filtering, *Mon. Not. R. Astron. Soc.*, **267**, 111

Pijpers F.P., Thompson M.J., 1994, The SOLA method for helioseismic inversion, *Astron. Astrophys.*, **281**, 231

Unno W., Osaki Y., Ando H., Shibahashi H., 1989, *Nonradial Oscillations of Stars*, 2nd Edition, University of Tokyo Press, Tokyo

Averaging kernels, error correlation functions and linear methods in helioseismology

Rachel Howe and Michael J. Thompson

Astronomy Unit, Queen Mary and Westfield College, London, UK

1 Introduction

Helioseismology is the study of the interior of the Sun using the observed properties of the Sun's global normal modes of oscillation. In particular, the frequencies
of the oscillations depend on the radial stratification of the solar interior and on
its rotation and can therefore be used in inverse analyses to make inferences
about these aspects of the Sun. General accounts of helioseismology may be
found in the review articles by Deubner & Gough (1984), Christensen-Dalsgaard
et al. (1985) and Gough & Toomre (1991). More details of helioseismic inverse
problems are presented by Gough (1985), Christensen-Dalsgaard et al. (1990),
Schou et al. (1994) and Thompson (1995).

Much use is made in helioseismology of linear inversion techniques. Various
tools – averaging kernels, inversion coefficients, error correlation functions – can
be used to understand the results, aid the choice of trade-off parameter values,
and compare methods. One frequently hears the argument that one should use
many different techniques to invert a given dataset, because they can reveal
different aspects of the true solution. Interestingly, in the helioseismic context,
it has been found using the above tools that seemingly quite different methods
are very similar in terms of their use of the data and the way they sample the
true solution (Christensen-Dalsgaard et al. 1990; Christensen-Dalsgaard et al.
1993)). In this paper, we discuss briefly some aspects of the linear inversions,
illustrated with a prototypical helioseismic application.

A global (so-called "spheroidal") mode of oscillation of a spherically symmetric star is described by three quantum numbers: the radial order n, the degree
l and the azimuthal order m. For such a mode, the horizontal dependence of
the vertical displacement (for example) is given by a surface harmonic function
$Y_l^m(\theta, \phi)$. Here (r, θ, ϕ) are spherical polar coordinates about the centre of the
star. In the spherically symmetric star, the frequencies ω_{nlm} of the oscillations
are independent of quantum number m. Departures from spherical symmetry,
such as are found in real stars, raise this degeneracy. The dominant symmetry-
breaking agent in the Sun is rotation, which leads to so-called rotational splitting
of frequencies with the same values of n and l but different values of m. We may

define the splitting to be

$$\Delta\omega_{nlm} = \omega_{nlm} - \omega_{nl0} . \tag{1}$$

The simplest case is that of a rotation rate $\Omega(r)$ which depends only on the radial coordinate r. This provides a useful prototypical helioseismic example with which to illustrate this paper. In this case, the splitting is simply proportional to m:

$$\Delta\omega_{nlm} = m \int_0^R K_{nl}(r)\Omega(r)\,\mathrm{d}r . \tag{2}$$

The mode kernels $K_{nl}(r)$ are functions that depend on the structure of the star, but will be assumed here to be known exactly; and R is the radius of the surface of the Sun. In this simplest case, we take out the factor of m to define a new quantity d_{nl}:

$$d_{nl} \equiv m^{-1}\Delta\omega_{nlm} = \int_0^R K_{nl}(r)\Omega(r)\,\mathrm{d}r . \tag{3}$$

These will be taken to be the basic data to which we apply our inversion techniques.

2 Linear inversion techniques

To simplify the notation, we replace the double suffix in equation (3) with a single index i running over the number M of data in our data set:

$$d_i = \int_0^R K_i(r)\Omega(r)\,\mathrm{d}r + \epsilon_i . \tag{4}$$

We have now explicitly acknowledged that the data contain errors, ϵ_i, which are presumed to be independent and Gaussian distributed, each with zero mean and standard deviation σ_i.

To be concrete, we consider the outcome of the inversion to be an estimate $\bar{\Omega}(r_0)$ of the rotation at selected radii r_0: we shall think of $\bar{\Omega}$ as the 'solution' of the inversion procedure. However, what we say would also be applicable to estimators of other linear functionals of the rotation rate.

A linear inversion is one in which the solution is a linear combination of the data, that is, explicitly or otherwise, there exist *inversion coefficients* $c_i(r_0)$ such that

$$\bar{\Omega}(r_0) = \sum_{i=1}^{M} c_i(r_0)d_i . \tag{5}$$

It follows from equations (4) and (5) that the solution and true rotation rate are linearly related according to

$$\bar{\Omega}(r_0) = \int_0^R \mathcal{K}(r_0,r)\Omega(r)\,\mathrm{d}r + \sum_{i=1}^{M} c_i(r_0)\epsilon_i , \tag{6}$$

where the *averaging kernel* $\mathcal{K}(r_0, r)$ is defined by

$$\mathcal{K}(r_0, r) = \sum_{i=1}^{M} c_i(r_0) K_i(r) . \tag{7}$$

The averaging kernel describes completely the resolution achieved by the inversion at radius r_0, while the last term in equation (6) shows how data errors propagate through to the solution. Indeed, we shall refer to this term as the error in the solution: it is the deviation of $\bar{\Omega}(r_0)$ from the value it would take in the absence of data errors. Of course, in general this error is not equal to $\Omega(r_0) - \bar{\Omega}(r_0)$ because of the limited resolution of the inversion – i.e. the averaging kernels are not delta functions.

The errors $\sum c_i \epsilon_i$ at two points r_0 and r_1 will not be independent, in general. The correlation between the errors in $\bar{\Omega}(r_0)$ and $\bar{\Omega}(r_1)$ can be described by the *normalized error correlation function* $C(r_0, r)$:

$$C(r_0, r_1) = \frac{E\left[\left(\sum c_i(r_0)\epsilon_i\right)\left(\sum c_j(r_1)\epsilon_j\right)\right]}{\left\{E\left[\left(\sum c_i(r_0)\epsilon_i\right)^2\right] E\left[\left(\sum c_j(r_1)\epsilon_j\right)^2\right]\right\}^{1/2}} , \tag{8}$$

where $E[\cdots]$ denotes the expectation value.

To date, all the inversion techniques commonly used in helioseismology have been linear methods. We select two methods to illustrate the above concepts. One is the Subtractive Optimally Localized Averages method (SOLA) of Pijpers & Thompson (1992), which is based on the work of Backus & Gilbert (1968, 1970). In SOLA, one explicitly constructs an averaging kernel to match a chosen target form $\mathcal{T}(r_0, r)$, by finding inversion coefficients c_i to minimize

$$\mathrm{R} \int_0^R [\mathcal{K}(r_0, r) - \mathcal{T}]^2 \, dr + \tan\theta \sum_i c_i^2 \sigma_i^2 / \bar{\sigma}^2 , \tag{9}$$

where $\bar{\sigma}$ is an average standard deviation defined by

$$\bar{\sigma}^2 = \frac{1}{M} \sum_{i=1}^{M} \sigma_i^2 . \tag{10}$$

The value of the trade-off parameter θ is chosen to weight the opposing aims of minimizing the propagated error and minimizing the mismatch between the kernel and the chosen target form. In this paper we use targets that are Gaussian:

$$\mathcal{T} \propto e^{-(r-r_0)^2/\Delta(r_0)^2} . \tag{11}$$

The width $\Delta(r_0)$ is another adjustable parameter, and the constant of proportionality is such as to make the target unimodular. Our second method is Regularized Least Squares (RLS), in which one makes a discretized approximation to the solution and then minimizes the chi-squared plus a penalty function.

Specifically, we shall use a discretized approximation to the second derivative in our penalty function, so we minimize

$$\sum_{i=1}^{M} \left(\frac{d_i - \int_0^R K_i \bar{\Omega}\, dr}{\sigma_i} \right)^2 + \lambda^2 \bar{\sigma}^{-2} R^3 \int_0^R \left(\frac{d^2 \bar{\Omega}}{dr^2} \right)^2 dr \,, \qquad (12)$$

where λ^2 is another trade-off parameter.

3 A helioseismic example

As an example, we apply the above ideas to a mode set (Set 1 of Christensen-Dalsgaard et al. 1990) comprising 834 p modes with frequencies in the range $2000 - 4000\,\mu$Hz and with degrees in the range $1 - 200$. This is a small mode set, but representative of the kind of modes that are observable on the Sun. For simplicity (but somewhat unrealistically), the standard deviations σ_i for all the data errors are taken to be equal. RLS solutions have been computed on a 100-point radial mesh. SOLA solutions have been computed at 50 evenly spaced target radii. To guide the eye, continuous curves have been used to join SOLA results at different radii.

Fig. 1 shows shows averaging kernels and error correlation functions at selected radii for a SOLA inversion. [Since the attainable resolution at different locations scales roughly with the adiabatic sound speed (Thompson 1993), we choose the target widths Δ at different radii by specifying the target width at $r = 0.5R$ and then scaling appropriately to other radii: thus the widths are smaller near the surface and larger near the centre.] Fig. 1a shows that except at the smallest radii (and in fact also close to the surface), this mode set permits one to construct fairly well-localized averaging kernels. The correlation functions in Fig. 1b indicate that the errors at neighbouring points are positively correlated, as one would expect. As remarked by Barrett (1994; see also Howe & Thompson 1996), the width of the region over which the errors are positively correlated is similar to the width of the main peak of the averaging kernel: thus propagated data errors and unresolved sharp features in the underlying rotation rate can give rise to signatures in the solution which possess similar length scales. Few of the modes in our set are sensitive to the region near the centre of the Sun, and hence they place only weak constraints on this region. This is reflected both in the poorer averaging kernel at $r_0 = 0.1R$ and in the high error correlation in the region $r_0 \lesssim 0.1R$.

In Fig. 2, we compare averaging kernels and error correlation functions at two selected radii for an RLS inversion and for two SOLA inversions with different trade-off parameters. The parameters in all cases were chosen so that the propagated error was similar at the two radii for all three inversions. At a given radius, the averaging kernels are similar in all three cases, confirming the finding of Christensen-Dalsgaard et al. (1990) that different inversions seem to make similar trade-offs between error and resolution. Characteristically, the RLS averaging kernels have negative sidelobes adjacent to the main positive peak (cf.

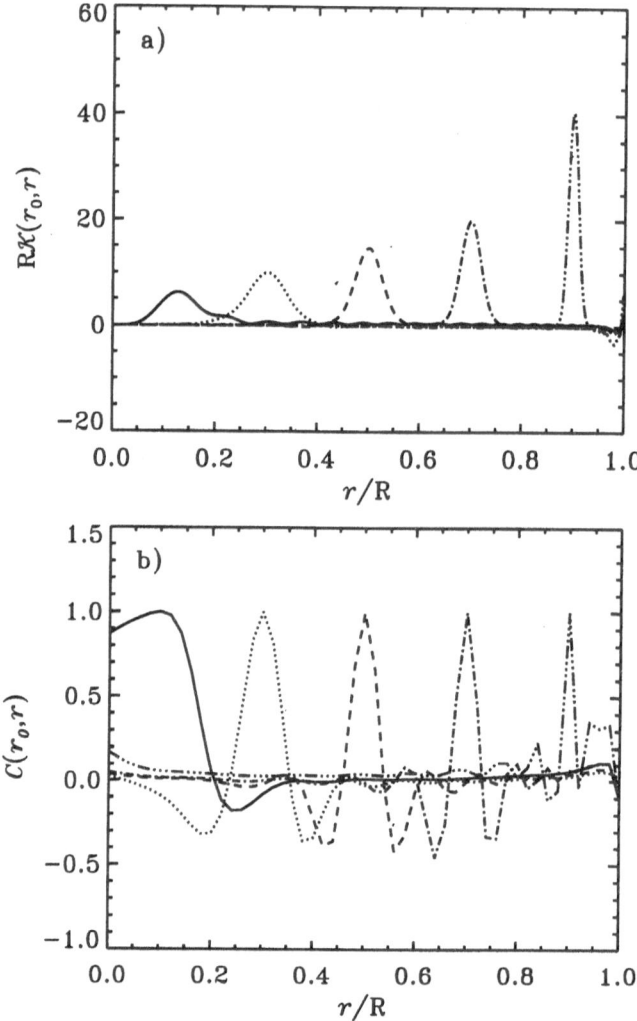

Fig. 1. (a) Averaging kernels at target radii $r_0/R = 0.1$, 0.3, 0.5, 0.7, 0.9, for a SOLA inversion with parameters $\theta = 0.3$, $\Delta(0.5R) = 0.035R$. (b) Normalized error correlation functions for the same SOLA inversion and radii r_0 as in panel (a).

Christensen-Dalsgaard et al. 1990, Thompson 1995). The situation is somewhat different with the error correlation functions. At the smaller radius, all three correlation functions have a positive peak with adjacent regions where the errors are negatively correlated. However, while for one choice of SOLA parameters the correlation function is quite similar to the RLS correlation function, for a less felicitous choice the SOLA correlation function has a substantial positive 'background component' that means that the error at points near the surface

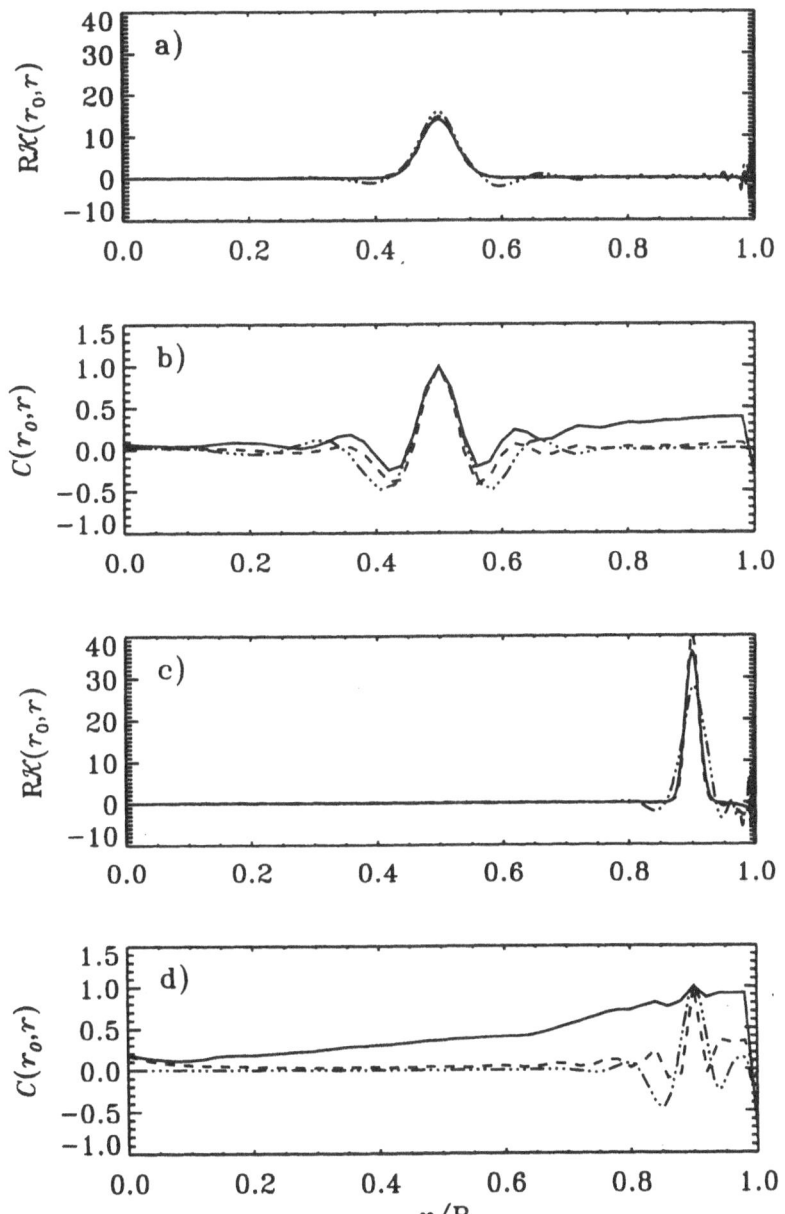

Fig. 2. Averaging kernels and normalized error correlation functions for two SOLA inversions with $\theta = 0.01$, $\Delta(0.5R) = 0.04R$ (solid) and with $\theta = 0.3$, $\Delta(0.5R) = 0.035R$ (dashed), and for an RLS inversion with $\lambda^2 = 5 \times 10^{-7}$ (triple-dot dashed). (a) Averaging kernels for $r_0 = 0.5R$. (b) Error correlation functions with $r_0 = 0.5R$. (c) Averaging kernels for $r_0 = 0.9R$. (d) Error correlation functions with $r_0 = 0.9R$.

is strongly correlated to the error at other radii. As we show elsewhere (Howe
& Thompson 1996), this comes about because at all target radii the inversion
is making similar use of shallowly penetrating high-degree modes to eliminate
near-surface structure in the averaging kernels. We conclude from this that it
is important to examine the error correlation function as well as the averaging
kernels when assessing the quality of an inversion.

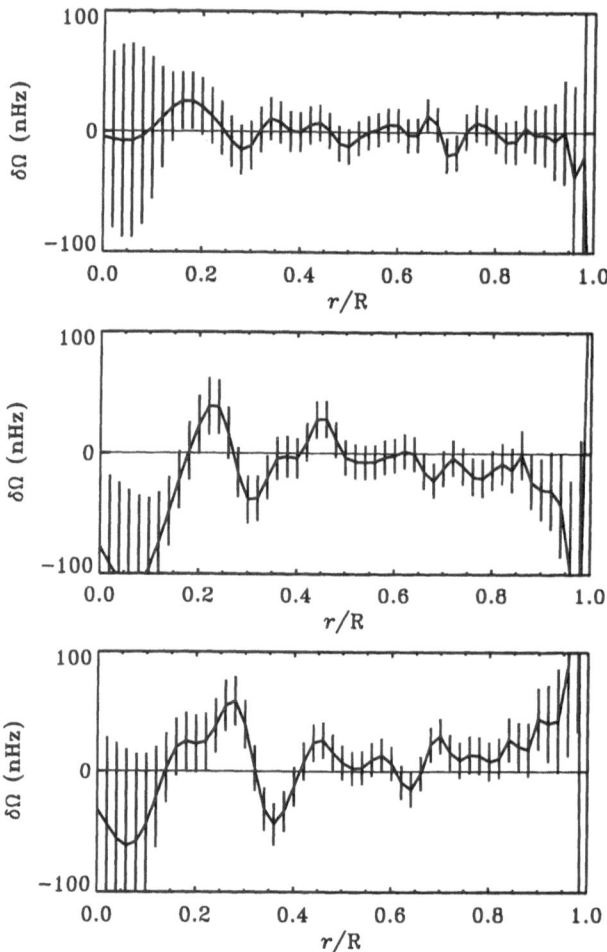

Fig. 3. The difference $\delta\Omega$ between a SOLA inversion of noisy data and the same inversion of noise-free data, for three realizations of data noise. The SOLA parameters were
$\theta = 0.01$, $\Delta(0.5\text{R}) = 0.04\text{R}$. The data errors were independent, Gaussian distributed
with zero mean and standard deviation $10\,\text{nHz}$.

Finally, we generate three random noise realizations (independent identically
distributed Gaussian data errors with zero mean and standard deviation $\sigma_i =
10\,\text{nHz}$) and compute the difference $\delta\Omega \equiv \sum c_i \epsilon_i$ between the inversion solution
for each noisy data set and the solution with noise-free data. The errors $\delta\Omega$ for

for the three realizations are illustrated in Fig. 3. Also shown are ± 1 σ formal error bars at each target r_0, the half-length of each bar being $\sqrt{(\sum c_i^2 \sigma_i^2)}$. The errors appear to be consistent with the error bars, as they should be. In some regions $\delta\Omega$ deviates systematically from zero, as indeed one expects for correlated solution errors: the length-scale of these regions is roughly the width of the positive peak in the correlation function. Clearly neighbouring points are highly correlated: this over-sampling not only increases the computational expense but may also lead to the over-interpretation of apparent features. Near the surface one sees systematic trends in the errors in two out of the three simulations: this observation is consistent with the correlation functions in Fig. 2 (solid lines).

Inversion coefficients and averaging kernels are widely used in helioseismology: we suggest that the error correlation function is also a useful diagnostic. In particular, it should ideally be taken into account when choosing trade-off parameters.

References

Backus, G. and Gilbert, F., 1968. The resolving power of gross Earth data. Geophysical Journal of the Royal Astronomical Society, 16, 169 – 205.

Backus, G. and Gilbert, F., 1970. Uniqueness in the inversion of inaccurate gross Earth data. Philosophical Transactions of the Royal Society of London, Series A, 266, 123 – 192.

Barrett, R.K., 1994. Ph.D. thesis, University of Glasgow.

Christensen-Dalsgaard, J., Gough, D.O. and Toomre, J., 1985. Seismology of the Sun. Science, 229, 923 – 931.

Christensen-Dalsgaard, J., Hansen, P.C. and Thompson, M.J., 1993. Generalized singular value decomposition analysis of helioseismic inversions. Monthly Notices of the Royal Astronomical Society, 264, 541 – 564.

Christensen-Dalsgaard, J., Schou, J. and Thompson, M.J., 1990. A comparison of methods for inverting helioseismic data. Monthly Notices of the Royal Astronomical Society, 242, 353 – 369.

Deubner, F.-L. and Gough, D.O., 1984. Helioseismology: Oscillations as a diagnostic of the solar interior. Annual Reviews of Astronomy & Astrophysics, 22, 593 – 619.

Gough, D.O., 1985. Inverting helioseismic data. Solar Physics, 100, 65 – 99.

Gough, D.O. and Toomre, J., 1991. Seismic observations of the solar interior. Annual Reviews of Astronomy & Astrophysics, 29, 627 – 685.

Howe, R. and Thompson, M.J., 1996. On the use of the error correlation function in helioseismic inversions. Monthly Notices of the Royal Astronomical Society, in press.

Pijpers, F.P. and Thompson, M.J., 1992. Faster formulations of the optimally localized averages method for helioseismic inversion. Astronomy & Astrophysics, 262, L33 – L36.

Schou, J., Christensen-Dalsgaard, J. and Thompson, M.J., 1994. On comparing helioseismic two-dimensional inversion methods. Astrophysical Journal, 433, 389 – 416.

Thompson, M.J., 1993. Seismic investigation of the Sun's internal structure and rotation. In: T.M.Brown (ed.), Proc. GONG 1992: Seismic investigation of the Sun and stars, Astronomical Society of the Pacific Conference Series vol. 42, 141 – 154.

Thompson, M.J., 1995. Linear inversions for the Sun's internal rotation. Inverse Problems, 11, 709 – 730.

Inversion for the Velocity Field in the Solar Interior

Jesper Schou

Stanford University, HEPL, California, USA

1 Introduction

The Sun is oscillating in a large number of normal modes. Most of the modes observed are acoustic waves, but surface gravity modes are also observed. These modes are described by 3 'quantum' numbers: the radial order n, the degree l and the azimuthal order m, with $|m| \leq l$. The angular dependence of the mode eigenfunctions is determined by l and m, while the radial dependence is determined by n and l. If the Sun were spherically symmetric the mode frequencies $\omega_{nlm} = 2\pi\nu_{nlm}$ would not depend on m. Asphericities such as rotation lift this degeneracy. Modes have been observed with l from 0 to several thousand, n from 0 to about 40 and frequencies between $\approx 1mHz$ and $\approx 10mHz$ with a peak at $\nu \approx 3mHz$. The modes have been observed with several different instruments. Most instruments are only able to observe a limited range of l and have observed for a limited time. In the future we expect high quality observations over a large range of l for long periods from instruments such as GONG (a ground based network) and SOI/MDI on the SOHO spacecraft launched in 1995.

Here I shall describe how inverse methods have been used to infer the rotation rate in the solar interior from mode frequencies.

2 The Forward Problem

The frequency shift caused by rotation is given by

$$2\pi\Delta\nu_{nlm} = \Delta\omega_{nlm} = \omega_{nlm} - \omega_{nl} = \int_0^\pi \int_0^1 K_{nlm}(r,\theta)\Omega(r,\theta)rdrd\theta \quad (1)$$

where r is the fractional radius, θ is the co-latitude, ω_{nl} is the unperturbed mode frequency and K_{nlm} is an (assumed) known non-negative function which can be found as described in Schou et al. 1994. For high-l

$$K_{nlm}(r,\theta) \simeq mP_l^m(\cos\theta)^2 \sin\theta\rho(r)rI_{nl}^{-1}\left(\xi_{nl}(r)^2 + \eta_{nl}(r)^2\right), \quad (2)$$

where P_l^m is an associated Legendre function and I_{nl} a normalization factor. Examples of the radial and latitude dependence are shown in Fig. 1, as can be seen the kernels vanish rapidly below the so-called radial turning point r_t and below the turning point in co-latitude θ_t and are oscillatory above these points.

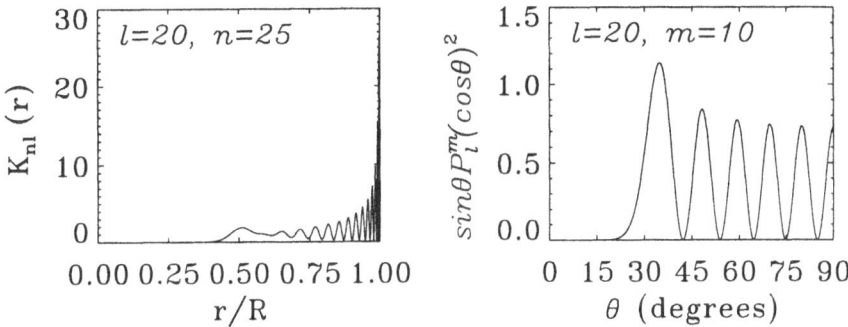

Fig. 1. Examples of kernels given by Eq. 2. The left plot shows an exact latitude averaged kernel used in Eq. 1. The right plot shows an example of $\sin\theta P_l^m(\cos\theta)^2$.

3 Solution of the Inverse Problem

In the 1D case (latitude independent rotation) the number of splittings is of the order a few thousand and the kernels can be described adequately using a discretization with, say 1000 points. These low numbers make it possible to use a number of inversion methods such as regularized least squares (RLS), Backus-Gilbert (known as (subtractive/multiplicative) optimally localized averages) and asymptotic methods. In the 2D case, the number of splittings is of the order 10^5 to 10^6 and the necessary number of mesh points is of the order 10^5. This and the fact that the problem can not be separated in radius and latitude means that a brute force implementation of most of the 1D methods is not feasible.

At least 3 different methods have been used for the 2D case:
1. Expand the latitude dependence of Ω in polynomials in $x = \cos\theta$ and the m dependence of the splittings in polynomials in m. These methods have been dubbed 1.5-dimensional methods (Brown, et al. 1989, Schou, et al. 1994).
2. So-called $1\otimes1$ methods (Sekii 1993) in which one separates radius and latitude in a way similar to Eq. 2.
3. A full two dimensional RLS inversion. (Schou et al., 1994.)

I will briefly describe method 3. In the current implementation we minimize

$$\sum_i \left(\frac{\int K_i\bar{\Omega}r\mathrm{d}r\mathrm{d}\theta - \Delta\omega_i}{\sigma_i}\right)^2 + \mu_r \int \left(\frac{\partial^2\bar{\Omega}}{\partial r^2}\right)^2 f_r\mathrm{d}\theta\mathrm{d}r + \mu_\theta \int \left(\frac{\partial^2\bar{\Omega}}{\partial\theta^2}\right)^2 f_\theta\mathrm{d}\theta\mathrm{d}r ,$$

$$(3)$$

here i stands for the combination nlm; σ_i is the standard deviation of $\Delta\omega_i$, $\bar{\Omega}$ is the inferred rotation rate, μ_r and μ_θ are trade-off parameters and f_r and f_θ are (non-negative) weight functions. The rotation rate is described as a bilinear function on a rectangular grid. Unfortunately it is not feasible to use a sufficient number of points to adequately resolve the kernels in both radius and latitude. Our solution to this problem has been to use a fine grid for calculating the integrals of the basis functions with the kernels and a coarser grid for the solution. The effect of the inadequate resolution is to add another regularization term to Eq. 3. So far, however, the signal to noise ratio has not been high enough to make use of the potential resolution of the kernels.

The choice of the trade-off parameters and weight functions determines the trade-off between the resolution and the effects of input errors on the inferred rotation rate. For the results shown here f_r has been chosen to be proportional to r and f_θ to be proportional to r^{-4}. I will not discuss how to choose the trade-off parameters in this poster. In addition to changing the regularization parameters it is possible to use other functional forms for the regularization such as first derivatives or mixed terms in radius and latitude.

Figure 2a shows a rotation rate used for generating artificial data. An inversion of data generated using this rotation rate is shown in Figs. 2b and 2c. The modeset used had modes with $1 \le l \le 200$ and frequencies between $1mHz$ and $4mHz$ representing what we expect to have in the near future from the GONG project. The errors were estimated theoretically assuming 1 year of continuous observations. This modeset had 153217 splittings and we used 50 intervals in radius and 24 in latitude to describe the inferred rotation rate (we have used more than twice the number of points in each direction for some inversions).

The difference between Figs. 2b and 2c indicates the magnitude of the errors, which have not been shown separately. These errors only represent the effect of the errors on the input splittings, they are *not* an estimate of the difference between the true and inferred rotation rates, which can be substantially larger due to rapid unresolved variations in the true rotation rate. Also the errors on different points of the inferred rotation rate can be highly correlated, which is generally ignored (see, however, Schou 1991).

As in any linear inversion method *averaging kernels* $\mathcal{K}$ exist such that

$$\bar{\Omega}(r_0, \theta_0) = \int_0^R \int_0^\pi \mathcal{K}(r_0, \theta_0; r, \theta) \Omega(r, \theta) r \mathrm{d}r \mathrm{d}\theta , \qquad (4)$$

plus an error term. These averaging kernels (the resolution matrix) describes how the inferred rotation rate is actually a weighted integral of the true rotation rate. Examples are shown in Fig. 2d.

4 Some Real Results

The result of an inversion of a real dataset is shown in Fig. 3. These results are from an instrument called LOWL (Tomczyk 1995), designed to cover all l's from 0 to 80. Most earlier instruments have not been able to cover both the lowest l's and higher l's, making it necessary to combine different datasets (possibly leading to systematic errors) to infer the rotation rate in the solar core. The inversion shown here used 2119 'splittings' from 716 (n, l) multiplets. Instead of using real

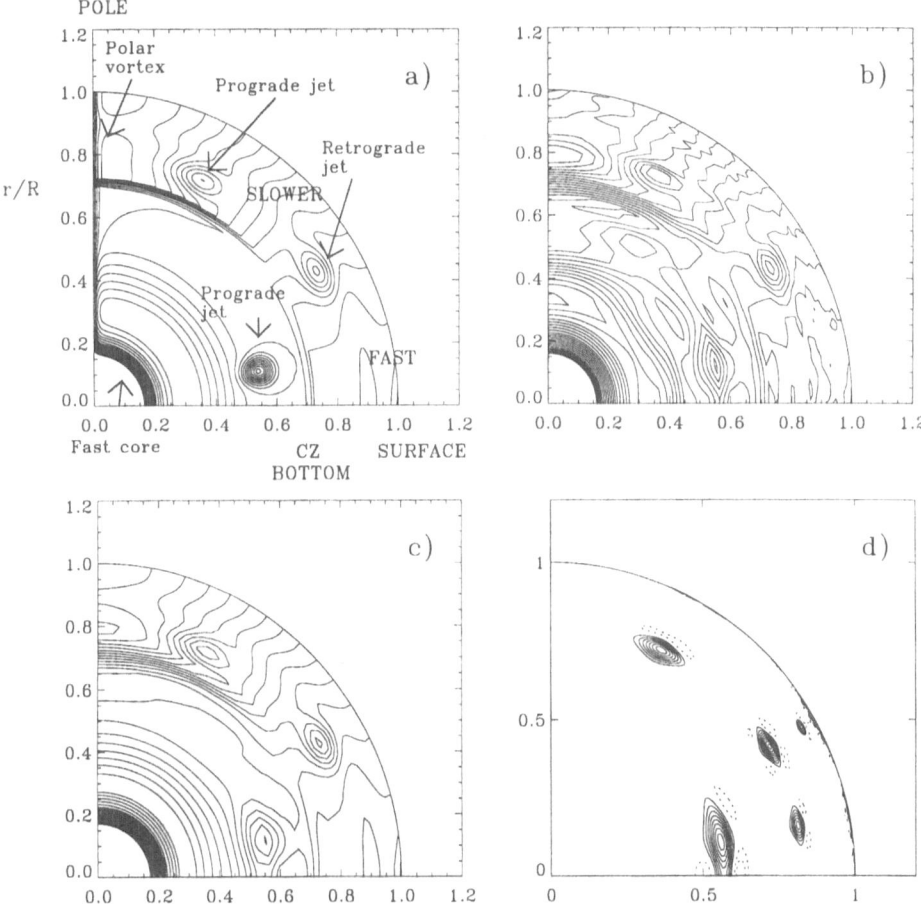

Fig. 2. (a) Artificial rotation profile. (b) Recovered solution with data containing random noise appropriate to one year's observations. (c) Recovered solution for noise-free data, but with the same data uncertainties assumed as for b. In a–c, the contour spacing is 10 nHz. (d) Averaging kernels at different target radii and latitudes. Solid contours are positive, broken contours are negative. For clarity the zero contour is omitted.

splittings the variation of mode frequency with m was expanded in polynomials and the inversion was performed using the coefficients in the polynomial fits. These results are in general agreement with those found previously, except that they go down to 0.2 unlike most earlier inversions which only went to 0.4.

The most notable features of the inferred rotation rate are the near discontinuity at the bottom of the convection zone (at ≈ 0.7) and a lack of significant radial variation of the rotation rate in the convection zone which is in sharp contrast to the results expected theoretically. In the radiative interior below the convection zone the Sun appears to rotate as a solid body.

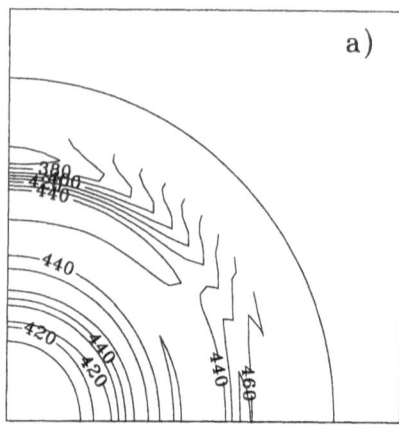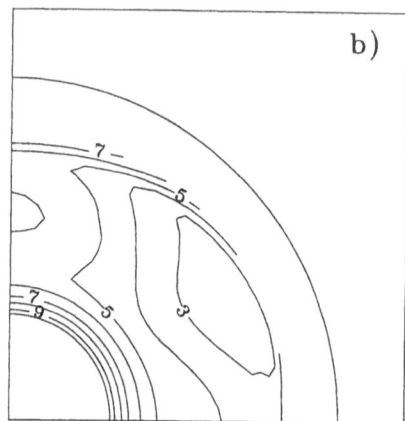

Fig. 3. (a) The solar rotation rate between $r = 0.2R_\odot$ and $r = 0.85R_\odot$, as inferred using the 2D RLS technique with $\mu_r = 10^{-7}$, $\mu_\theta = 10^{-3}$. The contour spacing is 10 nHz. (b) The corresponding 1-σ error levels (values are in nHz).

5 Discussion

As mentioned earlier the kernels K_{nlm} are assumed known. Fortunately the main effect of the uncertainties in the solar structure, from which the kernels are calculated, is to distort the radial scale slightly, which is fairly harmless.

Correlated errors are also a problem. Unfortunately little has been done to estimate the magnitude of the off diagonal elements of the covariance matrix (see Schou, et al. 1995 for some estimates) or their effect on the inversions.

The method described here is, as mentioned, only one of several that have been applied to helioseismic data and the estimation of the rotation rate is only one of several properties that have been investigated using inverse methods. Most of the methods applied have been rather 'traditional' in the sense that they have been RLS or Backus-Gilbert like. This has been due to a combination of computational problems, lack of good constraints (prior information) to apply and limited data quality. Some attempts to set bounds on the pointwise values of the rotation rate or certain averages have also been made. In particular the robustness of the conclusion that the rotation rate in the solar convection zone is constant on radii and that the latitude gradient disappears at the bottom of the convection zone has been the subject of several studies. (eg. Gough, et al. 1993, Schou and Brown 1994, and Genovese et al. 1995). A problem has been that we do not have tight limits on the rotational velocities based on physical considerations. Among the constraints imposed have been that the rotation speed should be less than the speed of light and (less strict) that the matter should be gravitationally bound. Another problem has been the expansion of the mode frequencies in polynomials which turns out to significantly limit how tightly the rotation rate can be constrained. In the future we expect to have significantly more terms in the polynomials or individual frequency splittings.

We are also working on developing other inversion methods and improving on the speed of some of the presently used methods. Despite the simplifications already made the algorithms are extremely memory and cpu time intensive. We are looking in to such options as preprocessing the data and using more complex discretizations than a rectangular grid.

When a new generation of instruments such as GONG and SOI/MDI start producing data we expect to obtain significantly more accurate splitting measurements over a substantial range in l and frequency, thereby vastly improving our knowledge of the solar interior.

A substantial part of the work described in this poster was done in collaboration with a number of people, in particular Jørgen Christensen-Dalsgaard, Mike Thompson and Steve Tomczyk.

References

Brown, T. M., Christensen-Dalsgaard, J., Dziembowski, W. A., Goode, P., Gough, D. O. and Morrow, C. A., 1989. Inferring the Sun's Internal Angular Velocity from Observed p-Mode Frequency Splittings. Astrophysical Journal, 343, 526-546.

Genovese, C. R., Stark, P. B. and Thompson, M. J., 1995. Uncertainties for Two-Dimensional Models of Solar Rotation from Helioseismic Eigenfrequency Splitting. Astrophysical Journal, 443, 843-854.

Gough, D. O., Kosovichev, A. G., Sekii, T., Libbrecht, K. G., and Woodard, M. R. 1993. The Form of the Angular Velocity in the Solar Convection Zone. In: T. M. Brown (ed.), GONG 1992: Seismic Investigation of the Sun and Stars, Astronomical Society of the Pacific Conference Series, Volume 42, San Francisco, 213.

Schou, J., 1991. On the 2-Dimensional Rotational Inversion Problem. In: D. Gough and J. Toomre (eds.), Challenges to Theories of the Structure of Moderate-Mass Stars, Lecture Notes in Physics volume 388, Springer-Verlag, Berlin, 93-100.

Schou, J. and Brown, T. M., 1994. On the Rotation Rate in the Solar Convection Zone. Astrophysical Journal, 434, 378-383.

Schou, J., Christensen-Dalsgaard, J. and Thompson, M. J., 1994. On Comparing Helioseismic Two-Dimensional Inversion Methods. Astrophysical Journal, 433, 389-416.

Schou, J., Christensen-Dalsgaard, J. and Thompson, M. J., 1995. Some Aspects of Helioseismic Time-Series Analysis. In: R. K. Ulrich, E. J. Rhodes, Jr. and W. Däppen (eds.), GONG '94: Helio- and Astero-Seismology From the Earth and Space, Astronomical Society of the Pacific Conference Series, Volume 76, San Francisco, 528-531.

Sekii, T., 1993. On an $1 \otimes 1$ Inversion Technique for Solar Rotation. In: T. M. Brown (ed.), GONG 1992: Seismic Investigation of the Sun and Stars, Astronomical Society of the Pacific Conference Series, Volume 42, San Francisco, 237-240.

Tomczyk, S., Streander, K., Card, G., Elmore, D., Hull, H. and Cacciani, A., 1995. An Instrument to Observe Low-Degree Solar Oscillations. Solar Physics, 159, 1-21.

On the choice of trade-off parameter in helioseismic SOLA inversion

A.A. Stepanov[1,2] and J. Christensen-Dalsgaard[1]

[1] Teoretisk Astrofysik Center, Danmarks Grundforskningsfond, and Institut for Fysik og Astronomi, Aarhus Universitet, Denmark
[2] Institute of Mathematics and Computer Science, University of Latvia, Riga, Latvia

1 Statement of the problem

A number of problems in geophysics, helioseismology, signal processing involve recovery of a function $\Omega(r)$ from measurements of the following form:

$$\Delta_i = \int_0^R K_i(r)\Omega(r)\,dr + \epsilon_i\,, \qquad i = 1, 2, \ldots, M\,, \tag{1}$$

where the kernels K_i are known functions, $K_i, \Omega \in L_2[0, R]$, and the errors ϵ_i are limited by

$$\left(M^{-1} \sum_{i=1}^M \epsilon_i^2 \right)^{1/2} \leq e$$

for some positive e. This inverse problem is ill-posed: for small error levels e there exist solutions $\tilde{\Omega}(r)$ reproducing the data Δ_i within the errors while having very large deviation from the true function $\Omega(r)$. Besides, the precise solution can be found only if the kernels $K_i, i = 1, 2, \ldots$ form an infinite complete set of functions (Xia & Nashed 1994).

Christensen-Dalsgaard *et al.* (1990) have compared methods suitable for the inverting helioseismic data. All of these numerical methods are linear, so that the approximating solution $\tilde{\Omega}(r)$ is the linear combination of the data Δ_i:

$$\tilde{\Omega}(r) = \sum_{i=1}^M c_i(r)\Delta_i\,. \tag{2}$$

We consider the optimally localized averages inversion method (Backus & Gilbert 1968) where the coefficients $c_i(r)$ are explicitly determined to control the resolution and error magnification. We consider the subtractive (SOLA) variant of the method, widely used for the inversion of helioseismic data (Jeffrey 1988; Pijpers & Thompson 1994) as well as for signal processing (Oldenburg 1981; Louis & Maaß 1991). Here the coefficients $c_i(r)$ minimize the functional

$$\int_0^R [A(r, r') - T(r, r')]^2\,dr' + \mu\Lambda^2(r)\,, \tag{3}$$

where

$$A(r, r') = \sum_{i=1}^{M} K_i(r')c_i(r)$$

is the averaging kernel characterizing the resolution of the method, $T(r, r')$ is a given target function, $\mu > 0$ is a trade-off parameter which must be chosen, and

$$\Lambda^2(r) = \sum_{i=1}^{M} c_i^2(r)$$

is the error-magnificatio· coefficient. For simplicity we have assumed that all data have the same stan‿ard deviation; this can always be achieved through suitable normalization of the data and kernels. We do not impose the condition

$$\int_0^R A(r, r') \, dr' = 1 \,, \quad r \in [0, R] \,, \tag{4}$$

on the coefficients $c_i(r)$ (Oldenburg 1981; Pijpers & Thompson 1994) but require this normalization for the target function. It is clear from equation (3) that the coefficients $c_i(r)$ and the solution $\tilde{\Omega}$ are functions of μ; thus we denote the approximating solution obtained by the SOLA method by $\tilde{\Omega}_\mu(r)$.

The purpose of present report is to consider the choice of trade-off parameter μ and the convergence of the SOLA solution as $e \to 0$.

2 The relation with Tikhonov regularization

The method of Tikhonov regularization can be successfully applied to the solution of the l·lioseismic inverse problem (Christensen-Dalsgaard *et al.* 1990). Due to discrete form of helioseismic data Δ_i, $i = 1, 2, \ldots, M$, the semi-continuous form of the smoothing functional (Wahba 1977) must be used. Then the regularized solution $\tilde{\Omega}_\mu^{\text{reg}}(r)$ minimizes in the space $L_2[0, R]$ the functional

$$M^{-1} \sum_{i=1}^{M} \left[\int_0^R K_i(r)\Omega(r) \, dr - \Delta_i \right]^2 + \mu \|\Omega\|_{L_2[0,R]}^2 \,,$$

and has the following form:

$$\tilde{\Omega}_\mu^{\text{reg}}(r) = K(r)(Q + \mu E)^{-1} \Delta \,, \tag{5}$$

where

$$K(r) = [K_1(r), K_2(r), \ldots, K_M(r)] \,,$$

Q is an $M \times M$ matrix with elements

$$q_{j,k} = \int_0^R K_j(r)K_k(r) \, dr \,, \quad j, k = 1, 2, \ldots, M \,;$$

E is the $M \times M$ unit matrix, and $\Delta = (\Delta_1, \Delta_2, \ldots, \Delta_M)^\mathsf{T}$.

The SOLA and Tikhonov techniques are related by

Theorem 1. Between SOLA solution and Tikhonov regularized one the following relation is valid:

$$\tilde{\Omega}_\mu(r) = \int_0^R T(r, r') \tilde{\Omega}_\mu^{\text{reg}}(r') \, dr' . \tag{6}$$

Hence the SOLA method yields the regularized solution smoothed by $T(r, r')$; the two solutions are identical if the target function is the Dirac δ function. A similar relation was found by Jeffrey (1988) between the SOLA solution for the δ-function target and the solution of Philips (1962) and Twomey (1963). Thus we may concentrate on the case $T(r, r') = \delta(r - r')$, results for different target functions following then from the transformation (6). This relation shows that in the SOLA method the trade-off parameter μ corresponds to the regularization parameter in the Tikhonov method. Therefore, the choice of μ may be based on methods well known in the regularization theory. We consider objective methods which guarantee the convergence of the approximating solution $\tilde{\Omega}_\mu(r)$ of the inverse problem to the true solution $\Omega(r)$ as $e \to 0$.

3 The discrepancy principle

We introduce the *rms* misfit $\rho(\mu)$ of the solution to the data by

$$\rho^2(\mu) = M^{-1} \sum_{i=1}^{M} \left[\int_0^R K_i(r) \tilde{\Omega}_\mu(r) \, dr - \Delta_i \right]^2 . \tag{7}$$

The discrepancy principle, commonly used in Tikhonov regularization, states that the optimal choice of the trade-off parameter μ is the solution of the equation $\rho(\mu) = e$. This definition satisfies the following

Theorem 2. Let $\tilde{\Omega}_\mu(r)$ be defined by equation (5) and satisfying the inequality $\rho(\mu) \leq e$. Then the discrepancy principle provides the smallest error magnification $\Lambda(\mu, r)$ and the maximal error in fitting the target δ function.

We illustrate the application of the discrepancy principle to the inversion of helioseismic data, using a set of 834 kernels $K_i(r)$ for modes of solar oscillation in the frequency range $2 - 4$ mHz and the artificial rotation law $\Omega(r)$ defined by Christensen-Dalsgaard *et al.* (1990). The dotted line in Fig.1 shows results of the inversion, including random errors in the artificial data with an assumed error level e of 0.1%. The solution recovers rather well the original function $\Omega(r)$ except at the centre and surface. The problem for small radius r is connected with the fact that at $r = 0$ all kernels K_i are zero. We show in section 4 that the problem near the surface can be solved by using a semi-optimal choice of trade-off parameter μ as well as a target function differing from the δ function.

Additional experiments with such data show that the discrepancy principle yields a good approximation only for small error levels (e less than 1%), which is rather smaller than the realistic measurements errors, whereas for larger values of e the solution is oversmoothed. The dashed line in Fig.1 shows results for an error of 3% which are not satisfactory.

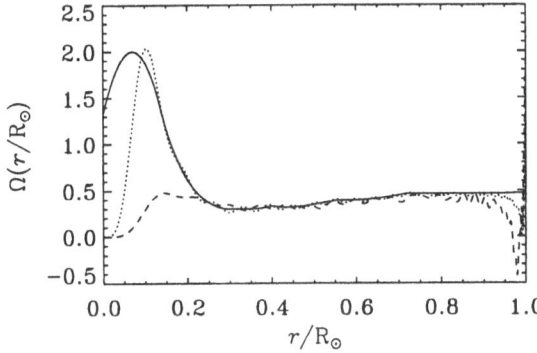

Fig. 1. Exact rotation rate (solid line) and solutions obtained with trade-off parameters determined from the discrepancy principle, for error levels e of 0.1% and 3% (dotted and dashed lines)

4 The choice of trade-off parameter

If the data $\{\Delta_i\}$ have no errors, i.e., $\epsilon_i = 0$, equation (5) shows that $\tilde{\Omega}_\mu^{\text{reg}}(r)$ converges to the function $\Omega^+(r) = KQ^+\Delta$ as $\mu \to 0$; here Q^+ is the generalized inverse matrix of Q. If the kernels $K_i(r)$ are linearly independent, Q^+ is simply Q^{-1}. From Nashed & Wahba (1974) it follows that in the space $L_2[0, R]$, $\Omega^+(r)$ is a least-squares solution of equation (1) with the minimal norm and as $M \to \infty$ converges to the exact solution $\Omega(r)$. To investigate the convergence of the approximating solution $\tilde{\Omega}_\mu(r)$ to the function $\Omega^+(r)$ as $e \to 0$ as well as to choose the trade-off parameter μ we have obtained the following estimate:

$$|\tilde{\Omega}_\mu^{\text{reg}}(r) - \Omega^+(r)| \le \Lambda(\mu, r)(e + \mu\|\omega\|_2), \quad r \in [0, R], \tag{8}$$

where the vector $\omega = Q^+\Delta$. Thus convergence will be achieved if $\mu \to 0$ as $e \to 0$ in such way that the right-hand side of equation (8) tends to zero, that is if $\Lambda(\mu, r) \to 0$ as well $\Lambda(\mu, r)\mu \to 0$. Thus the trade-off parameter μ cannot tend to 0 too quickly but must be consistent with the level e of errors. For example, the equality $\mu = e$ would guarantee the necessary convergence but such choice of μ is suitable only in the formal limit of $e \to 0$. For a given fixed level of errors $e > 0$ we have to obtain a constructive formulation of choice of μ. We do it on the basis of the estimate (8), assuming that the value e is known *a priori*.

The right-hand side of equation (8) shows that the error in the approximating solution consists of the resolution error $\Lambda(\mu, r)\mu\|\omega\|_2$ and the magnified error $\Lambda(\mu, r)e$. As we have previously noted, the discrepancy principle provides a minimum only for the second term; as a result, the resulting solution is as a rule oversmoothed and has insufficient resolution. The estimate (8) shows that in principle the overall error in the approximating solution can be reduced by increasing the magnified error and improving the resolution.

The methodology for the choice of the trade-off parameter μ depends on the available *a priori* information on the norm $\|\omega\|_2$.

A. Optimal choice of trade-off parameter:

We define the optimal value of μ_{opt} by minimizing the right-hand side of the inequality (8),

$$\mu_{opt} = \text{argmin } \Lambda(\mu, r)(e + \mu\|\omega\|_2), \quad r \in [0, R], \tag{9}$$

where argmin denotes the argument μ for which the functional is minimal. This local criterion defines a *trade-off function* $\mu_{opt} = \mu_{opt}(r)$ for $r \in [0, R]$.

B. Semi-optimal choice of trade-off parameter.

If the norm $\|\omega\|_2$ is unknown *a priori*, the choice

$$\mu(r) = \text{argmin } \{\Lambda(\mu, r)(\mu + e)\}, \quad r \in [0, R],$$

ensures convergence of the right-hand side of equation (8) to 0 as $e \to 0$. We should note that because the value $\|\omega\|_2$ in the space $L_2[0, R]$ may be arbitrarily large, convergence of the regularized solution to the true one may be arbitrarily slow, with arbitrarily large absolute errors in the approximating solution. Evidently, the error may be constrained if an estimate of $\|\omega\|_2$ can be obtained.

For all the solutions obtained with this semi-optimal trade-off parameter the approximating solution is significantly increased near the centre of the Sun, compared with the solution obtained using the discrepancy principle, bringing it in closer agreement with the exact solution; however, as in Fig. 1 there is still a deficiency near the surface. To avoid the latter discrepancy, we depart from the formally ideal resolution and use a Gaussian target function of the width σ instead of the δ-function target. This approach was used previously by Pijpers & Thompson (1994) (see also Oldenburg 1981). The parameters σ and $\|\omega\|_2$ were chosen such as to satisfy the condition $\rho(\mu) \leq e$ and to obtain well-localized averaging kernels $A(r, r')$. Specifically, we used $\sigma=0.03$ and $\|\omega\|_2=40$.

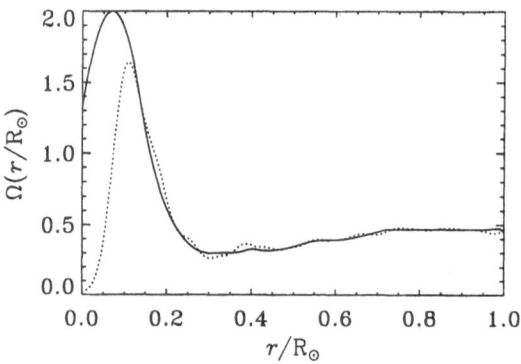

Fig. 2. Exact rotation law (solid line) and solution (dotted line) obtained with optimal choice of trade-off function and a Gaussian target function, for an error level e of 3%

The regularized solution with the semi-optimal value of μ recovers the true rotation function much better near the centre than does the solution obtained by the discrepancy principle, but it displays rapid oscillations similar to those obtained by Christensen-Dalsgaard *et al.* (1990) for the spectral-expansion method. However, the SOLA solution using Gaussian target function, shown in Fig. 2, quite well approximates the true function $\Omega(r)$, assuming an error level e representative of current observations.

Acknowledgements

This work was supported by the Danish National Research Foundation through its establishment of the Theoretical Astrophysics Center, through grant 93.606 of the Latvian Council of Science and with support from the Nordic Council of Ministers in the Nordic-Baltic-Scholarship Scheme.

References

Backus, G.E., Gilbert, J.F., 1968, The resolving power of gross Earth data, *Geophys. J. R. astr. Soc.* **16**, 169

Christensen-Dalsgaard, J., Schou, J., Thompson, M.J., 1990, A comparison of methods for inverting helioseismic data, *MNRAS* **242**, 353

Jeffrey, W., 1988, Inversion of helioseismic data, *ApJ* **327**, 987

Louis, A.K., Maaß, P., 1991, Smoothed projection methods for the moment problem, *Numer. Math.* **59**, 277

Nashed, M.Z., Wahba, G., 1974, Convergence rates of approximate least squares solutions of linear integral and operator equations of the first kind, *Math. Comp.*, **28**, 69

Oldenburg, D.W., 1981, A comprehensive solution to the linear deconvolution problem, *Geophys. J.R. astr. Soc.* **65**, 331

Philips, D.L., 1962, A technique for the numerical solution of certain integral equations of the first kind, *J. Assn. Comput. Mach.*, **9**, 84

Pijpers, F.P., Thompson, M.J., 1994, The SOLA method for helioseismic inversion, *A & A* **281**, 231

Twomey, S., 1963, On the numerical solution of Fredholm integral equations of the first kind by inversion of the linear system produced by quadrature, *J. Assn. Comput. Mach.*, **10**, 97

Wahba, G., 1977, Practical approximate solutions to linear operator equations when the data are noisy, *SIAM J. Numer. Anal.* **14**, 651

Xia, X.-G., Nashed, M.Z., 1994, The Backus-Gilbert method for signals in reproducing kernel Hilbert spaces and wavelet subspaces, *Inverse Problems* **10**, 785

Prospects for binary star research using disentangling

Pierre Maxted

NBIfAFG, Astronomisk Observatorium, Brorfelde, Denmark

1 Introduction

Binary stars are our most valuable source of accurate fundamental data for stars such as the masses and radii of individual stars. If masses and radii can be derived to an accuracy of better than 2%, we are then in a position to apply stringent tests to models of stellar structure and evolution (Andersen, 1991).

Disentangling is a new technique that applies sparse matrix algebra to the study of spectroscopic binary stars. The technique has been shown to provide reliable orbits for binary stars where the width of the spectral lines has hampered accurate measurements by other techniques (Simon et al.,1994, Sturm & Simon,1994). It also provides the separated ("disentangled") spectra of the individual components.

2 Formulation of the problem

The observed spectra of a spectroscopic binary star can be described as a matrix equation of the form $\mathbf{M} \cdot \mathbf{x} = \mathbf{b}$ where :

- $\mathbf{b}$ is the concatenation of the observed spectra $(\mathbf{b}_1, \mathbf{b}_2, \ldots, \mathbf{b}_n)$
- $\mathbf{x}$ is the concatenation of the individual spectra of the component stars $(\mathbf{x}_A, \mathbf{x}_B)$
- $\mathbf{M}$ is of the form:

$$\begin{pmatrix} N_{A,1} & N_{B,1} \\ N_{A,2} & N_{B,2} \\ \vdots & \vdots \\ N_{A,n} & N_{B,n} \end{pmatrix}$$

The structure of the submatrices depends upon the wavelength scales adopted for the observed spectra $(\mathbf{b}_1, \mathbf{b}_2, \ldots, \mathbf{b}_n)$. For studies of Doppler motion the natural wavelength scale is logarithmic, since the wavelength shift is then a constant function of wavelength. If we re-sample all the spectra onto a uniform heliocentric logarithmic wavelength scale, the submatrices contain two adjacent offset diagonals with values complementary to 1. The submatrix simply maps a section of $\mathbf{x}_A$ or $\mathbf{x}_B$ onto $\mathbf{b}_i$ using linear interpolation.

3 Solution of the problem

This system of linear equations is overdetermined (for $n > 2$) and rank deficient. To solve this equation we have used the LSQR algorithm (Paige & Saunders, 1982). LSQR uses an iterative method to approximate the solution of $\mathbf{M} \cdot \mathbf{x} = \mathbf{b}$ in the least squares sense when $\mathbf{M}$ is large and sparse. $\mathbf{M}$ need only be described by means of the operations $\mathbf{y} = \mathbf{y} + \mathbf{M} \cdot \mathbf{x}$ and $\mathbf{x} = \mathbf{x} + \mathbf{M}^{\mathrm{T}} \cdot \mathbf{y}$. We have written our own routines for performing these operations in order to take advantage of the simple structure of $\mathbf{M}$. The program runs on an HP 9000/735 workstation. The disentangling of 49 spectra of 4096 elements takes around 10s to converge to a precision of 10^{-6} in $r = \|M \cdot x - b\|$.

4 The clever bit

The accurate disentangling of the spectra clearly depends on having the correct spectroscopic orbit in order to define $\mathbf{M}$. Accurate spectroscopic orbits are precisely the quantities that are so hard to derive for some of the most interesting binary stars. We might then expect that for a poor spectroscopic orbit, the residual $r = \|M \cdot x - b\|$ will be large and that for the correct spectroscopic orbit r will reach a minimum. We can therefore, in principal, derive a spectroscopic orbit for the binary star by minimising r.

5 θ^2 Tau

θ^2 Tau is a bright binary star in the Hyades. The Hyades is a rich star cluster and one of the closest to Earth and is therefore one of the cornerstones of the astronomical distance scale. Since it is relatively close, θ^2 Tau can be resolved as a visual binary star. From the visual orbit (angular size) and the spectroscopic orbit (physical size) we can obtain an independent distance estimate for θ^2 Tau and thus to the cluster.

Although θ^2 Tau is bright and the secondary star contributes 30% of the total light, the secondary star rotates at over $100 \, \mathrm{km \, s}^{-1}$ at its equator, and this makes the absorption lines in its spectrum broad, shallow and difficult to measure.

We have applied the disentangling technique to spectra of θ^2 Tau obtained by Jocelyn Tomkin et al. (1995) The maximum radial velocities of the stars in their orbits are denoted K_1 and K_2. These are usually the least well determined parameters of a spectroscopic orbit, and have the largest influence on the masses derived. In Fig. 1 we see the value of $1/r$ as a function of K_1 and K_2 for θ^2 Tau based on a section of spectrum around 3950Å. We see a dependence of $1/r$ on K_2 that suggests a value of $K_2 \approx 30.5$. There are two peaks due to K1 at $34.6 \, \mathrm{km \, s}^{-1}$ and $36.2 \, \mathrm{km \, s}^{-1}$. Previous estimates of K_1 are some $5 \, \mathrm{km \, s}^{-1}$ lower. The difference may be the result of the broad lines of the secondary blending with the sharper lines of the primary and reducing the apparent amplitude of the Doppler shift.

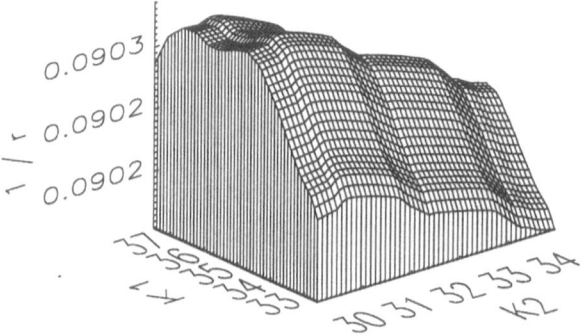

Fig. 1. $1/r$ versus K_1 and K_2 for θ^2 Tau based on spectra near 3950Å.

6 Massive stars

Herrero et al. (1992) have modelled the formation of spectral lines due to hydrogen and helium in the atmospheres of massive stars (10 to 60 Solar masses). This leads indirectly to estimates of the mass. These can be compared to those derived from the models of the evolution of massive stars and models of the stars' stellar winds. The agreement with models of stellar winds is good, but there is a discrepancy with the masses derived from evolutionary models. This is demonstrated in Fig. 2 where the mass derived from the analysis of the spectral lines ("Observed Mass") is compared to that predicted by models of stellar evolution based on the observed temperatures and luminosities of the stars ("Evolutionary Mass"). Despite the large errors involved, it is clear that the is a systematic difference between the two masses. Evolutionary models also fail to predict the observed pattern of helium abundances.

It is not clear which theory (if either) gives correct results, but rotation may be the key to the solution of this problem. Disentangling will lead to accurate orbits for massive stars, even with quite high rotation rates. The disentangled spectra can then be analysed in the same way as single star spectra. Some of the stars undergo mutual eclipses for which high quality data already exist. Combining the analysis of all these data will lead to accurate estimates of the masses, radii, luminosities and compositions of massive stars. We will then be in a position to test models of stellar evolution, spectral line formation and stellar winds.

Massive stars are also extremely luminous and can be observed in nearby galaxies. Massive eclipsing binaries are already being studied in the Large and

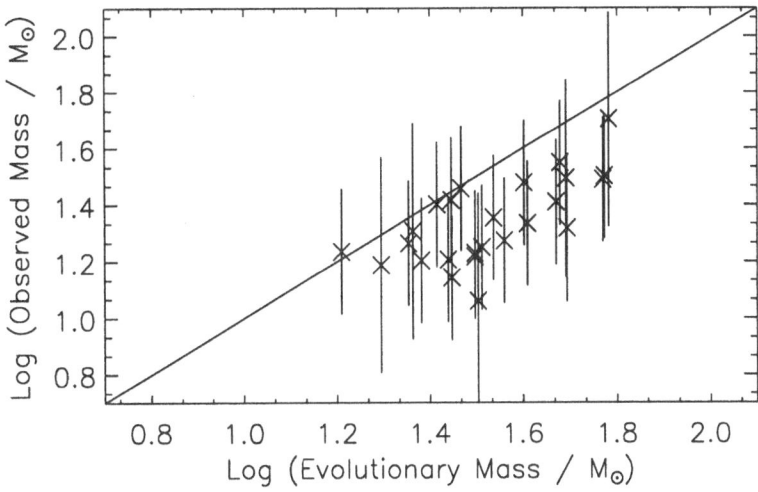

Fig. 2. The mass discrepancy for OB stars.

Small Magellanic Clouds, and the new generation of 8m class telescopes will enable us to extend these studies to few nearby galaxies such as the Andromeda Galaxy. Deriving accurate parameters for stars 1 million times fainter than can be seen with the naked eye is going to be hard work. The effort will be well spent because these stars will tell us for certain how far away these galaxies are and so may help to settle the debate over the value of the Hubble constant.

References

Andersen J.,1991, Accurate masses and radii of normal stars. Astronomy & Astrophysics Review, 3 ,91

Paige C.C., Saunders M.A., 1982, LSQR - An algorithm for sparse linear-equations and sparse least-squares, ACM Trans. Math. Softw., 8, 43

Simon K.P., Sturm E., Fiedler A., 1994, Spectroscopic analysis of hot binaries 2: The components of Y-Cygni, Astronomy & Astrophysics 292, 507

Sturm E., Simon K.P.,1994, Spectroscopic analysis of hot binaries 1: The components of DH Cephei, Astronomy & Astrophysics 282, 93

Tomkin J., Pan X., McCarthy J.K., 1995, Spectroscopic detection of the secondaries of the Hyades interferometric spectroscopic binary θ^2 Tauri and of the interferometric spectroscopic binary α Andromedae, Astronomical Journal 109, 780

Herrero A., Kudritzki J.M., Vilchez D., Kunze K., Butler K., Haser S., 1992, Intrinsic parameters of galactic luminous OB stars, Astronomy & Astrophysics 261, 209

Estimating Bidirectional Transport Parameters for the Blood-Retina Barrier - a Feasibility Study

Peter Dalgaard

Department of Biostatistics, Faculty of Health Sciences, University of Copenhagen

1 Introduction

Vitreous fluorometry is used to study the transport properties of the human retina and vitreous. A fluorescent tracer is injected into an arm vein and circulates in the bloodstream, from which it is slowly cleared by metabolism. The concentration in the bloodstream is monitored by taking blood samples. Small amounts of the tracer penetrates the blood-retina barrier and diffuses into the vitreous, where concentration profiles can be obtained by measuring blue-green fluorescence.

In the beginning (until about 20 minutes after the injection), most of the tracer is in a thin layer close to the retina. Subsequently, there is a phase where the profiles decrease towards a near-zero concentration in the mid-vitreous. Then, roughly 2 to 5 hours after injection, the mid-vitreous concentration rises and the profiles tend to flatten out. Finally, the profiles begin to decrease towards the retina.

A complication of the method is that there are actually two fluorescent substances involved. Fluorescein (F) is metabolized by the liver into fluorescein glucuronide (FG). FG is a weaker fluorophore but it also has slower clearance from the blood. The measured signal is a mixture of the fluorescence of F and FG. Partial separation of the substances is obtained by switching the excitation wavelength, utilizing the fact that the relative fluorescence intensity of FG compared to F is lower at higher wavelengths.

For further details of the experimental conditions, see Engler et al. (1994). It is desired to formulate a model which describes the pattern of profiles that is observed. The model contains parameters which describe properties of the eye, such as the leakiness of the blood-retina barrier, and the strength of the active transport in the opposite direction. These are scalar parameters, but, as shall be seen, it is also necessary to introduce unknown functions into the model.

Thus, we have an inverse problem involving a mixture of scalar and functional parameters. The purpose of the present paper is to evaluate an algorithm for the solution of the inverse problem and see if it yields satisfactory results on simulated data with realistic sampling characteristics.

2 Model considerations

A simple first-order model for the flux across a section of the blood-retina barrier is

$$\text{flux/area} = P_{in}c_+ - P_{out}c \ ,$$

in which P denotes *permeability*, c the concentration in the vitreous at the retinal interface, and c_+ external concentration (in the bloodstream). The two coefficients can be interpreted as the apparent permeabilities that would be observed if the concentration on one side of the barrier were fixed at zero. The coefficients are lumped constants, representing both active and passive mechanisms. Purely passive transport implies $P_{in} = P_{out}$.

P_{out} can only practically be estimated from long-term data when the flux is no longer dominated by the inward term.

The transport in the vitreous has the character of slow diffusion and is modeled using a standard isotropic diffusion model, although it is suspected that the diffusion is faster in the mid-vitreous than in the vicinity of the retina.

Current technology allows only measurements along the optical axis of the eye. Therefore, an assumption is imposed that the diffusion is spherically symmetric, i.e. of the form

$$\frac{\partial c}{\partial t} = \frac{1}{r^2}\frac{\partial}{\partial r}D(r)r^2\frac{\partial c}{\partial r}$$

with initial condition $c(r,0) = 0$ and boundary condition at the retina

$$D(R)\frac{\partial c}{\partial r}(R,t) = P_{in}c_+(t) - P_{out}c(R,t) \ .$$

The condition at the interior is a little tricky. Normally one takes a symmetry condition in r, but we wish to account for contributions from the anterior part of the eye, where the anatomy is different. This is accomplished by introducing a boundary condition, controlling the concentration $c_-(t)$ 4 mm from the midpoint of the vitreous, as an unknown parameter. This corresponds to modeling diffusion in a symmetric hollow sphere with boundary conditions on both the inside and the outside.

It is desired to study both the two fluorophores F and FG separately, so parameters appear for both of them. It is, however, difficult to detect differences between F and FG in their vitreous diffusion properties, so the model assumes that $D(r)$ is the same. All in all, the model contains four scalar parameters, P_{in}^F, P_{out}^F, P_{in}^{FG}, and P_{out}^{FG}, and three unknown functions, $D(r)$, $c_-^F(t)$, and $c_-^{FG}(t)$.

3 Numerical method

The equations of the model cannot in general be solved analytically. We use an adaptation of the well-known Crank-Nicolson finite-difference procedure. A conical section of the sphere is subdivided into a number of shell segments (Fig. 1).

The concentration in the innermost segment is $c_-(t)$ (we omit the F/FG superscript), and the mass fluxes across the shell interfaces can be approximated as

$$F = r^2 D(r) \frac{\Delta c}{\Delta r}$$

The rate of increase at the midpoint is then approximately

$$\frac{\partial c}{\partial t} = \frac{\Delta F}{v}$$

where v is the volume of the segment, roughly $r^2 \Delta r$ if r corresponds to the segment midpoint. (A proportionality factor is omitted on v and F.)

This converts the PDE to a system of ODEs which is solved implicitly by equating $\Delta c/\Delta t$ to the average of $\partial c/\partial t$ at times t and $t + \Delta t$, leading to the familiar tridiagonal systems of equations.

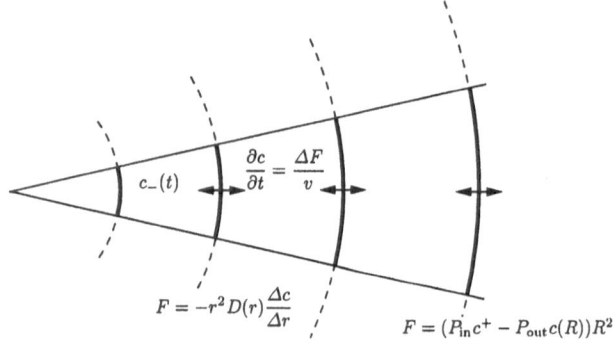

Fig. 1. Sketch of segmentation used in numerical method. Rightmost interface corresponds to blood-retina barrier. Fluxes across other interfaces follow Fick's law. The concentration in an inner shell is controlled by the parameter $c(t)$

4 General theory for discretized inverse problems

This was discussed at some length by Dalgaard (1993). Here we give a brief summary, with notes on the application to the present case.

The numerical method can be viewed as a single large matrix-vector equation, $G(\theta)c = h(\theta)$. In this equation, c is a vector containing the concentrations of F and FG on the entire finite-difference grid. The values are stored with the spatial index varying most rapidly, then the time index and lastly the fluorophore type. With this storage scheme, G is a gigantic but very sparse matrix consisting of two large diagonal blocks (corresponding to the separate solution of equations for F and FG), each being lower block bidiagonal with tridiagonal blocks (corresponding to the solution of successive tridiagonal equations in the Crank-Nicolson method). One would never actually store $\mathbf{G}$ in a computer program, its usage here is entirely notational. The vector $\mathbf{h}$ on the right hand side has to do with the external inputs to the system, it is zero, except at spatial indices corresponding to the boundaries, where it depends on $c_-(t)$ and $c_+(t)$ (the precise definition is unimportant here, it becomes apparent when the numerical method is spelled out).

The vector $\boldsymbol{\theta}$ contains the parameters. Of these the function parameters $D(r)$, $c_-^{\mathrm{F}}(t)$, and $c_-^{\mathrm{FG}}(t)$ are represented by their values at the grid-points (these are also the only values used in the numerical calculation of $\mathbf{c}$.)

The setup takes account of the fact that in general one does not observe $\mathbf{c}$ itself, but rather some affine mapping of it, $\boldsymbol{\eta} = \mathbf{Lc} + \mathbf{a}$. In the present case, the matrix $\mathbf{L}$ performs interpolation from the numerical grid to the observation points. It also takes care of instrument calibration and the mapping of concentrations of F and FG to the fluorescence measurements at 458 nm and 488 nm excitation obtained by a photon counter. The offset $\mathbf{a}$ models "dark counts" due to thermal noise in the photomultiplier tube.

However, it is not $\boldsymbol{\eta}$ that is observed either, but some noisy data, i.e. the realization $\mathbf{y}$ of a random vector $\mathbf{Y}$ with a statistical distribution relating to $\boldsymbol{\eta}$. For problems involving photon counting, such as the present one, it is realistic to assume $\mathbf{Y}$ Poisson distributed with mean $\boldsymbol{\eta}$, but the framework allows $\mathbf{Y}$ to have any of the distributions that are used in the family of generalized linear models (GLIM), cf. McCullagh and Nelder (1989), e.g. Normal, Gamma, and Binomial distribution. For some of these, it is convenient to allow a scale transformation of the response variable, as in the case of log-normal errors. This is done by introducing a "link" function and specifying $\boldsymbol{\eta} = \mathrm{link}(\mathbf{EY})$. Here, it is simply the identity mapping.

For GLIM models, the change in $-2 \log$ likelihood is called the *deviance*. This generalizes the sum of squares. A quadratic penalty $\boldsymbol{\theta}' \mathbf{Q}(\lambda) \boldsymbol{\theta}$ is added to the deviance and we seek to minimize this penalized deviance. The parameter λ controls the smoothness of the estimate.

This setup allows efficient formulas (exact up to round-off) for the gradient of the log-likelihood function and the Fisher information (the statistical expectation of the Hessian matrix). The latter in the sense that one can multiply vectors by it efficiently, the matrix itself is too large to be useful.

$$\gamma = [(\mathbf{y} - \boldsymbol{\mu})' \mathbf{DWLG}^{-1}](\mathbf{dh} - \mathbf{dGc}) \quad \text{and}$$
$$\mathbf{Sv} = (\mathbf{dGc} - \mathbf{dh})'[\mathbf{G}'^{-1} \mathbf{L}' \mathbf{WLG}^{-1}(\mathbf{dGc} - \mathbf{dh})\mathbf{v}]$$

(Terms in square brackets computed first.) 'd' denotes differentials with respect to $\boldsymbol{\theta}$. Note that the term $\mathbf{dG}$ is a three-way structure. In practice, one writes routines to compute $(\mathbf{dG}\mathbf{c})\mathbf{v}$ and $\mathbf{u}'(\mathbf{dG}\mathbf{c})$, the efficiency being obtained by utilizing the extreme sparseness of $\mathbf{dG}$. Note also that the calculations involve the solution of systems with coefficient matrix $\mathbf{G}'$. This is easily done in the manner of the Crank-Nicolson method, but since $\mathbf{G}'$ is *upper* block bidiagonal, the solution "runs in reverse time".

The matrices $\mathbf{D}$ and $\mathbf{W}$ come from standard GLIM theory. Both are diagonal, $\mathbf{D}$ contains the derivatives of the link function and $\mathbf{W}$ the so-called "iterative weights", proportional to the reciprocal variance of $\mathbf{DY}$. (The terminology stems from the standard method of iteratively using weighted linear regression to fit GLIMs).

Of course, the usefulness of this general framework cannot really be evaluated without seeing the details of the formulas and their computer implementation, which is not possible here. In my opinion it allows a well-structured top-down approach to a complex calculation that otherwise would become quite unwieldy.

5 Penalty functions

The classical choice of penalty for multi-function problems is

$$\sum_i \lambda_i \int f_i''(x)^2 \, dx$$

with the natural discretization of the integrals of the form $(\Delta x)^{-3} \sum_j (f(x_{j-1}) + f(x_{j+1}) - 2f(x_j))^2$.

Using matrix notation, this latter expression is $(\Delta x)^{-3} \mathbf{f}'(\mathbf{T}'\mathbf{T})\mathbf{f}$ where $\mathbf{f}' = (f(x_0), \ldots, f(x_N))$ and $\mathbf{T}$ is the second-order difference operator of dimension $(N-2) \times N$. Since $\mathbf{T}$ is triple-banded, $\mathbf{T}'\mathbf{T}$ is pentadiagonal and positive semi-definite with rank $N-2$.

However, it is often desired to impose boundary conditions on the functions, such as being "flat" at one end ($f' = 0$) or being "clamped" at zero ($f = f' = 0$).

For the vitreous diffusion problem, it is convenient to impose flatness on the diffusion coefficient $D(r)$ at the retinal interface $r = R$. Also, away from the boundary, concentrations rise very slowly from zero, so it is natural to clamp the mid-vitreous concentrations $c_-^F(t)$ and $c_-^{FG}(t)$.

These types of boundary conditions can be incorporated in the penalty by modifying the second-order difference operator $\mathbf{T}$ in the same way as is standard for finite-difference methods. The resulting $\mathbf{T}'\mathbf{T}$ matrices are still pentadiagonal, but have rank $N-1$ and N for "flat" and "clamped" boundary conditions, respectively.

For notational purposes, it is desired to have the penalty written in the form $\theta'\mathbf{Q}(\lambda)\theta$, with θ as described in Section 3. The matrix $\mathbf{Q}(\lambda)$ becomes block diagonal with the $\lambda_i(\Delta x_i)^{-3}\mathbf{T}_i'\mathbf{T}_i$ along the diagonal (Δx_i denoting Δr or Δt as appropriate). Diagonal elements corresponding to scalar parameters are zero. Note that it is possible to easily compute not only the penalty $\theta'\mathbf{Q}(\lambda)\theta$ itself, but also $\mathbf{Q}(\lambda)\mathbf{v}$ and $\mathbf{Q}(\lambda)^-\mathbf{v}$ (generalized inverse).

6 Algorithm

For large, nonlinear minimization problems with gradient information, it is attractive to use the conjugate gradient algorithm (CG), see e.g. Kennedy and Gentle (1980). However, the algorithm turns out to be very slow. This may seem surprising since truncated CG is a successful smoothing technique for linear ill-posed problems. However, it works because CG tends to give elements of the solution corresponding to large eigenvalues first. Regularization changes the problem into one whose eigenvalues increase with frequency.

Thus, to make CG work, it is necessary to precondition the system for better spectral properties. Tempering the large eigenvalues is the essential point, but also removing collinearity between the parameters is an issue.

A good preconditioner for CG must be positive definite and (at least in a neighborhood of the minimizer of the penalized deviance) similar to the inverse Hessian. The Hessian itself is not positive definite but its statistical expectation, $S + Q$ is. It is also somewhat easier to compute. However, neither the Hessian nor $S + Q$ is easily inverted, since both are large and non-sparse. Therefore, it is necessary to look for approximations of $(S + Q)^{-1}$, which can be computed using only inversion of lower-order matrices. One such formula is

$$X(X'SX)^{-1}X' + Q^{S-} - Q^{S-}SZ(Z'SQ^{S-}SZ + Z'SZ)^{-1}Z'SQ^{S-}$$

This involves the matrices X, spanning the null space of Q, and Z, spanning the "smooth functions". Z can be chosen with columns corresponding to the first terms of polynomial or trigonometric expansions of the functional parameters. $Q^{S-} = (I - P_X^S)Q^-(I - P_X^S)'$ with $P_X^S = X(X'SX)^{-1}X'S$ and Q^- any g-inverse. The idea of the approximation is to approximate $S + Q$ with Q on the non-smooth components of the parameter space. Ignore for simplicity the existence of the null space X (which leads to some technicalities) so that Q is invertible. Writing $S + Q = P_Z^{S'}SP_Z^S + (I - P_Z^S)'S(I - P_Z^S) + Q$, the second term is dominated by Q and is discarded (penalty dominates information for non-smooth functions). The remaining terms can be rewritten as $Q + SZ(Z'SZ)^{-1}Z'S$ which a standard formula allows to be inverted using only Q^{-1} and inverses of matrices the size of $Z'Z$. In principle, the higher dimension of Z the better approximation, but in practice, numerical cancelation sets in and makes the formula unstable. It is not unlikely that better formulas than the above can be found, but it has performed adequately for the present study.

The numerical method can become completely unstable if P_{in}^F, P_{out}^F, P_{in}^{FG}, P_{out}^{FG}, and $D(r)$ are negative. Also, if $D(r)$ becomes zero at either end of its domain, the boundary condition there no longer influences the observed data, leading to singularity problems. Therefore, it is necessary to design the algorithm so that it is ensured that the permeabilities are greater or equal to zero and $D(r) \geq 1 \times 10^{-6}\, \text{cm}^2/\text{s}$ throughout the iterations. Mid-vitreous concentrations are less of a concern, they can (and do) go negative without similar catastrophic effects on the calculations. Of course, negative concentrations give no physical meaning, but as long as the negative values are close to zero, there is no problem with the interpretation of the estimates and it is hardly worth the trouble to force them to be positive.

To enforce the restrictions, a variant of the constraint-set technique is incorporated in the CG algorithm. This technique works by constraining or unconstraining parameters one at a time according to the change implied by the

preconditioned gradient. Any change to the constraint pattern incurs a reset of
the CG algorithm and a renewed computation of the gradient, which makes it
a bit troublesome to use on restricted smooth-function parameters because it is
often so that the algorithm will e.g. try to set a function to zero over a substantial
part of its domain. In the present case, constraint violations for $D(r)$ fortunately
only happen occasionally in the first iterations.

It is a feature of the CG algorithm that it attempts to preserve the effect
of previous line minimizations. In nonlinear problems, this sometimes results in
an undesirable unwillingness to reconsider previously tried search directions. To
avoid this, a "nudge factor" has been incorporated in the algorithm, causing the
search direction to be reset periodically. Also the preconditioner itself requires
occasional recomputation, this is done at the same time and after large changes
in the penalized deviance.

7 Simulated data

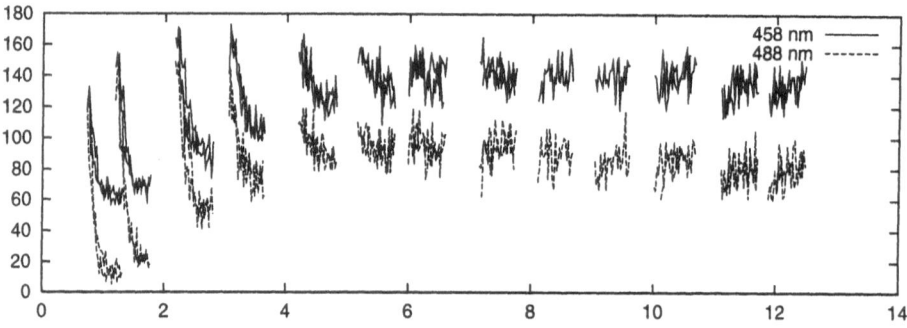

Fig. 2. Simulated data set. Pseudo-3D effect obtained by using time in hours + distance
from retina in cm as x-axis. Data for 458 nm excitation is offset by 50 counts for
separation from 488 nm data.

A set of simulated data was created as follows: The model was modified to full
spherical symmetry (no extra contributions from the anterior) and the numerical method was run with a fine grid-size. The diffusion coefficient was selected
as a quadratic spline, where a constant value of $6 \times 10^{-6}\,\mathrm{cm}^2/\mathrm{s}$ at the retina
rises to a constant level of $6 \times \times 10^{-5}\,\mathrm{cm}^2/\mathrm{s}$ in the mid-vitreous. The innermost sphere of radius 2 mm was assumed completely stirred. The permeabilities
were set to $P_{\mathrm{in}}^{\mathrm{F}} = P_{\mathrm{in}}^{\mathrm{FG}} = 10^{-7}\mathrm{cm}/\mathrm{s}$ and $P_{\mathrm{out}}^{\mathrm{F}} = P_{\mathrm{out}}^{\mathrm{FG}} = 10^{-5}\mathrm{cm}/\mathrm{s}$. Plasma
concentration curves were obtained by interpolating data from a volunteer. Cal-
ibration constants and position of points of measurement were available from the
same source. Data points between 1 and 8 mm from the retina were used. The
concentration values from the numerical calculations were interpolated to the

grid-points and converted to expected photon counts at 458 and 488 nm excitation wavelengths. Finally, independent Poisson random variables were generated with the calculated expectations. The resulting data are seen in Fig. 2.

8 Results

A set of fitted curves appear in Figs. 3 and 4, along with the true curves. The smoothing parameters were chosen by trial and error. The fitted permeabilities were $P_{in}^F = 9.18 \times 10^{-8}, P_{out}^F = 9.12 \times 10^{-6}$ and $P_{in}^{FG} = 8.88 \times 10^{-8}, P_{out}^{FG} = 1.00 \times 10^{-5}$. It should be noted that the simulations were made with somewhat pessimistic assumptions. Practical data are larger in magnitude and display a clearer decline towards the retina in the later phase. On the other hand real data are often disturbed by factors which are not completely controlled yet.

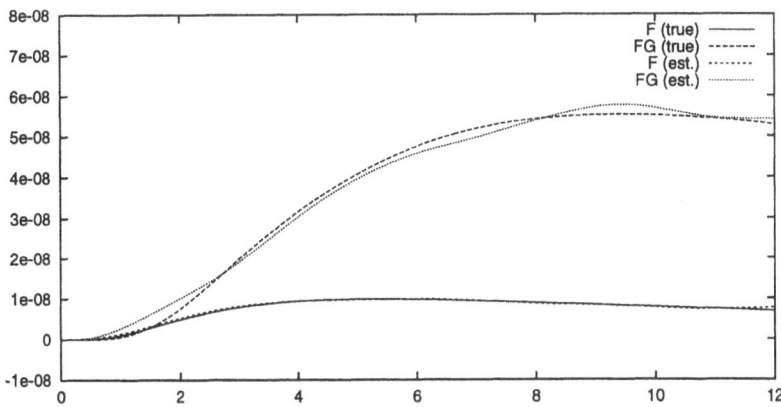

Fig. 3. Fitted and true mid-vitreous concentrations. Upper curves are FG, lower are F

Fig. 5 show plots of the $\log_{10}$ improvement as a function of CPU time on a 486DX2-80 PC clone system running the Linux operating system. The two overlaid plots differ in the **Z** matrix used in the preconditioner. It involves 5-term polynomial expansions in one plot and single-term in the other. It is seen that the added computational complexity is outweighed by faster convergence (slower but fewer iteration steps). Note that parts of the preconditioner can be computed once, whereafter the preconditioning is a relatively fast operation. The frequent re-precomputations due to initial large changes in the likelihood have a pronounced influence on the early part of the plots.

All in all, the algorithm is satisfactory. It converges in about two minutes, and though the numerical method is run on coarse grids, the exact calculation of derivative information avoids otherwise common problems where (e.g.) the approximate function does not increase along the approximate gradient, even very close to the optimum.

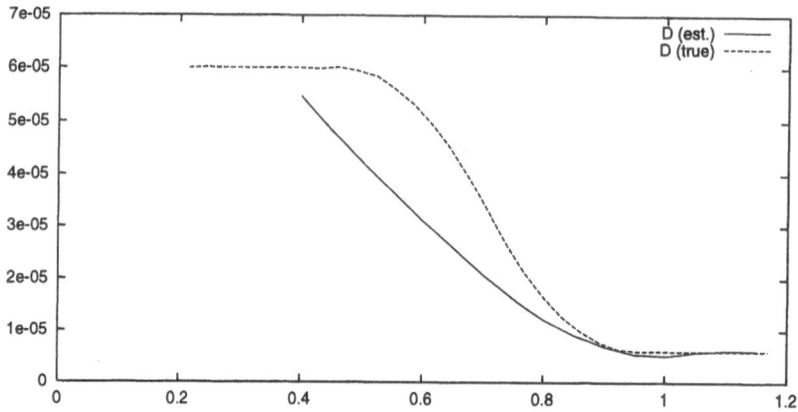

Fig. 4. Fitted and true diffusion coefficient $D(r)$. Retina at 1.2cm.

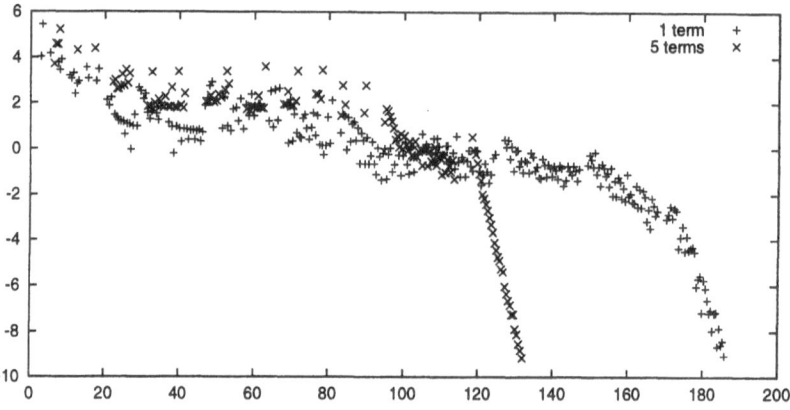

Fig. 5. Convergence plots. Log_{10} improvement vs. CPU seconds. +: One-term expansions in Z, ×: Five terms

References

Dalgaard, P., 1993. Estimation in large linear systems. In: K.Mosegaard (ed.), Proc.Interdisciplinary Inversion Workshop 2, The Niels Bohr Institute for Astronomy, Physics, and Geophysics, University of Copenhagen, 85–90.

Engler, C.B., Sander, B., Larsen, M. Dalgaard, P., Lund-Andersen, H., 1994. Fluorescein transport across the human blood-retina barrier in the direction vitreous to blood. Acta Ophthalmol., 72, 655–662.

Kennedy, W.J.Jr., Gentle J.E., 1980. Statistical Computing. Dekker.

McCullagh P., Nelder, J., 1989. Generalized Linear Models. Chapman and Hall.

Determination and variability of depth profiles from XPS backgrounds

Kamil L. Aminov[1,2], *Jørgen S. Jørgensen*[2], *and J. Boiden Pedersen*[2]

[1] On leave from: Kazan Phys.-Tech. Institute RAS, Kazan, Russia
[2] Fysisk Institut, Odense Universitet, Denmark

1 Introduction

X-ray photoelectron spectroscopy (XPS) is a non-destructive technique for probing the surface region of a solid. A typical XPS spectrum has a wide and rather structureless low energy background that contains information on the depth composition profile of the emitting atoms in the solid (see Fig. 1).

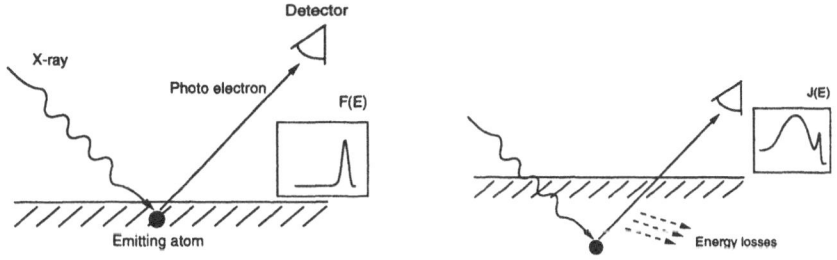

Fig. 1. Graphical illustration of XPS. Left: the intrinsic spectrum $F(E)$ is the initial energy distribution of the electrons and corresponds to the spectrum obtained from a thin layer close to the surface. Right: a typical observed XPS spectrum $J(E)$ with a large background due to energy losses during the flight to the surface.

When exposed to monoenergetic X-rays the atoms of the surface region emit photoelectrons with an energy distribution $F(E)$ (the intrinsic spectrum) that is assumed to be independent of the local environment of the atoms. Electrons originating from atoms below the surface must travel some distance inside the solid before escaping through the surface. During this path energy is lost via interactions with the solid (scattering events) and this results in a low energy background in the detected energy distribution $J(E)$ of the photoelectrons. The

---

[*] On leave from: Kazan Phys.-Tech. Institute RAS, 420029, Kazan, Russia

shape and size of this background depends on the depth profile $c(z)$, i.e. the concentration of emitting atoms at a depth z below the surface. The objective is to obtain an estimate of this profile.

The following discussion is based on a recent quantitative method (Aminov et al. (1995)). In many cases of practical interest the angular deflections for the scattering events are negligible and then the XPS spectrum is given by (Tougaard & Hansen (1989))

$$J(E) = \int_0^\infty dz\, c(z, \mathbf{q}) \int_{-\infty}^\infty dE'\, F(E') G(E' - E, z)$$

$$\text{where } G(E, z) = \int_{-\infty}^\infty ds\, \exp\left(2\pi i s E - z\sigma(s)/\mu\lambda\right)$$

$$\text{and } \quad \sigma(s) = 1 - \int_{-\infty}^\infty dE K(E) \exp\left(-2\pi i s E\right) \tag{1}$$

The parameters of the assumed profile model are contained in $\mathbf{q}$. The other functions and variables are assumed to be known. These are the inelastic scattering cross section $K(E)$, the mean free path of the electrons λ, and the direction cosine μ of the outgoing electrons measured from the surface normal.

2 Resolution estimates

For large depths $(z > 3\mu\lambda)$ the depth dependence of the electron energy distribution is well described by spectral diffusion. The average energy loss $E(z)$, and the average energy spread $\Delta(z)$ of an electron created at depth z is given by (Aminov et al. (1995)):

$$E(z) = (z/\mu\lambda)\sigma_1, \quad \Delta(z)^2 = (z/\mu\lambda)\sigma_2, \tag{2}$$

where σ_1 and σ_2 are the first two moments of the inelastic cross section:

$$\sigma_i = \int_{-\infty}^\infty dE\, K(E) E^i \tag{3}$$

From these expressions the maximum resolution Δz at depth z can be estimated by (Aminov et al. (1995))

$$\Delta z/\mu\lambda \sim \sqrt{z/\mu\lambda} \tag{4}$$

This shows that only features larger than about $\mu\lambda$ can be resolved and that the resolution falls off as the square root of the depth.

3 Algorithms for profile estimation

3.1 Least mean squares fitting

Using the described theory a profile restoration method was suggested (Tougaard & Hansen (1989)). This method, however, does not contain a quantitative measure of the goodness of the fit. We therefore reformulated the method into an optimization problem with a direct comparison to the measured data (Aminov et al. (1995)). Applying the Bayesian approach described in Jørgensen & Pedersen (1994) and using the fact that the statistics of the measurements are Poissonian (counting noise) the best fit is obtained by minimizing the log-likelihood function

$$\mathcal{L}(\mathbf{q}) = -2\sum_{j=1}^{N}\{J_j \ln J(E_j, \mathbf{q}) - J(E_j, \mathbf{q}) - \ln J_j!\} \tag{5}$$

where N is the number of measured data points J_j.

In the limit of large values of J_j this expression can be simplified by using Stirling's formula for $\ln J_j!$. Minimization of $\mathcal{L}(\mathbf{q})$ is then reduced to minimization of the usual mean squares measure

$$\chi^2(\mathbf{q}) \equiv \sum_{j=1}^{N}\left(\frac{J_j - J(E_j, \mathbf{q})}{\sigma_j}\right)^2 \tag{6}$$

where $\sigma_j \equiv \sqrt{J_j}$ is the estimated *rms* of the noise in channel j. In the following we will use only the latter measure and for simplicity use $\sigma_j = $ constant.

3.2 Maximum entropy estimation

As an alternative to assuming a definite profile model with adjustable parameters the surface profile can be discretized by n point c_i, i.e. $c(z_i) = c_i > 0$ for $z_i = i\Delta z, i = 1 \ldots n$. In that case regularization is needed, and a special choice is the maximum entropy method which can be argued to give the least biased solution consistent with the available information. The maximum entropy solution is determined by the optimization problem

$$\max\ (S - \lambda\chi^2), \quad S \equiv -\sum_{i=1}^{n} f_i \ln f_i, \quad f_j \equiv \frac{c_j}{\sum_i c_i} \tag{7}$$

where λ is a Lagrange multiplier which is tuned to satisfy a χ^2 constraint. A typical choice for the constraint is $\chi^2 \sim n$ (Gull & Daniell (1978)). It has also been suggested (Skilling & Gull (1985)) to use $\chi^2 \sim n + 3.29\sqrt{n}$ which is the upper 99% confidence limit of a χ_n^2 distribution.

4 Simulated XPS data

Simulated XPS spectra were generated from Eq. (1) using an analytical background correction method (Aminov & Pedersen (1995)). The intrinsic spectrum $F(E)$ was modeled by a single normalized Gaussian with mean 0 and standard deviation 0.05. The inelastic cross section $K(E)$ was constructed as a sum of two Gaussians with means 0.5 and 1.5, and standard deviations 0.1 and 0.3, respectively and it was normalized to give a value of $\alpha = 0.95$. Gaussian noise with standard deviation 0.01 was added to all spectra. These spectra were analyzed by application of various models and techniques and the results were compared with the true profile used to construct the spectra.

5 Variability of model parameters

When calculating the variability of the model parameters two cases must be considered. The simples case arise if the model is linear in the parameters, or at least linear in a neighborhood around the best fit values. In that case the covariance matrix V of the parameters is well approximated by

$$V = 2H^{-1}, \tag{8}$$

where H is the Hessian in the best fit point, and the variability is then easily obtained.

In the more general non-linear case, i.e. where linearization cannot be applied, the full parameter distribution is needed. This can be calculated by use of Bayes' theorem (Jørgensen & Pedersen (1994)). For the special case of Gaussian noise the following expression can be found

$$p(\mathbf{q}) \propto \exp\left(-\chi^2(\mathbf{q})/2\right) \tag{9}$$

Figure 2 demonstrates the latter approach for two spectra generated from rectangular profiles positioned at different depths. It is seen that the distribution

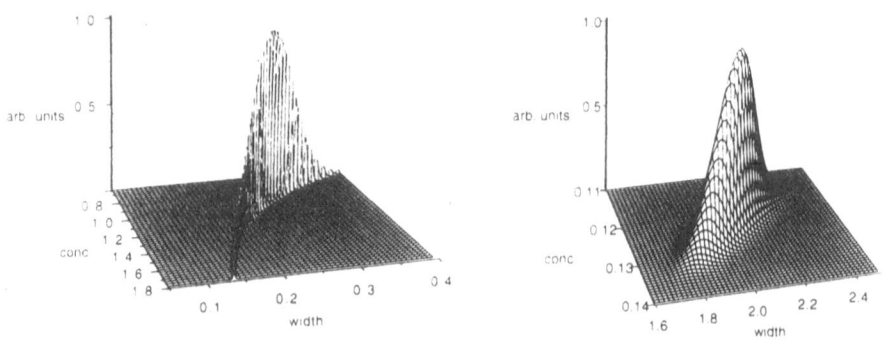

Fig. 2. Two simulated XPS spectra were generated from rectangular profiles. Both rectangles were positioned at depth $0.5\mu\lambda$. The figures show the probability distribution Eq. (9) for the concentration and the width assuming that the depth is known. The true parameters are (left) width $= 0.25\mu\lambda$, conc. $= 1$, and (right) width $= 2\mu\lambda$, conc. $= 0.125$. (©John Wiley & sons, Ltd., 1995. Reprinted from Aminov et al. (1995) by permission of John Wiley & Sons.)

for the narrow rectangle is thin and curved showing that the two parameters are extremely correlated: only the product of the parameters is well defined. The explanation is that the width of the rectangle is smaller than the maximum resolution at this depth and that only the product of the width and the concentration is defined. The distribution for the broad rectangle shows that the parameters are correlated but otherwise quite well determined. This is an indication that the width of the rectangle is about the same as the resolution limit at this depth. It is evident that linearization would work in the latter case, but would fail for the former.

6 Estimation of profiles

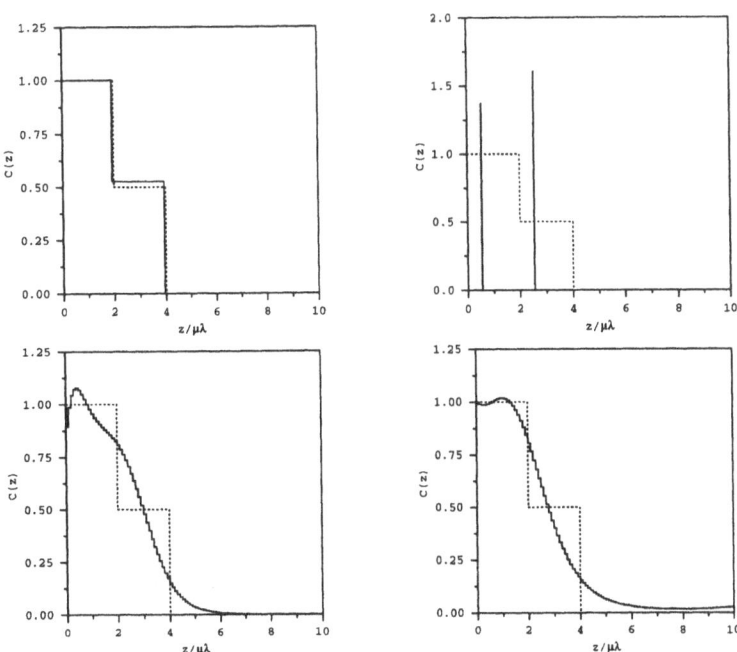

Fig. 3. Different profile estimations using a simulated XPS spectrum generated from the Stransky-Krastanow profile shown with a dashed line on the figures above. The number of data points were $n = 256$. The pictures show: Stransky-Krastanow fit, $\chi^2 = 254$ (top left), fit by two delta layers, $\chi^2 = 261$ (top right), maximum entropy, $\chi^2 = 256$ (bottom left), and maximum entropy, $\chi^2 = 309$ (bottom right)

To illustrate the inverse nature of the estimation problem, a simulated XPS spectrum was generated from the Stransky-Krastanow profile defined by the dashed line in Figure 3. The figure shows four different profiles that are consistent with the spectrum, i.e. they all have a statistically acceptable χ^2 value. The top figures show the results of a least mean squares fitting to two different models. As expected, the Stransky-Krastanow model gives a very good fit. More surprisingly, a model consisting of two delta functions fit the spectrum equally well. This illustrates that the class of feasible models is very wide and that prior information about the model type is needed.

The bottom figures show the maximum entropy estimations corresponding to the two different aim values of the χ^2 value mentioned in section 3.2. It appears that the maximum entropy estimation is quite insensitive to the specific choice of aim value. The left figure shows slightly more structure due to the harder constraint, but the general features of both profiles are that the first rectangle is reasonably well resolved while the second one is buried in a broad tail. Furthermore, both profiles are consistent with the resolution estimate, Eq. (4).

One may argue that the Stransky-Krastanow fit is probably not robust since the spectrum is equally well described by a model consisting of two delta functions. But this is not necessarily the case. What *can* be concluded is that the choice of model function (prior information) is crucial and that very different model functions describe the observed data equally well. If the model function is not confirmed by other observations then the significance of the estimated profiles is rather low. On the other hand the maximum entropy method produces solutions that look "reasonable" and is consistent with the resolution estimates. Furthermore, since the two lower figures can be considered estimates corresponding to two different noise levels the method is seen to be quite robust.

7 Main findings

1. The resolution of the profile decreases as the square root of the depth and the maximum resolution is of order one effective scattering length $\mu\lambda$.
2. Typically the parameters are strongly correlated in a non-linear way.
3. Radically different models fit the spectra equally well.
4. The maximum entropy method gives "reasonable" solutions for all cases studied and the solutions are consistent with the resolution estimates.

References

Aminov, K. L., Jørgensen, J. S., and Pedersen, J. B., 1995. Estimation of depth profiles from inelastic backgrounds in XPS. I. General relationships. Surf. Interf. Anal., 23, 753–763.

Aminov, K. L. and Pedersen, J. B., 1995. Analytical background correction in numerical Fourier transform procedures. Application to electron spectroscopy. Surf. Interf. Anal., 23, 717–722.

Gull, S. F. and Daniell, G. J., 1978. Image reconstruction from incomplete and noisy data. Nature, 272, 686–690.

Jørgensen, J. S. and Pedersen, J. B., 1994. Calculation of the variability of model parameters. Chemom. Intell. Lab. Syst., 22, 25–35.

Skilling, J. and Gull, S. F., 1985. Algorithms and applications. In: Smith, C. R. and Grandy, Jr., W. (eds.), Maximum-entropy and Bayesian methods in inverse problems, 83–132. Reidel.

Tougaard, S. and Hansen, H. S., 1989. Non-destructive depth profiling through quantitative analysis of surface electron spectra. Surf. Interf. Anal., 14, 730–738.

Uncertainties in Seismic Inverse Calculations

John A. Scales

Department of Geophysics, Center for Wave Phenomena, Colorado School of Mines, Golden, Colorado, USA

Introduction

Solving an inverse problem means making inferences about physical systems from real data. This requires taking into account three different kinds of information or uncertainty.

- What is known about the parameters independently of the data? In other words, what does it mean for a model to be reasonable or unreasonable?
- How accurately are the data known? That is, what does it mean to "fit the data"?
- How accurately is the physical system modeled? Does the model include all the physical effects that contribute significantly to the data?

Exploration seismologists have a large amount of information available that could be used to refine inferences about the subsurface. Some of this information could readily be encapsulated as hard constraints, while in other cases, probabilities, whether derived directly from data or personal prejudices, seem more appropriate. In this paper I will discuss how such information can be used to quantify the uncertainties in inverse calculations.

A Simple Example

To motivate the discussion, let us consider first a problem of linear tomographic inversion.

The Forward Problem Acoustic sources and receivers are lowered into two vertical boreholes as shown in Figure 1. Sources are excited in one borehole, seismograms are recorded in the other, and the goal is to use the time-of-flight of waves to infer the elastic properties of the medium between the boreholes. To some approximation the wave propagation can be treated kinematically, so the

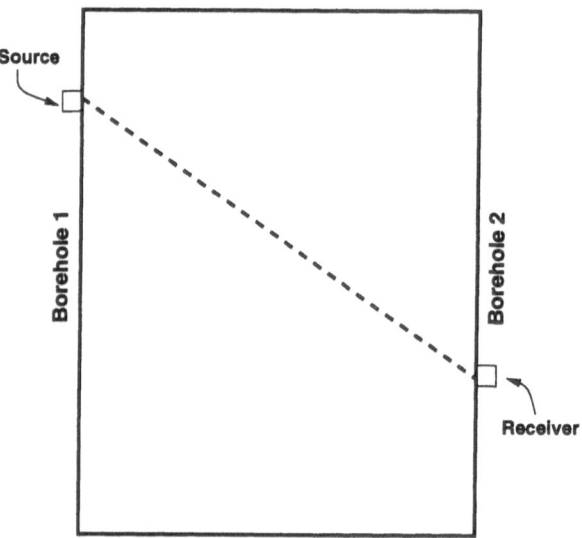

Fig. 1. A simple model of cross-hole or x-ray tomography

travel time of a wave (e.g., the direct arrival) is related to the unknown slowness between the boreholes by

$$t(s(x,y,z)) = \int_{\mathrm{ray}(s(x,y,z))} s(x,y,z)\,d\ell. \tag{1}$$

The details of the calculation are not important here, they are described in various places (Bording *et al.* (1987), for example). The point is that within certain approximations (small perturbations, high frequency, etc.) the forward operator mapping model parameters to data can, when discretized, be represented as a linear system of equations:

$$\delta \mathbf{t} = J \cdot \delta \mathbf{s} \tag{2}$$

where δt is the vector containing the differences between the observed travel times and those computed using an approximate slowness model $\mathbf{s_0}$, $\delta \mathbf{s}$ is a vector of perturbations to $\mathbf{s_0}$ (what we're seeking), and J is the Jacobian derivative matrix of the forward operator.

The Inverse Problem Now that we have a tractable, linear forward problem mapping model parameters (the components of $\delta \mathbf{s}$) into predicted data, we can consider how to use the travel time observations to make inferences on these parameters. The first step is to decide upon a criterion for measuring the degree of data fit. This is called the likelihood function. A widely used likelihood function is

$$\frac{1}{\sqrt{(2\pi)^N \det C_d}} \exp\left[-\frac{1}{2}(\mathbf{d}_{\mathrm{computed}} - \mathbf{d}_{\mathrm{obs}})^T C_d^{-1}(\mathbf{d}_{\mathrm{computed}} - \mathbf{d}_{\mathrm{obs}})\right]$$

where $\mathbf{d}_{obs}$ and $\mathbf{d}_{computed}(\mathbf{m})$ are the observed and computed (i.e., predicted) data, and C_d is the covariance matrix of data uncertainties. The argument of this exponential defines an N−dimensional quadratic form, the level surfaces of which are ellipsoidal error surfaces. For any given degree of data fit, the corresponding level surface tells us the range of model parameters which "fit the data" at that tolerance. For simplicity, let us suppose that the data (travel times in this case) are subject to known independent Gaussian uncertainties. Then the solution of the inverse problem would seem to be: find slowness values δs such that

$$\frac{1}{N} \sum_{i=1}^{N} \left(\frac{t_i^c(\delta \mathbf{s}) - t_i^o}{\sigma_i} \right)^2 \qquad (3)$$

is approximately equal to 1, where N is the number of observations, t_i^o is the i-th observed travel time, t_i^c is the i-th travel time calculated through the current slowness model, and σ_i is the standard deviation of the i-th datum. Making Equation 3 approximately equal to one corresponds to fitting the data on average to one standard deviation. Of course, other choices could be made.

Let's see how this works in practice. Figure 2 shows an example of such a tomography problem. The model consists of a homogeneous background with slowness equal to 2, and an embedded inclusion shaped like an "E" having slowness equal to 1.5. At the right we see the illumination of the rays. There are 10 sources, 20 receivers and 20^2 cells within which the slowness is assumed to be constant.

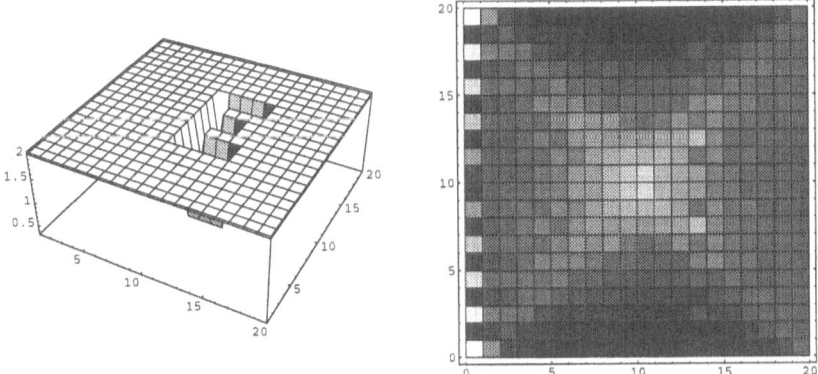

Fig. 2. Straight rays are traced between 10 sources and 20 receivers. The goal is to be able to characterize the unknown inclusion (left). On the right, the ray "illumination" plot is shown.

Next add uncorrelated Gaussian noise, so that the standard deviations σ_i are constant and known. The textbook solution to the problem is to do a singular value decomposition (SVD) of the Jacobian matrix. Then, build up the solution one singular vector at a time, starting with the largest singular values and moving down the spectrum, until the solution *just* manages to fit the data.

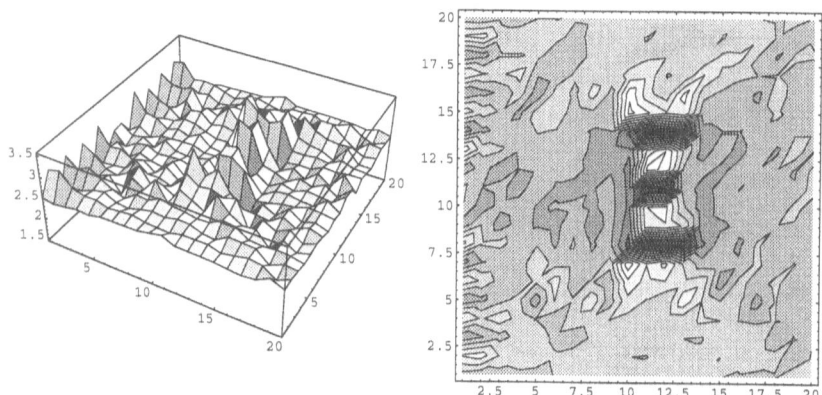

Fig. 3. A perspective (left) and contour (right) plot of the truncated SVD solution which *just* fits the data in the sense that its normalized χ^2 is 1.

Figure 3 shows this truncated SVD solution as both a perspective (left) and contour (right) plot. Although we can see the inclusion in the contour plot, the question we have to address is whether exhibiting this model really solves an inverse problem. The model fits the data, undeniably, but are the complicated looking features really required to fit the data or are they merely numerical artifacts? Is this model realistic? Are there other realistic models that would fit the data just as well?

What's the Point? One conclusion to be drawn from this simple experiment is that data fit alone is *never* enough to judge the solution of an inverse calculation. At best we can determine ranges of parameter values that are consistent with the data, for example, by looking at the confidence intervals defined by the level surfaces of the likelihood function. But since unrealistic models can fit the data too, one must balance the extent to which a model fits the data against the extent to which it conforms to our knowledge about what makes a model realistic or not in a particular case. This knowledge takes many different forms. It might be as simple as saying that a particular parameter must be positive, or that it must lie in some interval. Or it might be more complex. We might have previous measurements of this parameter or measurements of a related parameter.

Now it could be argued that in the tomography calculation described above, it would make more sense to use a different kind of regularization than SVD truncation. A finite-difference penalty function would also regularize the inversion and assure a smooth solution (e.g., Constable *et al.* (1987) or Scales *et al.*

(1990)).[1] Certainly this is a conservative strategy for problems where there is little *a priori* information. But for least squares, such a regularization scheme is equivalent to choosing a particular *a priori* relationship among the model parameters: a model covariance matrix, in fact. For example, using an anisotropic Laplacian to regularize the calculation would be equivalent to making different assumptions about the degree of horizontal and vertical smoothness of the velocity model; probably a reasonable thing to do in sedimentary basins. So this begs the question of whether we could make a more systematic characterization of such covariance matrices from other geological or geophysical observations.

Further, even in this toy problem we have introduced a number of assumptions and approximations which must ultimately be quantified in order that the inferences made be meaningful. For example, what errors have been introduced by neglecting refraction, by assuming ray theory was valid in the first place, by discretizing the model into a certain number of cells of constant slowness, by using a least squares criterion to measure data fit, by assuming that the data errors were i.i.d. Gaussian random variables, and on and on.

Information and Uncertainty

Even from this simple example it is clear that to be able to assign meaningful uncertainties to the results of an inverse calculation one must quantify what is known about the parameters of the physical system before the experiment, and what is known about the uncertainties in the data.

Information About Models

In exploration seismology there is a large amount of *a priori* information that could be used to influence inverse calculations. Here, *a priori* refers to the assumption that this information is known independently of the data. Plausible geologic models can be based on rock outcrops, models of sedimentological deposition, subsidence, etc. There are often *in situ* and laboratory measurements of rock properties that have a direct bearing on macroscopic seismic observations, such as porosity, permeability, crack orientation, etc. There are other, less quantitative, forms of information as well, the knowledge of experts for instance.

There is a simple conceptual model that can be used to visualize the application of this diverse *a priori* information. Suppose we have a black box into which we put all sorts of information about our problem. We can turn a crank or push a button and out pops a model that is consistent with whatever information that is put in. If this information consists of an expert's belief that, say, a certain sand/shale sequence is 90% likely to appear in a given location, then we must

---

[1] Assuming, of course, that there are no smooth model eigenvectors in the null-space of the forward problem. This need not be the case for multi-parameter inverse problems (Scales *et al.*, 1990). As a result, many of the standard "regularization" schemes may not in fact result in invertible operators.

ensure that 90% of the models generated by the black box have this particular sequence. One may repeat the process indefinitely, producing a collection of models that have one thing in common: they are all consistent with the information put into the black box.

Let us assume that a particular layered Earth description consists of the normal-incidence P-wave reflection coefficient r at 1000 different depth locations in some well-studied sedimentary basin. Suppose, further, that we know from *in situ* measurements that r in this particular sedimentary basin almost never exceeds .1. What does it mean for a model to be consistent with this information? We can push the button on the black box and generate models which satisfy this requirement. Figure 4 shows some examples.

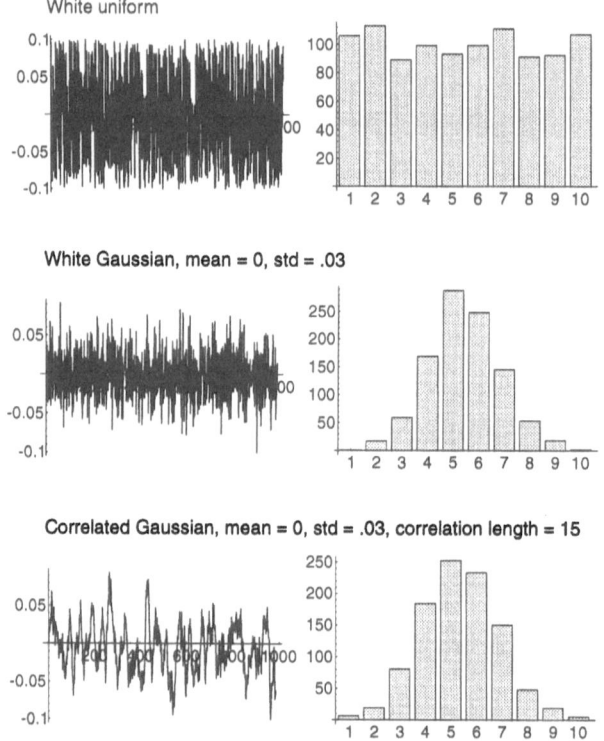

Fig. 4. Three models of reflectivity as a function of depth which are consistent with the information that the absolute value of the reflection coefficient must be less than .1. On the right is shown the histogram of values for each model. The top two models are uncorrelated, while the bottom model has a correlation length of 15 samples.

The three models shown satisfy the hard constraint that $|r| \leq .1$ but they look completely different. The question is, which is most consistent with our assumed prior information? And how do we measure this consistency? If we make a histogram of the *in situ* observations of r and it shows a nice bell-shaped curve, are we justified in assuming a Gaussian prior distribution? On the other hand, if we do not have histograms of r but only extreme values, so that all we really know is that $|r| \leq .1$, are we justified in thinking of this information probabilistically?

If we accept for the moment that our information is best described probabilistically, then a plausible strategy for solving the inverse problem would be to generate a sequence of models according to the prior information and see which ones fit the data. In the case of the reflectivity sequence, imagine that we have surface seismic data to be inverted. So for each model generated by the black box, compute synthetic seismograms, compare them to the data and decide whether they fit the data well enough to be acceptable. If so, the models are saved; if not, they are rejected. Repeating this procedure many times results in a collection of models that are by definition *a priori* reasonable and fit the data. If the models in this collection all look alike, then the features the models show are well-constrained by the combination of data fit and *a priori* information. If, on the other hand, the models show a diversity of features, then these features cannot be well-resolved by the surface data.

Should Hard Constraints Be "Softened"? It is plausible that any probability distribution consistent with a given hard constraint is more informative than the constraint itself. To say, for instance, that any model with $|r| \leq .1$ is feasible is certainly not the same thing as saying that all models with $|r| \leq .1$ are equally likely. And while we could look for the most conservative or least favorable such probabilistic assignment, Backus (1988) makes an interesting argument against any such probabilistic replacement in high- or infinite-dimensional model spaces. His point can be illustrated with a simple example. Suppose that all we know about an n−dimensional model vector $\mathbf{r}$ (now no longer the reflection coefficient, but any model vector) is that its length is less than some particular value–unity for the sake of definiteness. In other words, suppose we know *a priori* that $\mathbf{r}$ is constrained to be within the n−dimensional unit ball B_n. Backus considers various probabilistic replacements of this hard constraint; this is called "softening" the constraint. For instance, allow each component of $\mathbf{r}$ to be uniformly distributed on $[-1, 1]$. This particular choice turns out to be untenable since the expectation of r^2 goes to infinity as the dimension of the model space does, but this can be overcome without too much difficulty. We could choose a prior probability on $\mathbf{r}$ which is uniform on B_n. Namely, the probability that $\mathbf{r}$ will lie in some small volume $\delta V \in B_n$ shall be equal to δV divided by the volume of B_n. Choosing this uniform prior on the ball, it is not difficult to show that the expectation of r^2 for an n−dimensional $\mathbf{r}$ is

$$\langle r^2 \rangle = \frac{n}{n+2}$$

which converges to 1 as n increases. Unfortunately this does not overcome Backus' basic objection, since the variance of r^2 goes as $1/n$ for large n, and thus we seem to have introduced a piece of information that was not implied by the original constraint; namely that for large n, the only likely vectors $\mathbf{r}$ will have length equal to one. The reason for this apparently strange behavior has to do with the way volumes behave in high dimensional spaces. If we compute the volume of an n–dimensional shell of thickness ϵ just inside an R–diameter ball we can see that:

$$V_\epsilon \equiv V(R) - V(R - \epsilon) = C_n(R^n - (R - \epsilon)^n) = V(R)\left(1 - \left(1 - \frac{\epsilon}{R}\right)^n\right)$$

where C_n depends only on the dimension. Now for $\epsilon/R \ll 1$ and $n \gg 1$ we have

$$V_\epsilon \approx V(R)\left(1 - e^{-n\epsilon/R}\right).$$

This says that as n gets large, nearly all of the volume of the ball is compressed into a thin shell just inside the radius.

But even this objection can be overcome with a different choice of probability distribution to soften the constraint. For example, choose $r \equiv |\mathbf{r}|$ to be uniformly distributed on $[0, 1]$ and choose the $n-1$ spherical polar angles uniformly on their respective domains. This probability is uniform on $|\mathbf{r}|$, but non-uniform on the ball. However it is consistent with the constraint and has the property that the mean and variance of r^2 is independent of the dimension of the space. A different approach altogether, which I will not discuss, is the use of physical invariances associated with the model to derive so-called noninformative priors (Tarantola, 1987).

So, as Backus has said, we must be very careful in replacing a hard constraint with a probability distribution, especially in a high-dimensional model space. Apparently innocent choices may lead to unexpected behavior. One might well ask whether it was possible to measure the extra information, if any, implied by the softened version of a constraint. As we have seen, for specific choices it is possible to quantify certain aspects of this extra information. But perhaps it is best to not pose this question from the point of view of information, but rather to ask what are the consequences of various choices on the inferences we make. In other words, by computing the *a posteriori* resolution of the individual parameters (whether from a Bayesian viewpoint or via an approach such as confidence set inference (Stark, 1992)), we should be able to determine precisely the consequences of any given assumption about the information we use in an inverse calculation.

Information Naturally Suited to a Probabilistic Treatment In the last section we looked at information that took the form of a hard constraint on the model vector or some function of the model vector, such as its length. Some information comes to us in a form that seems naturally expressed probabilistically. For example, Figure 5 shows an example of a sonic log,[2] which can be thought of

---

[2] That is, an *in situ* estimate of acoustic wave speed as a function of depth inferred from travel times between a source and receiver in a measuring tool lowered into a well. NB, that the "data" are in fact the result of an inference. This is a common situation in which there is no fundamental distinction between the input and the output of an inverse calculation.

as characterizing the short vertical wavelength (1 m or less) material fluctuations at a given location in the earth. Perhaps we can extract the statistical features of these data and use them to refine inferences made with other data sets from the same area, such as surface seismic data. In order to focus on the statistical aspects of the data it is useful to subtract the long-wavelength trend, which is likely the result of more gradual, deterministic changes such as compaction. The trend and the resulting fluctuations are shown in Figures 6 and 7.

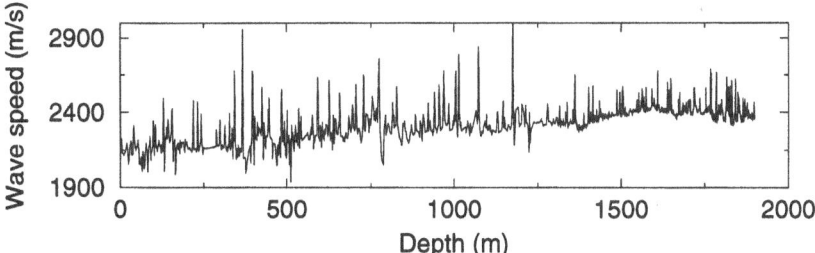

Fig. 5. Estimates of P and S wave velocity are obtained from the travel times of waves propagating through the formation between the source and receiver on a tool lowered into the borehole.

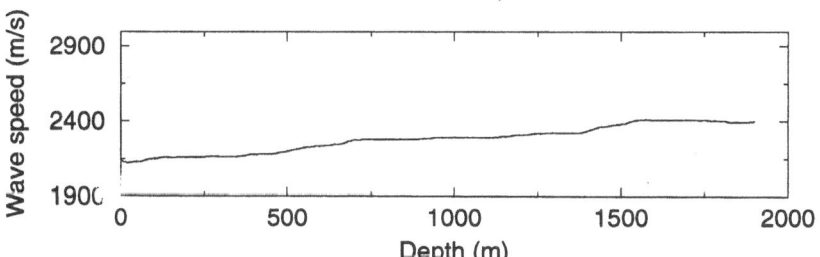

Fig. 6. Trend of Figure 5 obtained with a 150 sample running average.

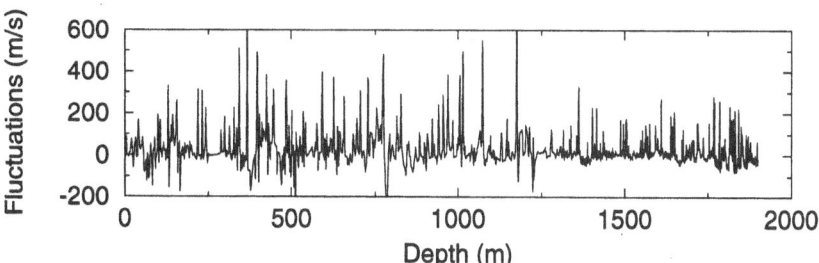

Fig. 7. Fluctuating part of the log obtained by subtracting the trend from the log itself.

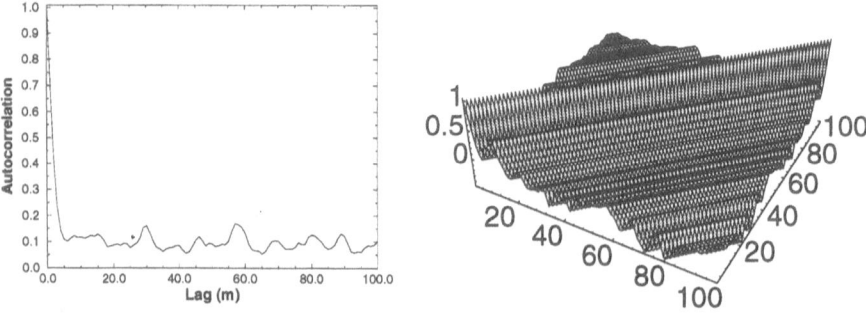

Fig. 8. Autocorrelation and approximate covariance matrix (windowed to the first 100 lags) for the well log.

To go further, we need an expression for the ultimate (so-called *a posteriori*) probability on the space of models as a function of the prior information and data uncertainties. It is

$$\sigma(\mathbf{m}) = k\rho_M(\mathbf{m})L(\mathbf{m}) \tag{4}$$

where σ is the posterior probability, which assimilates all the available information, ρ_M is what we know about the models *a priori*, L is the likelihood function (which depends implicitly on the data), and k is the normalization. This equation is derived in the appendix for completely general kinds of information. For now let us assume that the fluctuations of the well log are stationary and Gaussian, so that ρ_M and L are multi-dimensional Gaussian distributions. In this case it is relatively straightforward to compute a covariance matrix associated with the observations. For example, Figure 8 shows the first 100 lags (1 meter per lag) of the correlation function for the fluctuating part of Figure 7, as well as an approximate covariance matrix. Then, the Bayesian prior probability $\sigma(\mathbf{m})$, is proportional to

$$\exp\left[-\frac{1}{2}(\mathbf{m} - \mathbf{m}_{\text{prior}})^T C_M^{-1}(\mathbf{m} - \mathbf{m}_{\text{prior}})\right] \tag{5}$$

where C_M is the covariance matrix describing uncertainties in the models (as in Figure 8) and $\mathbf{m}_{\text{prior}}$ is the center of this distribution. Examples of such Gaussian calculations and their relation to Tikhonov regularization are shown in Gouveia (1996) and Gouveia and Scales (1996).

But the question remains as to whether the assumptions of stationarity and Gaussianity are reasonable. These two effects can trade off against one another. Figure 9 shows the standard deviation of the log as a function of depth. The non-stationarity is apparent and is probably ubiquitous in practice. One simple technique to overcome it is to divide the log into windows within which stationarity can be safely assumed. There is some preliminary evidence on the question of Gaussianity in Figure 10. This shows the third order cumulant and bispectrum of a small window (75 meters) of the fluctuating part of the data. As described in Subba Rao and Gabr (1984), a rigorous test for Gaussianity involves showing that the bispectrum is zero at all frequencies by looking at the estimated values over an appropriate grid of points.

In any case, if one were to accept the Gaussian hypothesis, then it would be reasonable to compute a covariance matrix for the *a priori* model distribution from widely available information. If the Gaussian hypothesis is rejected, then one needs to incorporate the higher order statistical information into the inverse calculation. How to do this is not obvious. If some number of moments of the unknown distribution can be estimated accurately from the data, then a plausibly conservative strategy would be to compute the maximum entropy distribution subject to these moments as constraints. If the first and second moments are used, this results in a Gaussian distribution; i.e., the Gaussian is the maximum-entropy distribution subject to fixed first and second moments. In Gouveia, Moraes & Scales (1996), we show examples of the estimation of 1D maximum-entropy prior distributions using up to four sample moments as constraints. Unfortunately, these techniques are not straightforward to apply for n-dimensional distributions, even assuming that one could accurately estimates moments beyond the second, and this is a big assumption.

Another possibility is to try to approximate directly the joint distribution function of the unknown process by making histograms of its low order marginal distributions. Here we rely on the fact that the n-dimensional distribution can be expressed in terms of the marginal distributions of the process. For example, the three dimensional joint distribution $f(x, y, z)$ can be written as $f(x|y, z)f(y, z)$, which can in turn be written as $f(x|y, z)f(y|z)f(z)$ and so on by induction for the general n-dimensional distribution. In principle, the marginal distributions can be estimated by making histograms. Here is a simple example of how such a procedure might be used in practice. Using the fluctuating part of the well log, as shown in Figure 7, we begin by constructing histograms to estimate the marginal distributions of the process. The histogram of nearest-neighbor values along the log is shown in Figure 11. If the process were Markovian, this would be sufficient to completely characterize the fluctuations of the log. If not, then we would need to compute histograms of non-nearest-neighbor pairs, histograms of triples, and so on. The problem now is when to stop? One possibility is to use a statistical hypothesis test such as Kolmogorov-Smirnov to decide when models pseudo-randomly sampled from a given set of histograms are sufficiently close to the observations. Another possibility would be to argue on *a priori* grounds which marginals are needed to capture the important information in the process. But I have no idea how one would do this.

If we can compute a non-parametric estimate of the joint distribution function of some piece of information, then this information can readily be incorporated into a Monte Carlo importance sampling scheme of the type proposed by Mosegaard and Tarantola (1995). In their approach, it is not necessary to assume that the prior information is expressed in terms of analytic functions; all that is required is a means of generating models in proportion to their *a priori* probability (i.e., importance sampling the prior distribution). These models are then

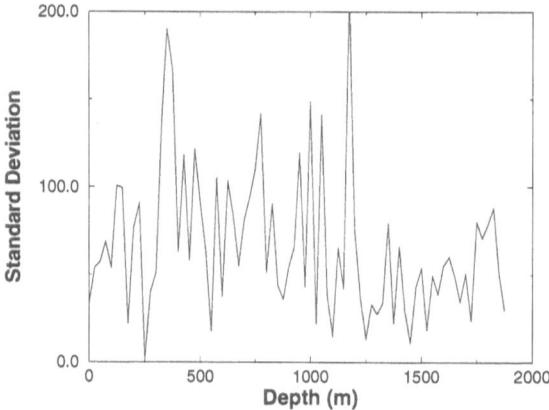

Fig. 9. The standard deviation as a function of depth of Figure 7.

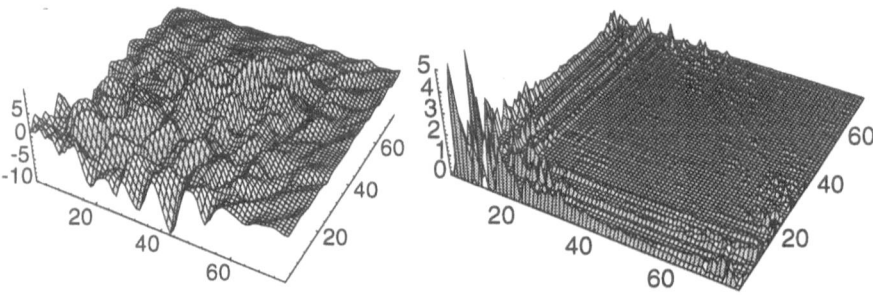

Fig. 10. The normalized third order cumulant and bispectrum of the fluctuating part of the data in Figure 7, computed for a short window of 75 lags. To account for non-stationarity, we can slide this window down the full length, repeating the estimates in each window.

accepted or rejected in a Metropolis-like procedure according to how well they fit the data, which guarantees that the collection of models that are generated do indeed sample the full *a posteriori* probability distribution. The key advance of of Mosegaard and Tarantola (1995) is that we no longer need to generate models that are *a priori* unrealistic–we only compute the response (predicted data) of those models which have been sampled from the prior. By making the prior sufficiently informative, it may be possible to overcome the profound cost of global sampling in high dimensional spaces.

Uncertainties in the data

In the last section we looked at probabilistic characterizations of models of the Earth's upper crust from commonly available measurements. Next we look at

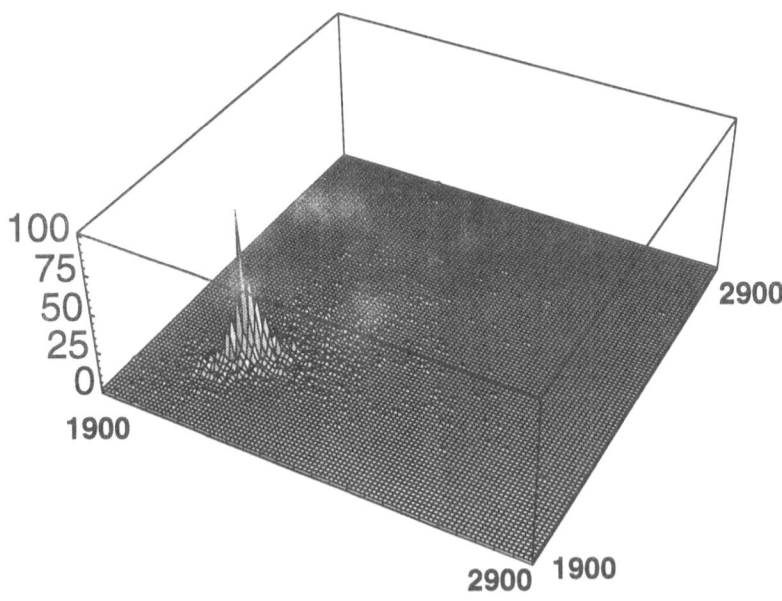

Fig. 11. Histogram of two-point transitions of length 1 (i.e., pairwise transitions for adjacent sites along the log) for the P-wave sonic log. The x and y axes correspond to measured values of velocity, while the z axis is the number of samples. If for a given value of x and y axes, say (v_i, v_j), the count is n, that means that there are n sites along the log where the site itself and its neighbor have the quantized values v_i and v_j respectively.

estimating uncertainties in the seismic data themselves. This is less controversial than the assignment of prior probabilities, but no less important since a *sine qua non* for any inference calculation is a specification of just what it means to fit the data.

We are used to thinking of seismic experiments as being non-repeatable. While this is true in earthquake seismology, it is not necessarily true in controlled-source seismology. Most land reflection data recorded nowadays uses a vibroseis source, i.e., a swept continuous wave signal. By careful stacking and cross-correlation with the known input signal, such data can approximate the results of explosive sources with high signal to noise ratio. But usually the raw, unstacked data are not saved; once the stacking and correlation have been done, the raw data are usually discarded. Figure 12 shows an example in which the raw traces have been saved. There are 25 unstacked, uncorrelated vibroseis traces recorded by the Reservoir Characterization Project of the Colorado School of Mines. Each trace corresponds to a separate vibroseis sweep at a single far-offset source location. All 25 traces were recorded within a span of about 10 minutes. On the left is shown the full 7 second sweep, while on the right the data have been windowed to just the first second, so as to exclude any source generated signal.

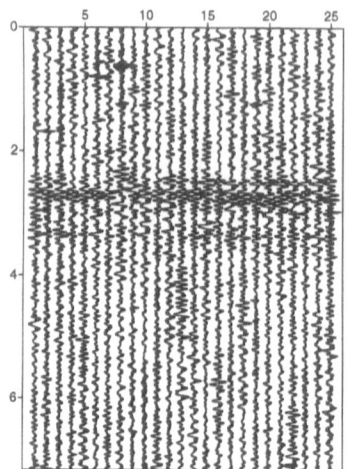
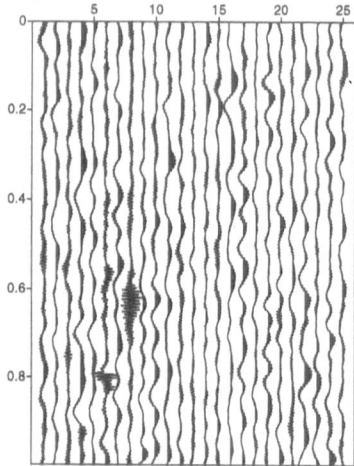

Fig. 12. 25 unstacked, uncorrelated vibroseis traces. Each sweep is 7 seconds long. Each raw trace can be regarded as a realization of a controlled source reflection seismic experiment. These are far offset sweeps, thus by isolating the first second of data (right), source generated energy is removed.

Since we now have repeated realizations of the experiment, it is possible to directly estimate a data covariance matrix. This is shown in Figure 13. Once we have the data covariance matrix C_d, then we can compute the Bayesian likelihood function

$$\frac{1}{\sqrt{(2\pi)^N \det C_d}} \exp\left[-\frac{1}{2}(g(\mathbf{m}) - \mathbf{d}_{\mathrm{obs}})^T C_d^{-1}(g(\mathbf{m}) - \mathbf{d}_{\mathrm{obs}}) \right]. \qquad (6)$$

Well, almost. The problem is that we have been implicitly thinking about the data covariance matrix as representing random errors or noise, whereas there is often no clear distinction between random noise and unmodeled physics. If we neglect to include anisotropy in the forward model, for instance, then energy in the data due to anisotropy will be erroneously mapped into isotropic features. Sometimes the distinction is largely a matter of taste. In the cross-hole tomography example, suppose we attempted to repeat the experiment in order to gather statistical information. Perhaps we cannot reposition the sources and receivers in exactly the right places, so small shifts in the–assumed known–source and receiver locations will introduce errors into the problem. Do we regard these as being essentially random fluctuations which broaden the distribution of the data, or do we look at

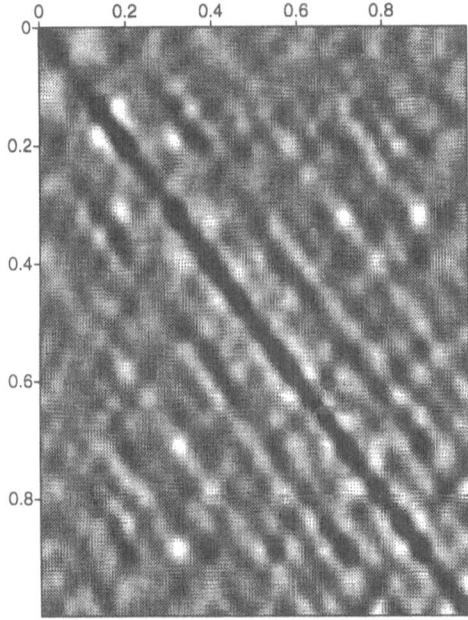

Fig. 13. Sample covariance matrix for the ambient noise present in a reflection seismic experiment. The ensemble consists of the repeated sweeps of the vibrator (before stacking and correlation) and has been windowed to the first second of data, before any source generated energy has arrived at the receivers.

the shifts in source and receiver locations as being model parameters (like station corrections) to be inferred from the data?

Tarantola (1987) shows that as long as the uncertainties are Gaussian, we can still use Equation 6 to describe the likelihood function, provided we replace the data covariance matrix C_d, with a combined covariance matrix

$$C_D = C_d + C_T$$

where C_T represents the contribution due to theoretical errors.

More often than not, theoretical errors are neglected in real applications. And the reason is plain, for it is not usually possible to account for a particular kind of unmodeled physics without incurring the additional expense that led to the neglect of this physics in the first place. How can one quantify the effects of neglecting anisotropy without doing the complete calculation anisotropically from the beginning? In some cases it is possible to do just that. For example, looking at well logs, which exhibit heterogeneity on the 1 meter or smaller length scale, it is reasonable that multiple-scattering dispersion and attenuation will be a non-negligible effect in typical sedimentary rocks. The magnitude of the dispersive effects can be estimated by comparing the dispersion free ray-theoretic speed with the Backus long-wavelength effective speed; total dispersion of 5% or more is common (Scales, 1993). So in this case, one can estimate the data error associated with the neglect of dispersion or simply correct for it, so as to eliminate the error altogether (Scales & Van Vleck, 1996). Similar effects which are potentially as large or larger include azimuthal anisotropy due to aligned systems of cracks within reservoirs, long-wavelength effective transverse isotropy due to fine layering, anelasticity associated with fluid-filled pores, 3D wave propagation effects (at least in structurally complex areas), and many others. A careful treatment of these modeling errors is beyond the scope of this paper, but it seems clear that there are a number of such effects which must be taken into account in even the simplest problems of real exploration interest.

Conclusions

As in all such calculations, seismic inversion requires careful estimation of prior information and data uncertainties in order to be able to make quantitative inferences about the Earth. The prior information comes from diverse sources, some of which may best be incorporated as hard constraints on parameters or functions of parameters, and some which may best be described probabilistically. There are also sources of subjective prior information from experts. It is important that we assimilate such information consistently and quantitatively. In the event that all the uncertainties can be treated as being Gaussian random variables, this information conveniently takes the form of model and data covariance matrices. In that case, the computational formalism applied is superficially similar to Tikhonov regularization, or constrained least squares, although the goals and conclusions are potentially rather different. But in some cases such Gaussian assumptions may not be justified. In the general case, in which the probability distribution associated with the prior information can only be sampled point-wise, it may be necessary to use an importance sampling procedure based on Monte Carlo methods. But this begs the question of how one estimates such a general prior probability. I have given some suggestions as to how these probabilities may be estimated, but I suspect we are a long way from a definitive answer to this question.

Acknowledgements

This work was inspired by my many discussions over the past few years with Albert Tarantola. Also, I learned a lot, as well as how much more I need to learn, from several stimulating conversations in Aarhus and Golden with Philip Stark. I would like to thank Tom Davis and Steve Roche of the Reservoir Characterization Project at the Colorado School of Mines for the vibroseis data, and Wences Gouveia and Fernando Moraes for useful discussions. Wences Gouveia wrote the code that was used to compute Figure 10. This work was partially supported by the sponsors of the Consortium Project on Seismic Inverse Methods for Complex Structures at the Center for Wave Phenomena, Colorado School of Mines, the Shell Foundation and the Army Research Office.

References

Backus, G. 1988. Hard and soft prior bounds in geophysical inverse problems. *Geophysical Journal*, **94**, 249–261.

Bording, R.P., Gersztenkorn, A., Lines, L.R., Scales, J.A., & Treitel, S. 1987. Applications of seismic travel time tomography. *Geophysical Journal of the Royal Astronomical Society*, **90**, 285–303.

Constable, S.C., Parker, R.L., & Constable, C.G. 1987. Occam's inversion: a practical algorithm for generating smooth models from electromagnetic sounding data. *Geophysics*, **52**, 289–300.

Duijndam, A. J. 1987. *Detailed Bayesian inversion of seismic data*. Ph.D. thesis, Technical University of Delft.

Gouveia, W. 1996. A study of model covariances in amplitude seismic inversion. *In: Proceedings of Interdisciplinary Inversion Conference on Methodology, Computation and Integrated Applications*. Springer-Verlag.

Gouveia, W., & Scales, J.A. 1996. Bayesian seismic waveform inversion. *Preprint*.

Gouveia, W., Moraes, F., & Scales, J.A. 1996. Entropy, Information and Inversion. *Preprint*.

Mosegaard, K., & Tarantola, A. 1995. Monte Carlo sampling of solutions to inverse problems. *JGR*, **100**, 12431–12447.

Scales, J. A. 1993. On the use of localization theory to characterize elastic wave propagation in randomly stratified 1-D media. *Geophysics*, **58**, 177–179.

Scales, J.A., & Van Vleck, E.S. 1996. Lyapunov exponents and localization in randomly layered elastic media. *Preprint*.

Scales, J.A., Docherty, P., & Gersztenkorn, A. 1990. Regularization of nonlinear inverse problems: imaging the near-surface weathering layer. *Inverse Problems*, **6**, 115–131.

Stark, P.B. 1992. Minimax confidence intervals in geomagnetism. *Geophysical Journal International*, **108**, 329–338.

Subba Rao, T., & Gabr, M.M. 1984. *An introduction to bispectral analysis and bilinear time series models*. Springer-Verlag.

Tarantola, A. 1987. *Inverse Problem Theory*. New York: Elsevier.

A The Bayesian Posterior Probability

The Bayesian posterior probability density on the space of models must be the product of two terms: a term which involves the *a priori* probability on the space of models and a term which measures the extent of data fit

$$\sigma(\mathbf{m}) = k\rho(\mathbf{m})\, L(\mathbf{m}) \qquad (7)$$

where k is the normalization constant and L is the likelihood function, which depends implicitly on the data.

We now show how Equation (7) follows logically from Bayes' theorem provided we generalize our notion of "data" to allow for the possibility that the data might be specified by probability distributions (Tarantola, 1987). Following Duijndam (1987), we begin by using the notation common amongst Bayesians, then we show how this relates to the more standard inverse-theoretic notation in Tarantola (1987).

In this approach we assume that we have some prior joint distribution $p_0(\mathbf{m}, \mathbf{d})$. Further, we suppose that as the result of some observation, the marginal pdf of $\mathbf{d}$ changes to $p_1(\mathbf{d})$. We regard $p_1(\mathbf{d})$ as being the "data" in the sense that we often know the data only as a distribution, not exact numbers. In the special case where the data are exactly known, $p_1(\mathbf{d})$ reduces to a delta function $\delta(\mathbf{d} - \mathbf{d}_{obs})$.

How do we use this new information in the solution of the inverse problem? The answer is based upon the following assumption: whereas the information on $\mathbf{d}$ has changed as a result of the experiment, there is no reason to think that the conditional degree of belief of $\mathbf{m}$ on $\mathbf{d}$ has. I.e.,

$$p_1(\mathbf{m}|\mathbf{d}) = p_0(\mathbf{m}|\mathbf{d}). \tag{8}$$

From this one can derive the posterior marginal $p_1(\mathbf{m})$:

$$p_1(\mathbf{m}) \equiv \int_D p_1(\mathbf{m}, \mathbf{d}) \, d\mathbf{d} \tag{9}$$

$$= \int_D p_1(\mathbf{m}|\mathbf{d}) p_1(\mathbf{d}) \, d\mathbf{d} \tag{10}$$

$$= \int_D p_0(\mathbf{m}|\mathbf{d}) p_1(\mathbf{d}) \, d\mathbf{d} \tag{11}$$

$$= \int_D \frac{p_0(\mathbf{d}|\mathbf{m}) p_0(\mathbf{m})}{p_0(\mathbf{d})} p_1(\mathbf{d}) \, d\mathbf{d} \tag{12}$$

$$= p_0(\mathbf{m}) \int_D \frac{p_0(\mathbf{d}|\mathbf{m})}{p_0(\mathbf{d})} p_1(\mathbf{d}) \, d\mathbf{d}, \tag{13}$$

where D denotes the data space.

Switching now to the inverse-theoretic notation, let us regard $p_0(\mathbf{d})$ as being the prior distribution on data $\mu_D(\mathbf{d})$: this is what we know about the data **before** we've actually done this particular experiment. Further, we identify $p_1(\mathbf{m})$ as the posterior distribution on the space of models, $p_1(\mathbf{d})$ as the data errors, $p_0(\mathbf{m})$ as the prior distribution on the space of models, and $p_0(\mathbf{d}|\mathbf{m})$ as the modeling uncertainties.

$$p_1(\mathbf{m}) \equiv \sigma(\mathbf{m}) \tag{14}$$

$$p_1(\mathbf{d}) \equiv \rho_D(\mathbf{d}) \tag{15}$$

$$p_0(\mathbf{m}) \equiv \rho_M(\mathbf{m}) \tag{16}$$

$$p_0(\mathbf{d}|\mathbf{m}) \equiv \Theta(\mathbf{d}|\mathbf{m}), \tag{17}$$

then we arrive at essentially Equation (1.65) of Tarantola (1987), except that there is no explicit reference to the non-informative prior. The prior distributions simply represent what we know about the problem–models and data–before the data have been collected.

$$\sigma(\mathbf{m}) = \rho_M(\mathbf{m}) \int_D \frac{\rho_D(\mathbf{d})\Theta(\mathbf{d}|\mathbf{m})}{\mu_D(\mathbf{d})} \, d\mathbf{d} \tag{18}$$

An important special case occurs when the modeling errors are negligible, i.e., we have a perfect theory. Then the conditional distribution $\Theta(\mathbf{d}|\mathbf{m})$ reduces to a delta function $\delta(\mathbf{d} - g(\mathbf{m}))$ where g is the forward operator. In this case, the posterior is simply

$$\sigma(\mathbf{m}) = \rho_M(\mathbf{m}) \left[\frac{\rho_D(\mathbf{d})}{\mu_D(\mathbf{d})}\right]_{\mathbf{d}=g(\mathbf{m})} \tag{19}$$

Constrained Waveform Inversion of Seismic Well Data

Marwan Charara[1], *Christophe Barnes*[1,2] *and Albert Tarantola*[1]

[1] Institut de Physique du Globe de Paris, France
[2] École des Mines de Paris, France

1 An underdetermined problem

The inversion of seismic data allows one to obtain an image of the subsurface through the determination of a certain number of physical parameters. Borehole data (VSP) are of special interest because of the geometry of acquisition which provides more informative signals on the medium properties than surface seismic data (high frequency signal, energetic P-S conversions). However, the lack of seismic data redundancy (one shot) renders the inversion problem underdetermined. Classical least squares inversions (without constraints) fail to give a sensible image of the subsurface when the medium has a complex geological structure. This arises because of the non-uniqueness related to the degrees of freedom in the inversion problem.

One way to overcome the lack of redundancy in the data is to extract as much information as possible from them. Therefore a fine forward problem such as finite difference method, when solving the elastic wave equation, is needed. The possibility of explaining various seismic events contained in the data (transmitted, reflected, multiple reflected, converted P and S waves) requires one to find an Earth model that can explain all these events, thus removing part of the underdetermination of the problem. We will show in this paper that the other way to reduce significantly the degrees of freedom in the inverse problem is to incorporate constraints applied to the data and the model space. Using the least squares criterion, these constraints can be expressed simply through covariance matrices.

2 Least squares inversion

Using a probabilistic formalism (Tarantola, 1987), we can rewrite the misfit function as the limit of a series of convergent functions S_n linearized in the vicinity of $\mathbf{m}_n$:

$$\begin{aligned}
S_n(\mathbf{m}_n + \delta\mathbf{m}) = {} & (\delta\mathbf{d}_n - \mathbf{G}_n\delta\mathbf{m})^t \, \mathbf{C}_D^{-1}(\delta\mathbf{d}_n - \mathbf{G}_n\delta\mathbf{m}) \\
& + (\Delta\mathbf{m}_n - \delta\mathbf{m})^t \, \mathbf{C}_m^{-1}(\Delta\mathbf{m}_n - \delta\mathbf{m}) + (\delta\mathbf{m})^t \, \mathbf{C}_a^{-1}(\delta\mathbf{m}),
\end{aligned} \quad (1)$$

where

- $\delta\mathbf{m}$ is the perturbation in the vicinity of the current model $\mathbf{m}_n$
- $\delta\mathbf{d}_n$ are the data residuals for the model $\mathbf{m}_n$
- $\Delta\mathbf{m}_n$ is the difference between the current model $\mathbf{m}_n$ and the *a priori* model $\mathbf{m}_{prior}$
- $\mathbf{C}_D$ is the covariance matrix over the data space (which combines experimental and theoretical uncertainties)
- $\mathbf{C}_m$ is the covariance matrix over the model space (defining the *a priori* probability density, here gaussian)
- $\mathbf{C}_a$ is a covariance matrix over the model space. This term will disappear at the convergence.

Applied to waveform inversion problem of estimating the density and the elastic parameters of the earth, the minimization of the misfit function can be solved by iterative gradient methods (Tarantola, 1988). Each iteration consists of the propagation of the actual source in the current medium and the propagation of residuals as if they were sources acting backward in time. The time correlation of these two fields yields the perturbation of the medium parameters. In principle, all the terms in equation (1) listed above in the misfit function can be incorporated. In practice, for reasons of simplicity, covariance matrices are set to identity (Kolb et al., 1986; Pica et al., 1990; Crase et al., 1990), although if introduced properly they can be a powerful tool, as we will show below, to constrain the inverse problem.

3   Constraints on the data space

Quantification of uncertainties in seismic data is not done systematically, because in many cases it is a difficult task. For instance, data errors generated by receiver response are correlated in time, which means accounting for them is not straightforward. To illustrate the possible contribution of constraints provided by the analysis of uncertainties of data, we have performed two inversions from the synthetic data experiment (Figure 1). The first inversion is done without any data constraints, whereas in the second one we have incorporated polarisation wave analysis in the covariance matrix $\mathbf{C}_D$. This matrix treats the uncertainties on displacement according to eigenbasis of the particle motion. The results of these inversions (Figure 2) show that constraints on the wavefront incidence to receivers contribute to recovering the sharp model interfaces at the receivers level.

4   Constraints on the model space

One often finds in the literature that the covariance on the model space, which should in principle describe the geological information, is used as an algorithmic tool. To avoid this confusion, we have explicitly separated them.

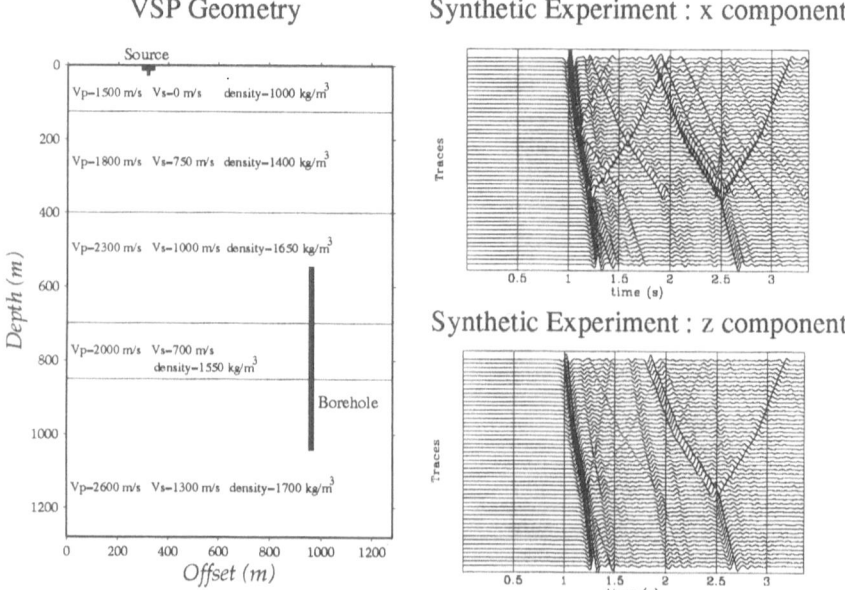

Fig. 1. Left: Synthetic experiment configuration where the cross near the surface denotes the source position, whereas the solid line denotes the location of the two-component receivers. Right: the two displacement components seismograms obtained for this experiment using a finite difference approximation to the wave equation.

4.1 Algorithmic constraints

We have added to the classical formulation of the misfit function (Tarantola, 1987) a covariance matrix C_a, to take algorithmic constraints into account. Stated in this way, this matrix is a generalization of linear damping which is usually taken as a constant diagonal matrix. This C_a term forces the algorithm to stay in regions where the model is, a priori, sensible, and when the convergence is reached its effect disappears.

4.2 Constraints on relationships among parameters

Statistical studies of parameters can be obtained from well logs, or in a general case from laboratory studies such as the well known relation of Gardner et al. (1974), which links P-wave velocity to density. This available information, for the case of a statistical study, can be simplified to obtain a Gaussian probability density or in the case of geological laws, linearized around a mean model. Then, the result will be introduced in the covariance matrix C_m. If we invert the same synthetic data (Figure 1), but this time putting a constraint on the correlation between parameters (here arbitrarily we have taken a 0.5 correlation between the three parameters), we can see on Figure 3 that compared to non-constrained inversion, a perturbation on one parameter (here on P-velocity) will lead to a perturbation on the other parameters.

Whithout constraint   **With polarization constraint**

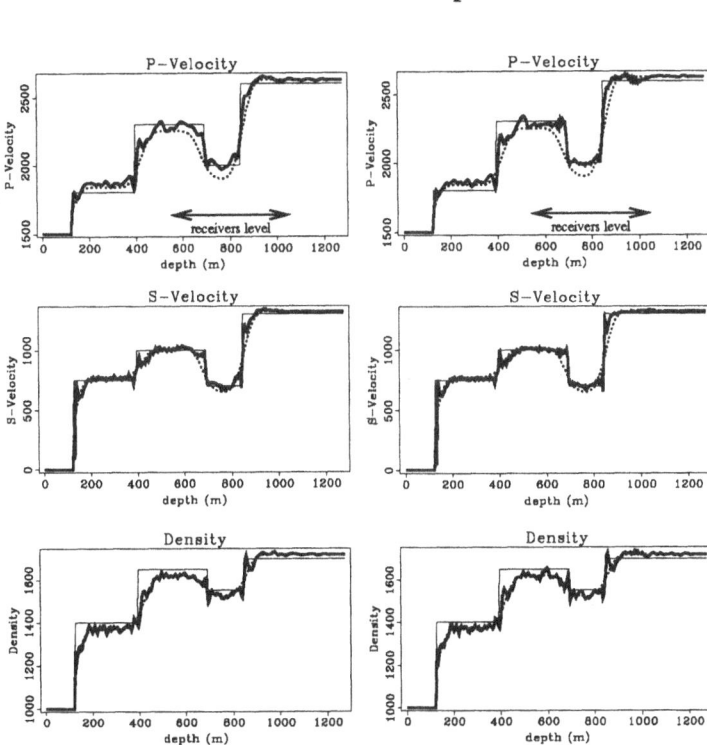

Fig. 2. Two inversion experiments: On the left without constraints, on the right with polarisation constraints on particle motion. Dashed lines for the initial model, thick solid line for the parameter model at the current iteration (here after 20 iterations) and the thin solid line for the true model. At the level of the receivers, the constrained inversion contributed to the recovery of the sharp interfaces.

4.3 Constraints on spatial distributions of parameters

We can distinguish two kinds of geological a priori information that can be incorporated in an inversion process. On the one hand information obtained by measurements (well logging), that we consider as "objective" information, allows the definition of an a priori model at the well vicinity. On the other hand, information coming from geological interpretations such as geological layer dips, that we qualify as "subjective" information, can be simplified and introduced into the covariance matrix $\mathbf{C}_m$.

5 Quantification of number of degrees of freedom

In order to evaluate the importance of constraints on an inversion, it is necessary to quantify the number of degrees of freedom. This number can be estimated from the set of parameters as a function of existing correlations among them and

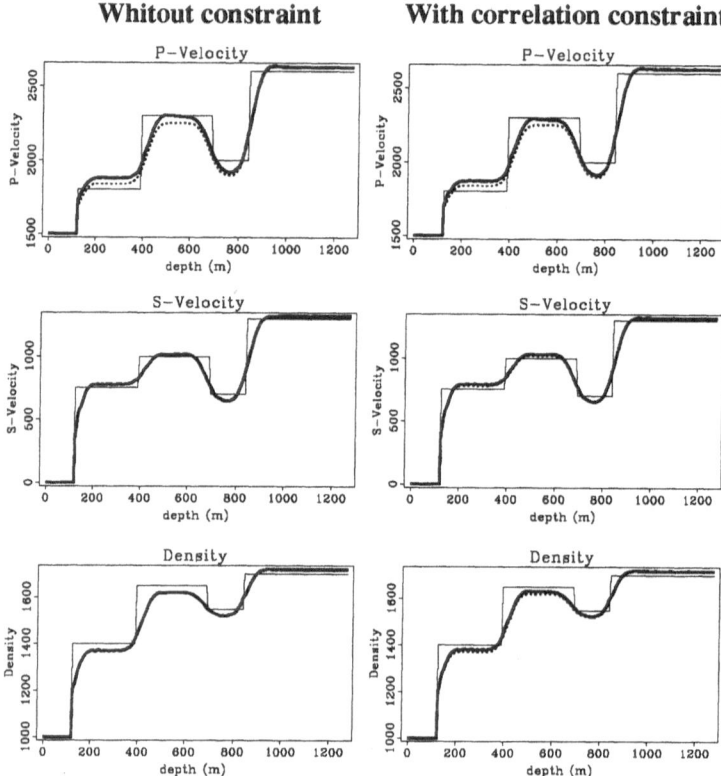

Fig. 3. Two inversion experiments: on the left without constraints, on the right with correlation between parameters. Dashed lines is the initial model, thick solid line is the parameter model at the current iteration (here first iteration) and the thin solid line is the true model. A perturbation on a parameter (here P-wave velocity) is propagating on other parameters.

independently of their standard deviations. The number N of free parameters is equal to the number n of parameters in the absence of correlations, and it is reduced to 1 in the case of perfect correlations.

5.1 General case

Let us consider a 1D Gaussian random field $\mathbf{X}$, the domain is discrete and finite consisting of a set of n points; this field is equivalent to the set of random variables $(X_1, X_2, \ldots X_n)$. It can be characterized by the following probability density function (or p.d.f.):

$$f(\mathbf{x}) = \frac{1}{(2\pi)^{n/2} \det(\mathbf{C})^{1/2}} \exp\left(-\frac{1}{2}(\mathbf{x} - \mathbf{m})^{\top} \mathbf{C}^{-1}(\mathbf{x} - \mathbf{m})\right), \qquad (2)$$

where: $\mathbf{x}$ is a realization, $\mathbf{m}$ is the expectation of the field $(\mathrm{E}(\mathbf{X}) = \mathbf{m})$, and $\mathbf{C}$ is the covariance matrix of the field.

The covariance matrix can be rewritten with the following form $C = DRD$, where R is the correlation matrix and D the diagonal matrix of the standard deviations.

If there is no correlation among the random variables X_i, then, $R = I$ and the covariance matrix is diagonal, the number of free parameters is then n. If a correlation exists, we then use the LU decomposition method to estimate the number of free parameters (see appendix A for more details). We can write $R = MM^\top$ where M is the left lower triangle matrix of the LU decomposition of the matrix R. The number of free parameters is then reduced to:

$$N = trace(M)$$

5.2 Spatial correlation on the model space

For illustration purposes, let us consider the special case where the spatial correlations are described by an exponential model. Thus, for any pair of points M_1 and M_2 (belonging to the same horizontal line), the correlation ρ between the pair of random variables X_1 and X_2 corresponding to the points M_i can be expressed as a function of the distance:

$$\rho(X_1, X_2) = \exp\left(-\frac{d(M_1, M_2)}{r}\right),$$

where $d(.)$ is a distance and r is the range of the correlation.

By setting the variance for each variable to be constant $\sigma_i = \sigma$, the p.d.f of equation 2 reduces to,

$$f(x) = \frac{1}{(2\pi)^{n/2} \cdot \sigma^n \cdot \sqrt{\det(R)}} \exp\left(-\frac{1}{2}\left(\frac{x-m}{\sigma}\right)^\top R^{-1}\left(\frac{x-m}{\sigma}\right)\right),$$

simplifying, as we will show below, the study of the relation between the matrix R and the degrees of freedom using an exponential correlation function.

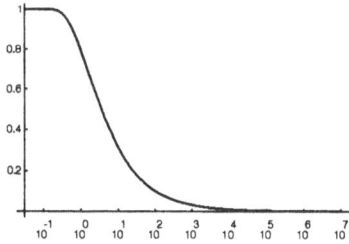

Fig. 4. Degree of freedom per point (for a regular grid) as a function of the range (expressed in multiples of the grid step) of the correlation for an exponential correlation function.

Let us consider the same Gaussian field X, but this time defined on a finite $1D$ domain regularly sampled (as for a grid) containing n points. The distance between points are multiples of the step Δx and the correlation matrix has the following form (for ordered points):

$$R = \begin{pmatrix} 1 & a & a^2 & a^3 & a^4 & \cdots & a^n \\ a & 1 & a & a^2 & a^3 & \cdots & a^{n-1} \\ \vdots & \vdots & \vdots & \vdots & & \ddots & \vdots \\ a^n & a^{n-1} & a^{n-2} & a^{n-3} & a^{n-4} & \cdots & 1 \end{pmatrix},$$

where

$$a = \exp\left(-\frac{\Delta x}{r}\right) \le 1.$$

One interesting property of the R matrix is that the M matrix of the LU decomposition has a simple form

$$M = \begin{pmatrix} 1 & 0 & 0 & 0 & 0 & \cdots & 0 \\ a & b & 0 & 0 & 0 & \cdots & 0 \\ a^2 & ab & b & 0 & 0 & \cdots & 0 \\ \vdots & \vdots & \vdots & \vdots & & \ddots & \vdots \\ a^n & a^{n-1}b & a^{n-2}b & a^{n-3}b & a^{n-4}b & \cdots & b \end{pmatrix},$$

with $b = \sqrt{1-a^2}$. This means that, in terms of a sequential simulation, the $n-1$ variables, X_2 to X_n, have the same degree of freedom: b. The total degrees of freedom N of the field X is then:

$$N = 1 + (n-1) \cdot b,$$

i.e., for n parameters (or n grid points), the number of free parameters, considered as the degrees of freedom, of the inversion is N. Figure 4 shows the relation between the range of the correlation (expressed in number of steps, Δx) and the values of b.

5.3 Real data example

We have performed two constrained inversions on filtered borehole data obtained from the North Sea. The geometry of acquisition can be seen from Figure 5. The spatially discretized window of parameters to invert consists of 230 horizontal elements and 456 vertical elements. The choice of a 1D model is motivated by a surface seismic migration image of the region, which shows an almost 1D earth model down to a depth of 3km. For the first inversion, we have applied a classical 1D inversion. The number of parameters involved corresponds to a column (456 points) times the three physical parameters to invert (P-wave velocity, S-wave velocity and density): in total 1368 free parameters. The second inversion is a 2D inversion (230x456 elements) assuming a lateral stationarity of relation among points that have been described by the exponential model of correlation function; the details are shown in table (1). As we have three physical parameters to invert, the total number of free parameters is $17940 \cdot 3 \approx 53800$, i.e. 5.8 times lesser than a non constrained inversion.

depth(m)	vertical number of grid points	horizontal range of the exponential model	degree of freedom per point	degrees of freedom of the area
0-2390	240	100	10	5520
2400-3090	70	30	5.4	2980
3100-4190	110	10	3.2	7910
4200-4550	36	30	5.4	1530
0-4550	456		5.8	17940

Table 1. The table shows the characteristics of the four homogeneous areas in term of correlation we have chosen and the number of free parameters per area. The values of the fourth column can be estimated from the figure 4. The last raw gives the values for the window of parameters to invert, i.e. the union of the four areas.

The 1D inversion converged after 45 iterations and was unable to explain the data, leaving 30 % of coherent residuals (Figure 6). The constrained 2D inversion explains most seismic phases, leaving 12.8 % of residuals that cannot be explained by our forward problem (3D effects) and some small reflected phase residuals that could have been explained at the 230th iteration due to slowness of convergence. Although our a priori knowledge of the geological structure of the region was weak, we have been able to impose sensible constraints, reducing by a factor of 5.8 the number of parameters and successfully explaining the data with a model (Figure 5) which is sensible from the geological point of view.

6 Conclusions

In this paper, we have illustrated how to introduce in the inverse problem, constraints consistent with the least squares formalism (through covariance matrices) on both the data and the model space. The incorporation of all our a priori knowledge of the parameters and all statistical studies on data, allows not only the algorithmic stabilization of the inversion process, but also the reduction of the solution set for an underdetermined problem.The purpose being not necessarily to converge quickly towards a good model (in term of residuals), but to prospect regions of the model space populated by models that are sensible, a priori, and also yielding the lowest possible misfit.

7 Acknowledgments

We thank Alessandro Forte for his appreciable remarks and comments. This research has been sponsored in part by the sponsors of the Groupe de Tomographie Géophysique and by the INSU (CNRS). We thank ELF for providing us the VSP data and the CNCPST for using their computational resources.

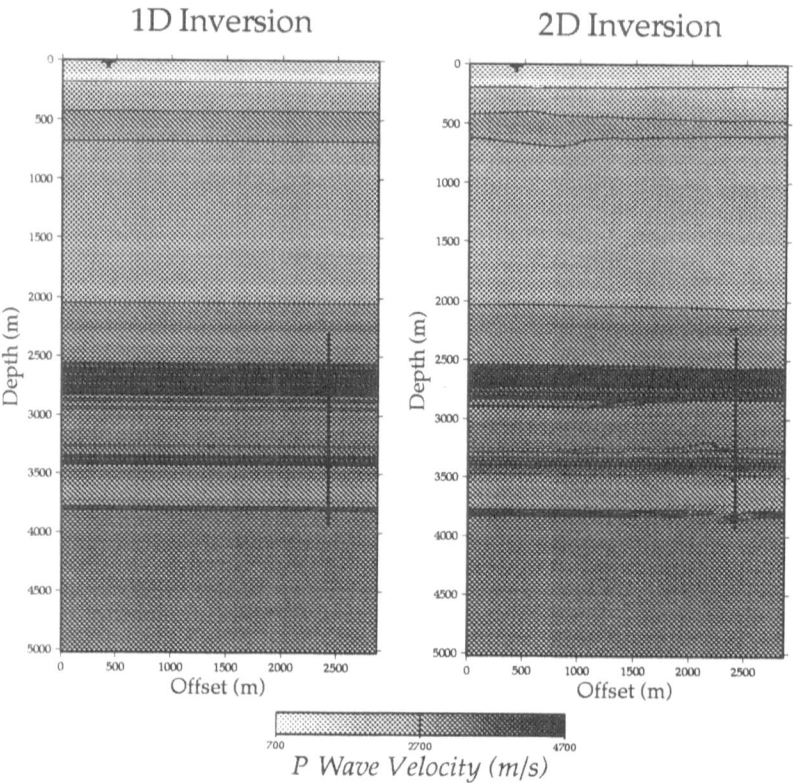

Fig. 5. On the left the result of a 1D inversion model, on the right the result of a 2D inversion model with horizontal correlation. Even with horizontal correlation constraints applied to the 2D inversion, the earth model obtained shows dipping layers at the well bottom which explains the associated reflected wave observed in data.

8 References

Crase, E., Pica, A., Noble, M., McDonald, J., Tarantola, A, 1990: Nonlinear inversion: application to real data. Geophysics, **55**, 527–538.

Gardner, G. H. F., Gardner, L. W., Gregory, A. R., 1974: Formation velocity and density–the diagnostic basis of stratigraphic traps. Geophysics, **39**, 770–780.

Kolb, P., Collino, F., Lailly, P., 1986: Pre-stack inversion of a 1D medium. Proceedings of the IEEE, **74**, 498–508.

Pica, A., Diet, J. P., Tarantola, A., 1990: Practice of nonlinear inversion of seismic reflection data in a laterally invariant medium. Geophysics, **55**, 284–292.

Tarantola, A., 1987: Inverse Problem Theory: methods for data fitting and model parameter estimation. Elsevier Science Publishing Co.

Tarantola, A., 1988: Theoratical background for the inversion of seismic waveforms, including elasticity and attenuation. Pure Appl. Geophys., **128**, 365–399.

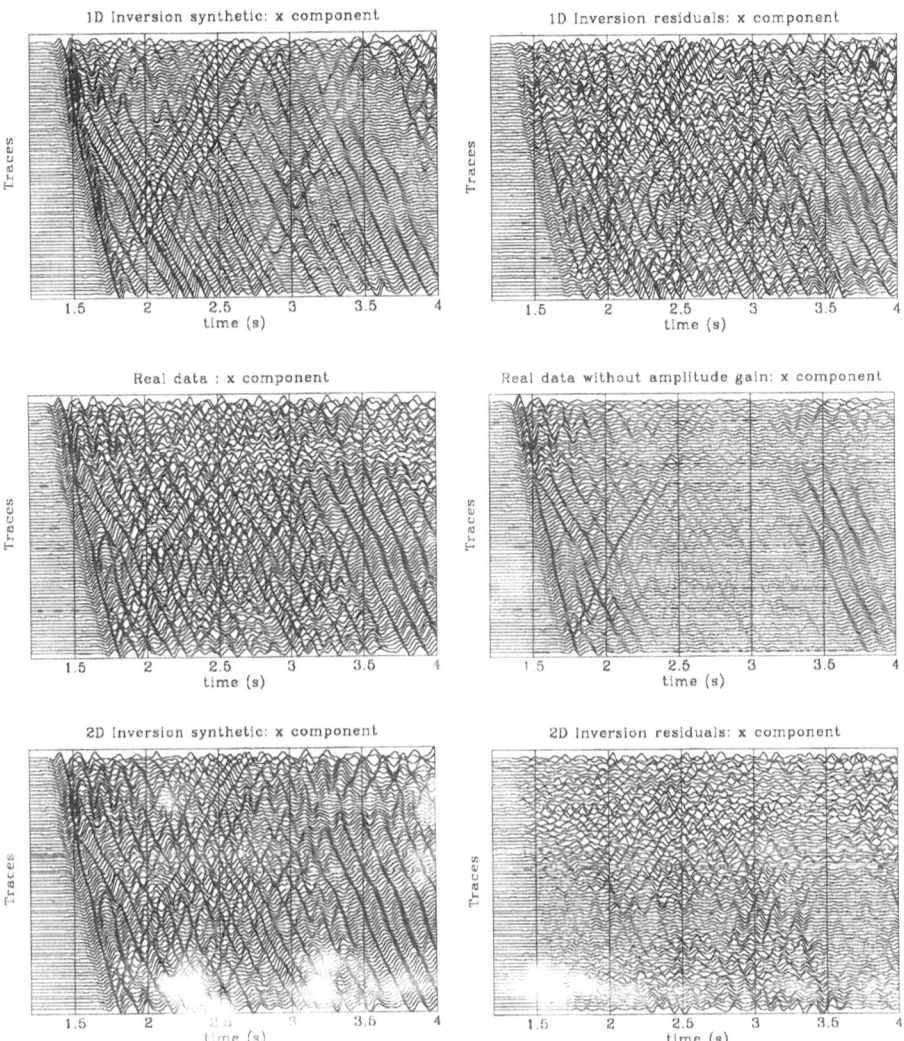

Fig. 6. Top: synthetic seismograms on the left and the corresponding residuals on the right for 1D inversion; Middle: low pass filtered real data (0-12Hz), with amplitude gain on the left and without on the right; Bottom, synthetic on the left and on the right residuals seismograms for 2D constrained inversion. The coefficients of the gain (used for display purposes) has been computed from real data, and applied to all seismograms except real data at the right center. The 1D inversion was unable to explain correctly the data, the residuals (upper right) are still important and coherent. This not the case for the 2D horizontally constrained inversion, the residuals (bottom right) are mainly unstructured and the phases still unexplained could be due to 3D effects. The obtained seismograms are very close to the real data, reproducing its complexity.

A Correlation and degrees of freedom

A.1 Generalities

Let us consider a 1D Gaussian random field $\mathbf{X}$ defined in section 5 and characterised by its p.d.f. given by equation 2.

If the covariance matrix $\mathbf{C}$ is diagonal, (i.e. no correlation between the random variables X_i), equation (2) can be rewritten:

$$f\left(\mathbf{x}\right) = \frac{1}{\left(2\pi\right)^{n/2} \prod_i \sigma_i} \exp\left(-\frac{1}{2} \sum_i \left(\frac{x_i - m_i}{\sigma_i}\right)^2\right)$$

$$= \prod_i \frac{1}{\sqrt{2\pi}\sigma_i} \exp\left(-\frac{1}{2}\left(\frac{x_i - m_i}{\sigma_i}\right)^2\right) = \prod_i f_i\left(x_i\right),$$

(3)

where σ_i is the standard deviation of the Gaussian random variable X_i for the ith point, and $f_i\left(x_i\right)$ the p.d.f. associated with this variable X_i. For a Gaussian field, when correlation is inexistent, the variables X_i are independent and as a consequence the p.d.f. of the field is the product of the marginal p.d.f. of the variables, the f_i.

Assuming for simplicity that the m_i and the σ_i are constant, the field is made of n independent variables, identically distributed. Then, considering we have in this case n free parameters, we will study the effect of the non independence of variables on the number of the free parameters, incorporating correlation between variables through a non diagonal covariance matrix.

A.2 Sequential simulation

Whatever the discrete random field, the knowledge of the joint p.d.f. (like f defined in equation 2 for the Gaussian field $\mathbf{X}$) allows to simulate a realisation of the field sequentially using the marginal-conditional decomposition (correct whatever the order of the indexes):

$$f\left(x_1, x_2, ...x_n\right) = g_1\left(x_1\right) \cdot g_2\left(x_2|x_1\right) \cdot g_3\left(x_3|x_1, x_2\right) \cdot ... g_n\left(x_n|x_1, x_2, ...x_{n-1}\right), \quad (4)$$

where g_i is the marginal-conditional p.d.f. of the random variable X_i independently of the random variables X_j for $j > i$, and conditionally to the random variables X_k for $k < i$. More precisely, g_i is defined by the relation:

$$g_i\left(x_i|x_1, x_2, ...x_{i-1}\right) = \frac{\int dx_{i+1} \int dx_{i+2} ... \int dx_n f\left(x_1, x_2, ...x_n\right)}{\int dx_i \int dx_{i+1} ... \int dx_n f\left(x_1, x_2, ...x_n\right)}.$$

Using equation 4, it is possible to simulate a realisation of the field sequentially. First, we use $g_1\left(x_1\right)$ to generate a realisation of the random variable X_1 independently of all other variables. Once $x_1 = x_1^0$ is determined, we use $g_2\left(x_2|x_1^0\right)$

to generate a realisation x_2^0 of the random variable X_2 dependent to the value x_1^0, but independent of the other random variables X_3 to X_n. We repeat this procedure using g_3, then g_4 and so on, until the last value x_n^0 is determined, thereby obtaining a realisation $\mathbf{x}^0$ of the field $\mathbf{X}$.

We can link the degree of freedom of the variable X_i, when the values x_1 to x_{i-1} are known, to the standard deviation σ_i^c of the Gaussian defined by the p.d.f. $g_i(x_i|x_1, x_2, ...x_{i-1})$. To account for the correlation between the variables X_i independently of the values of the standard deviations σ_i of the marginal p.d.f. $f_i(x_i)$, we can then define the degrees of freedom of the variable X_i (when x_1 to x_{i-1} are known) as the following ratio:

$$d = \frac{\sigma_i^c}{\sigma_i},$$

which equals one when X_i is not correlated to the variables $(X_1, ...X_{i-1})$ and zero when the correlation is perfect.

A.3 The LU simulation technique

In order to simplify the calculation of the degrees of freedom, we introduce here the **LU** decomposition technique for Gaussian field simulation as this method can be interpreted in term of sequential simulation using equation 4.

A way to simulate a Gaussian discrete random field $\mathbf{X}$ defined by the p.d.f. in equation 2 (n points with expectation $\mathbf{m}$ and covariance matrix $\mathbf{C}$) is the use of the **LU** decomposition of the covariance matrix $\mathbf{C}$:

$$\mathbf{C} = \mathbf{LU} = \mathbf{LL}^\mathsf{T}, \tag{5}$$

where $\mathbf{L}$ denotes the left lower triangle matrix and $\mathbf{U}$ the upper right one.

Considering a random field $\mathbf{E}$ constituted by n random variables, independently and identically distributed following a Gaussian law with zero mean and unit standard deviation; the field $\mathbf{m} + \mathbf{LE}$ is then equivalent to $\mathbf{X}$. Therefore, to simulate a realisation $\mathbf{x}$ of the field $\mathbf{X}$, we can use a realisation $\mathbf{e}$ of the field $\mathbf{E}$ and write:

$$\mathbf{x} = \mathbf{m} + \mathbf{Le}. \tag{6}$$

A.4 Interpretation of the LU decomposition of the covariance matrix

Let l_{ij} be the elements of the matrix $\mathbf{L}$. The relation given by equation 6 can be interpreted in term of sequential simulation: the first line gives $x_1 = m_1 + l_{11}e_1$, meaning that we simulate the realization of X_1 independently of the other variables, as if we used the $1D$ marginal p.d.f. $g_1(x_1)$ defined in equation 4:

$$g_1(x_1) = \frac{1}{\sqrt{2\pi}\sigma_1} \exp\left(-\frac{1}{2}\left(\frac{x_1 - m_1}{\sigma_1}\right)^2\right) = f_1(x_1). \tag{7}$$

Once the value of x_1 is fixed, the second line of the relation 6 means that the second random variable X_2 can be written:

$$X_2 = m_2 + l_{21}e_1 + l_{22}E_2 = m_2 + l_{21}\left(\frac{x_1 - m_1}{l_{11}}\right) + l_{22}E_2,$$

meaning that we simulate the realisation of X_2 independently of $X_3, ...X_n$ but dependent on the value $X_1 = x_1$, as if we used the conditional marginal p.d.f. $g_2(x_2|x_1)$(see equation 4). A given line i of the relation 6 means we can simulate a random variable X_i conditionally to the determined values of the preceding variables x_1 to x_{i-1}, as if we used the proper marginal-conditional p.d.f. $g_i(x_i \mid x_1, x_2, ...x_{i-1})$ (see equation 4).

Let us come back to the estimation of the degrees of freedom. The standard deviations σ_i^c are given here by the ith diagonal element L_{ii} of the matrix $\mathbf{L}$. This standard deviation is lesser than the standard deviation of the variable X_i independently of the other variables, (i.e. the standard deviation σ_i of the Gaussian defined by the marginal p.d.f. $f_i(x_i)$). When correlation between X_i and the variables X_1 to X_{i-1} is strong, the ratio $\dfrac{L_{ii}}{\sigma_i}$ is weak. When X_i is independent of the variables X_1 to X_{i-1} the ratio equals 1, meaning the freedom of the variable is total. We can say that when variables are independent (no correlation) the number of the free parameters is $n = \sum_i \frac{1}{1}$ and when correlation exist it is reduced to:

$$N = \sum_i \frac{l_{ii}}{\sigma_i}.$$

A.5 The LU decomposition of the correlation matrix R

The covariance matrix can be written with the following form:

$$\mathbf{C} = \mathbf{DRD},$$

where $\mathbf{R}$ is the correlation matrix and $\mathbf{D}$ the diagonal matrix of the standard deviations, i.e.:

$$\mathbf{D} = \begin{pmatrix} \sigma_1 & 0 & \cdots & 0 \\ 0 & \sigma_2 & \cdots & 0 \\ \vdots & \vdots & \ddots & \vdots \\ 0 & 0 & \cdots & \sigma_n \end{pmatrix}.$$

The **LU** decomposition of $\mathbf{R}$ gives: $\mathbf{R} = \mathbf{MM}^\top$, with $\mathbf{L} = \mathbf{DM}$. Therefore, the field $\mathbf{X}$ can be written (see equation 6): $\mathbf{X} = \mathbf{m} + \mathbf{LE} = \mathbf{m} + \mathbf{DME}$.

The number of free parameters is then:

$$N = \sum_i \frac{l_{ii}}{\sigma_i} = trace(\mathbf{M}).$$

B Degrees of freedom for a pair of Gaussian random variables (X, Y)

Let us consider a pair of random variables (X, Y) characterized by the following p.d.f.:

$$f(x, y) = \frac{1}{2\pi\sqrt{\det(\mathbf{C})}} \cdot \exp\left(-\frac{1}{2}\begin{pmatrix} x - m_x \\ y - m_y \end{pmatrix}^{\top} \mathbf{C}^{-1} \begin{pmatrix} x - m_x \\ y - m_y \end{pmatrix}\right),$$

where $\mathbf{C}$ is the covariance matrix, we have:

$$\mathbf{C} = \begin{pmatrix} \sigma_x^2 & \rho\sigma_x\sigma_y \\ \rho\sigma_x\sigma_y & \sigma_y^2 \end{pmatrix}, \det(\mathbf{C}) = \sigma_x^2\sigma_y^2(1 - \rho^2), \mathbf{C}^{-1} = \frac{1}{1 - \rho^2}\begin{pmatrix} \frac{1}{\sigma_x^2} & \frac{-\rho}{\sigma_x\sigma_y} \\ \frac{-\rho}{\sigma_x\sigma_y} & \frac{1}{\sigma_y^2} \end{pmatrix},$$

where ρ is the correlation between the random variables X and Y. An example is given on the Figure 7, with the following values: $m_x = m_y = 0$, $\sigma_x = \sigma_y = 1$, $\rho = 4/5$.
Equation 4 of the decomposition of the p.d.f. is written:

$$f(x, y) = f_1(x) \cdot f_2(y \mid x),$$

where:

$$f_x(x) = \frac{1}{\sqrt{2\pi}\sigma_x} \cdot \exp\left(-\frac{1}{2}\left(\frac{x - m_x}{\sigma_x}\right)^2\right),$$

$$f_{y|x}(y \mid x) = \frac{1}{\sqrt{2\pi}\sigma_y\sqrt{1 - \rho^2}} \cdot \exp\left(-\frac{1}{2}\left(\frac{y - m_y - \rho\sigma_y\left(\frac{x - m_x}{\sigma_x}\right)}{\sigma_y\sqrt{1 - \rho^2}}\right)^2\right).$$

We can note that the characteristics of these p.d.f. are given by the elements of the left triangle matrix of the covariance $\mathbf{LU}$ decomposition:

$$\mathbf{L} = \begin{pmatrix} \sigma_x & 0 \\ \rho\sigma_y & \sigma_y\sqrt{1 - \rho^2} \end{pmatrix}.$$

Indeed, the first line corresponds to the standard deviation of the p.d.f. $f_x(x)$. The conditional p.d.f. $f_{y|x}(y \mid x)$ is plotted on Figure 7. It is characterised by its mean $m_c = m_y + \rho\sigma_y\left(\frac{x - m_x}{\sigma_x}\right)$, and its standard deviation $\sigma_c = \sigma_y\sqrt{1 - \rho^2}$. In the second row, the first element corresponds to the scale factor of the correction of the mean of the conditional p.d.f. $f_{y|x}$, and the second element to the standard deviation of the conditional p.d.f.. In summary, the diagonal elements are the standard deviations of the random variables X and $Y \mid X$. The effect of the correlation ρ is to reduce, by the factor $\sqrt{1 - \rho^2}$, the degree of freedom of the second variable Y when x is known, as we see in Figure 7. We then note that

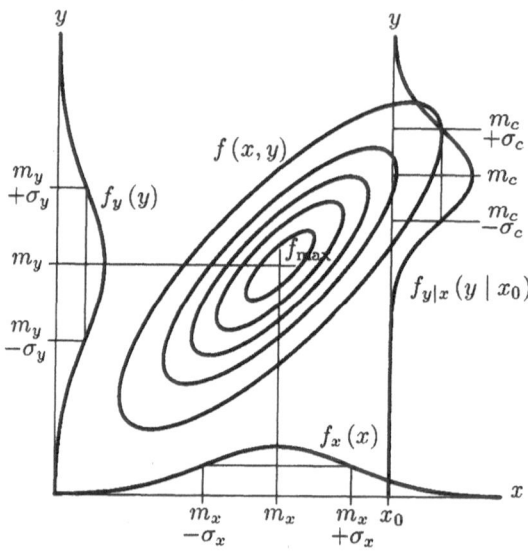

Fig. 7. The joint probability density $f(x, y)$ of the couple of random variable (X, Y) is plotted using contour levels (at level $0.9 \cdot f_{max}$, $0.7 \cdot f_{max}$, $0.5 \cdot f_{max}$, $0.3 \cdot f_{max}$ and $0.1 \cdot f_{max}$), the marginal probability densities for the variables themselves ($f_x(x)$ for X and $f_y(y)$ for Y) are plotted on the x and the y axes. The value $x_0 = 1.5$ has been fixed and the conditional probability density $f_c(y) = f_{y|x}(y \mid x_0)$ is plotted, the mean $m_x = m_y = 0$ and m_c has been reported as the standard deviations $\sigma_x = \sigma_y = 1$ and σ_c, the correlation $\rho = 0.8$ appears only in the ellipticity of the ellipses for the joint p.d.f. f. As equation 8 shows, the number of free parameters of the couple (X, Y) is $1 + \dfrac{\sigma_c}{\sigma_y} = 1 + \sqrt{1 - \rho^2} = 1.6$.

the total degree of freedom of this pair is (taking as a reference the uncorrelated pair characterized by the diagonal covariance matrix, $\rho = 0$):

$$d = \frac{\sigma_x}{\sigma_x} + \frac{\sigma_y \sqrt{1 - \rho^2}}{\sigma_y} = 1 + \sqrt{1 - \rho^2}. \tag{8}$$

We can note that $d = 1$ when the correlation is perfect ($\rho = \pm 1$) and $d = 2$ when the correlation is inexistent ($\rho = 0$).

Geological Information and the Inversion of Seismic Data

Christophe Barnes[1,2], *Marwan Charara*[1] *and Albert Tarantola*[1]

[1] Institut de Physique du Globe de Paris, France
[2] École des Mines de Paris, France

Introduction

In the inversion of seismic data, the incorporation of a priori information of geological nature requires, for theoretical reasons, the use of Monte-Carlo algorithms. Reciprocally, pragmatic reasons of efficiency involve the introduction of a priori information, which, in our case, can be geological information.

Realistic geological information can not be correctly described by well-shaped probability densities like those which are imposed by gradient algorithms. These algorithms are limited by their mathematical basis to probability densities related to norms (normed vectorial space structure is necessary for these algorithms), for example, Gaussian probability densities for the L2 norm (least squares).

In fact, only algorithms based on a probabilistic formalism (measurable space structure) are able to take into account complex information and thus, correctly incorporate realistic geological information. This fundamental remark explains why we have chosen to develop Monte-Carlo methods for inverse problems.

We can restate the incorporation of a priori information for an inverse problem by using the information theory formalism. The general solution of the inverse problem can be considered as a conjunction of states of information, which means in our case, the product of the probability densities representing the geological and the geophysical information (we can note the symmetrical role of both informations). These two kinds of information do not constrain the parameters of the models in the same way. Therefore, the geological information, when it is available, can drastically reduce the number of freedom degrees in the inversion process and then increases significantly the efficiency of Monte-Carlo algorithms. If we now express that in terms of realization of random fields, we can say that the introduction of constraints during the pseudo-random generation of models enables us to reduce the volume of the model space that needs to be explored. This overcomes the most important problem of Monte-Carlo methods: their relative inefficiency (calculation cost).

1 The Monte-Carlo Method

1.1 The Metropolis Algorithm

Let ρ be the the a priori probability density describing a priori information, let L be the likelihood measuring the misfit between real data and synthetic data, and let σ be the a posteriori probability density or, in other words, the general solution of the inverse problem.

Then, if $\mathbf{m}$ denotes any model of the model space, we have to sample the probability density defined by (cf. Tarantola and Valette (1982)):

$$\sigma(\mathbf{m}) \ = \ K \cdot \rho(\mathbf{m}) \cdot L(\mathbf{m}) \tag{1}$$

The Metropolis algorithm first introduced by Metropolis and al. (1953) can be used to sample σ. Assuming that we know how to sample the a priori probability density ρ: let $\mathbf{m}_i$ be the current model, $\mathbf{m}_{i+1}$ be the tested model, we now calculate $L(\mathbf{m}_{i+1})$.

• If $L(\mathbf{m}_{i+1}) > L(\mathbf{m}_i)$, we take the tested model as the new current model,
• and if $L(\mathbf{m}_{i+1}) < L(\mathbf{m}_i)$, we take the tested model with the probability:

$$P = \frac{L(\mathbf{m}_{i+1})}{L(\mathbf{m}_i)}.$$

This procedure ensures that the set of all retained models is representative of the a posteriori density σ.

1.2 Remarks for an efficient sampling of probability densities

In general, the Metropolis algorithm can be used in order to sample any probability distribution. This means it can be used, for example, to sample the a priori probability density ρ (if the generation of models according to the a priori probability density ρ can not be done directly).

First remark: the choice of the perturbation law for the random walk. The most important feature we have to define in the Metropolis algorithm is the graph of the possible perturbations (that is, the different laws we give to the algorithm to define how to draw $\mathbf{m}_{i+1}$ when we have $\mathbf{m}_i$). First, the graph must have some properties to ensure the convergence of the algorithm (see Mosegaard and Tarantola (1995)). Moreover, the choice of the graph determines drastically the efficiency of the algorithm. A simple rule to keep in mind is: to sample a probability density with a high efficiency, it is better to walk on its level lines (meaning that we prefer the biggest model parameters perturbations that keep the computed data as unchanged as possible).

Second remark: How to decrease the number of forward problem to be resolved. When the resolution of the forward problem requires several computations, we can partition the data space. Moreover, if the data (moreprecisely the uncertainties of observed data) from the different groups are independent, the likelihood can be written as a product of sub-likelihoods,

$$L(\mathbf{m}) = \prod_i (L_i(\mathbf{m})). \tag{2}$$

Thus, the synthetic data of the different groups can be computed separately as well as the different sub-likelihoods. It is then straightforward to apply the Metropolis algorithm successively for each group of data. If the order of the group of data to be tested by Metropolis is well chosen, this procedure can thus be more efficient than the evaluation of the global $L(\mathbf{m})$.

Third remark: approximation of the forward problem. Another way to make Monte-Carlo more efficient is to introduce a well chosen weighting function $h(\mathbf{m})$ in order to modify the coefficients of the graph defined by $L(\mathbf{m})$. This implies that we can rewrite equation (1) as:

$$\sigma(\mathbf{m}) = K \cdot \rho(\mathbf{m}) \cdot h(\mathbf{m}) \cdot \left(\frac{L(\mathbf{m})}{h(\mathbf{m})}\right). \tag{3}$$

For example, $h(\mathbf{m})$ can correspond to a simple and approximative calculation of synthetic data (in order to save time). Naturally, $h(\mathbf{m})$ must be consistent with $L(\mathbf{m})$ (i.e., give high probabilities where $L(\mathbf{m})$ is high) and be less informative than $L(\mathbf{m})$, in order to have a well defined quotient everywhere in the model space.

2 1D Inversion of Arrival Times

The following is an illustration of incorporation of geological information in a Monte-Carlo process applied to a simple problem of arrival-time inversion for a set of well seismic data (Offset VSP data). In order to estimate the general solution of the inverse problem (i.e., to retrieve the set of all likely P and S velocity models, we take into account as many arrivals as possible (e.g. transmitted P wave, multiples, converted waves P-S, etc).

2.1 Geological Information

The geological information is of a different nature. The sonic log (see figure 1) provides information about the vertical variations of the P wave velocities in the well vicinity, whereas non geological information from other geophysical experiments (e.g., surface seismic survey) ensures that the sub-surface in this region is almost flat, down to approximately 3000 meters depth. This allows us to make, in a first step, a 1D inversion of the data. Moreover, geological surveys in the region provide information on statistical relationships between S wave velocities and P wave velocities (only indirect informations on S wave velocities are available).

For the a priori information, we have defined positions of interfaces on the sonic log and we have characterized these depths by non-symmetric triangular shaped probability distributions (quite arbitrary). Outside the log limits we infer the existence and the depth of other interfaces from migrated surface seismic

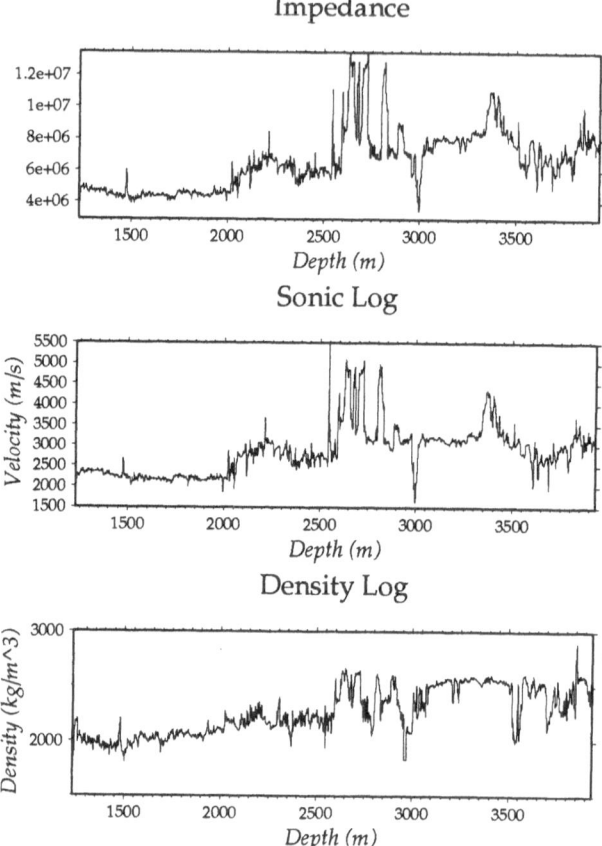

Fig. 1. The sonic log and the density log are measurements made in the well. The impedance log is only their product. Note that we do not have information in the upper part of the well, nor direct information on S velocities.

data (with more uncertainties). Although the depths are well known in the log vicinity, 2D structures exist below 3000m depth, therefore we increase artificially the uncertainties for the positions of these interfaces. We have estimated 1D distributions for the P wave velocity in each layer assuming that the random process within each layer is stationary. Then, even if the amount of available data is weak, we can state that in some layers the probability density tends to be Normal whereas it is probably log-Normal in others. Lastly, using petrological surveys, we have inferred correlations between the S-wave and P-wave velocity distributions.

2.2 The Observed and Computed Data

The picking of the arrival times have been made for 14 different phases (see figure 2). The computed times are obtained applying Fermat principle on a hor-

Arrival time

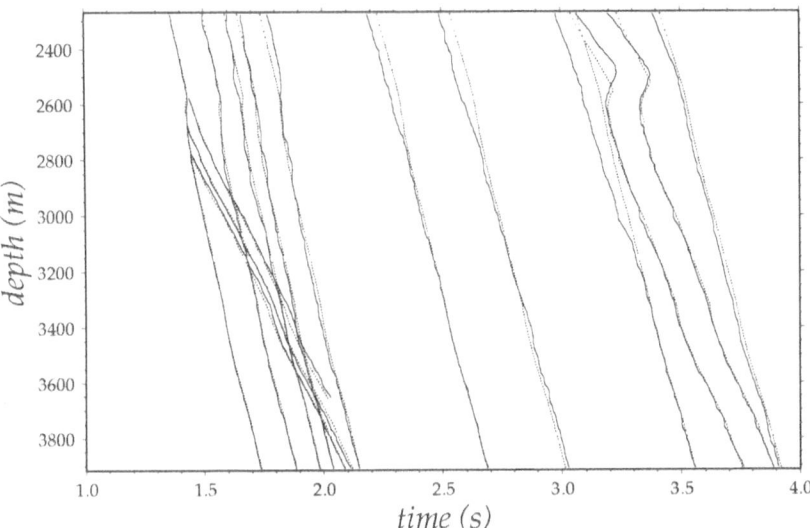

Fig. 2. Fourteen picked phases used in the inversion, in solid lines (P direct arrival, P multiples between the free surface and reflectors, converted P-S waves). The picked phases have to be associated with a path in the model in order to model them. The corresponding computed arrival times are shown in dashed lines (computed from a model sampled from the a posteriori probability density). Note that some systematic errors exist, they are typically one period length, the L_1 norm and the non uniform uncertainties associated to each picked events imply that part of the computed data does not fit.

izontally invariant velocity model. The more important practical problem is to associate picked phases to a defined path in the 1D Earth model. A wrong association could severely bias the solution and, to avoid this we have first used the "trial and error" method, verifying with finite difference simulation that our interpretation was not too wrong. The L_1 norm is used to compare calculated and observed arrival times. Moreover, we have associated a confidence uncertainty for each picked arrival.

2.3 The Algorithm

Once the a priori information was defined, the Metropolis algorithm has been used to sample the a priori probability density $\rho(\mathbf{m})$ (see subsection 1.2). An example of a random walk characterizing this probability density is shown on figure 3.

As the different phases can be computed separately and their uncertainties considered as independent, equation (2) has been used (decomposition of the likelihood in several sub-likelihood). Since the uncertainties associated with the

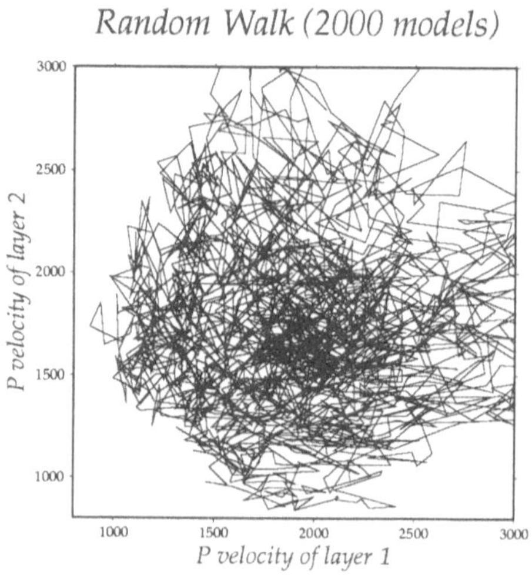

Random Walk (2000 models)

Fig. 3. Example of random walk, here it defines the a priori probability density for the parameters of the layers 1 and 2 of the P-wave velocity model.

selection of the first P arrivals are smaller than the uncertainties for other phases, the first sub-likelihood we test in the successive Metropolis processes is the sub-likelihood associated with this phase (as it is more selective in terms of rejected models, then for a large part of the tested models we just have to calculate the arrival times for this phase). Finally, we have introduced a precomputation of arrival time based on root-mean-square P and S velocities, in order to rapidly reject many of the unconsistant models. This procedure defines a weighting function $h(\mathbf{m})$ (as explain in subsection 1.2, see the equation (3)), which modifies the coefficients of the graph.

2.4 Results of the inversion

The set of all retained models is the solution of our inverse problem. In order to show the information given by the data, the a priori and the a posteriori informations are shown. Two representations are used in order to show in detail the relationship between two given parameters of the model (2D-histograms are plotted), and also to show the variability of the models sampling the a priori and a posteriori probability densities (i.e., the models kept during the corresponding random walk). See figure 4-5-6 for the discussion of the results.

Although this representations are not necessary for such a simple case (the a posteriori histograms can be described by a simple statistical model, as Gaussian, for example), a more complicated case will probably show interesting features in these representations. A complete description of the solution would then requires an accurate study of these features.

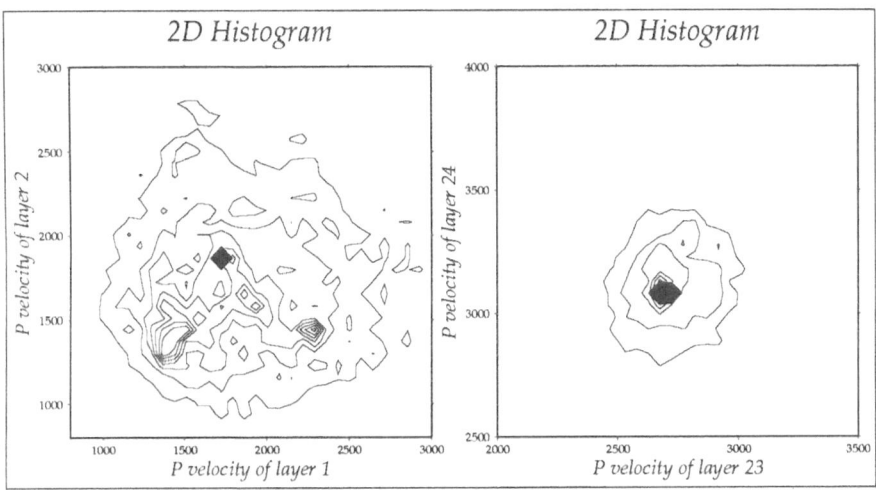

Fig. 4. Histograms representing different 2D marginals of the a priori probability density (the thin level lines) and the a posteriori probability density (the thick level lines) when the 14 phases arrival times are inverted. The 2D marginals are projections of the (3×34)D probability density (P wave velocity, S wave velocity and depth of the interface for each of the 34 layers). In other words, we show here the relationship between two given parameters of the inversion. On the left histogram, the considered layers correspond to a shallow depth (above the region where log measurement are available), therefore the a priori histogram is spread, containing few information. On the right, the a priori histogram is sharper containing information from the sonic log, we can notice that the a posteriori and the a priori histograms are centered on the same range of velocities, meaning that the information contained in the data is consistant with the values estim. d from the sonic log. Nevertheless, the data have brought information in regards to the a priori information as the a posteriori histogram is sharper than the a priori histogram.

Conclusion

The preceding example, even simple, demonstrates the power and utility of Monte-Carlo methods. There is no theoretical limitation for applying this method to a more complicated problem, e.g., incorporation of complex geological information for the model space or, in the data space, introduction of probabilities to characterize the association between phases and paths in the models (see section 2.2).

The main difficulty remains the cost of the resolution of the forward problem. But, it is always possible (if information is available) to reduce the size of the volume of the model space to be investigated (it is unavoidable if the dimension of the model space is too high). A critical point is also to define an appropriate graph in the model space, and an optimal circulation in this graph to make Monte-Carlo methods efficient.

On the other hand, if the a priori information is correctly introduced and the statistics of the uncertainties on the data space are well described, then Monte-

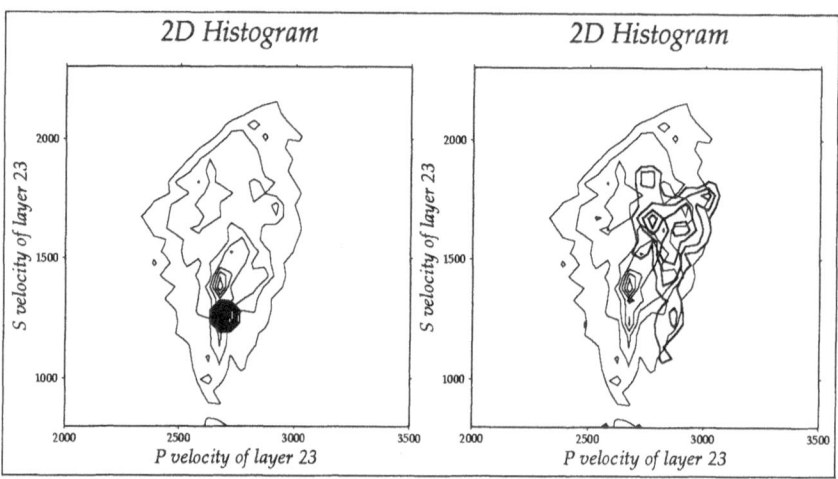

Fig. 5. Histograms representing relationship between the P wave velocity and the S wave velocity of the 23th layer when the 14 phases are inverted (including converted P-S waves) on the left and when only the first P-wave arrival times is inverted on the right. The thin level lines denote the a priori probability density and the thick level lines the a posteriori probability density. Note that the addition of S-wave phases in the inversion process increases the resolution on the S wave velocity parameter. In a similar way, we have notice that the addition of multiple phases tends to increase the a posteriori resolution. Moreover, on the left histogram, we can notice that the estimated S velocity for the a priori information (provided by geological regional studies) has a high value compared to the inversion results meaning the regional studies provide approximative information for this borehole.

Carlo methods enable direct access to the (possibly non-linear) resolution of the inversion solution.

Acknowledgments

We thank Alessandro Forte for the clarifying discussions we had. This research has been supported in part by the sponsors of the Groupe de Tomographie Géophysique and by the INSU (CNRS). We thank ELF for providing us the VSP data and the CNCPST for using their computational ressources.

References

Metropolis, N., Rosenbluth, A.W., Rosenbluth, M.N., Teller, A.H., and Teller, E., 1953, Equations of state calculations by fast computing machines, *Journal of Chemical Physics*, **21**, 1087–1092.

Mosegaard, K., and Tarantola, A., 1995, Monte Carlo Sampling of Solutions to Inverse Problems, *Journal of Geophysical Research*, **100**, **B7**, 12431–12447.

Tarantola, A., and Valette, B., 1982, Inverse Problems = Quest for Information, *J. Geophys*, **50**, 159–170.

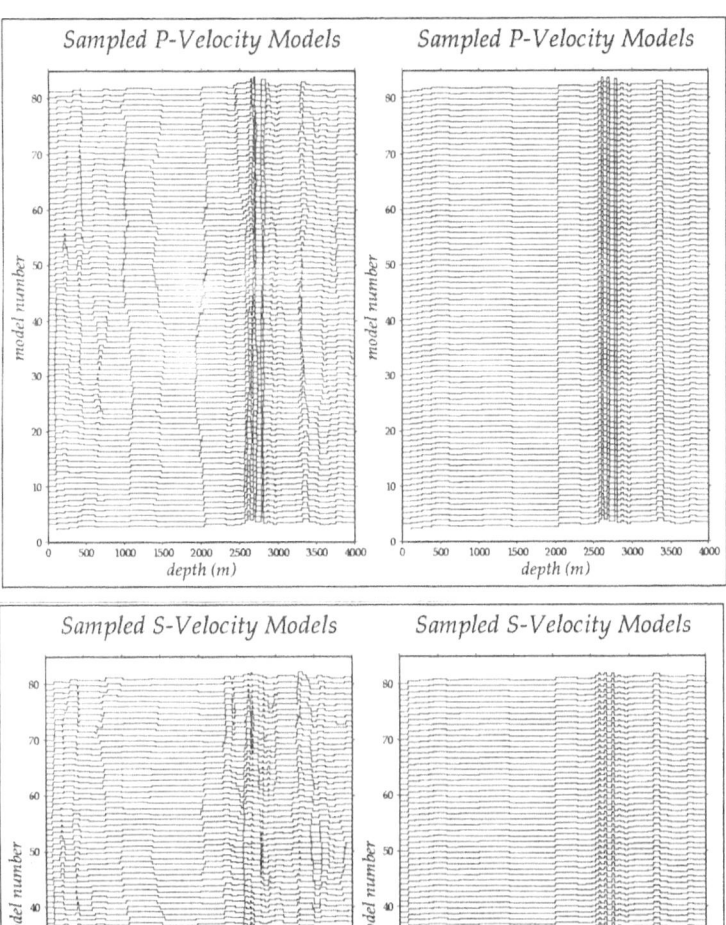

Fig. 6. 80 models (1 every 100 kept models during the random walk), P-velocity models above, and S-velocity models bellow, on the left for the a priori probability density, and on the right for the a posteriori probability density. Note that, as the weak variability of a the posteriori models shows, the a posteriori probability density can be well described by a central model and uncertainties around it (using a Gaussian distribution for example). Then, if the a posteriori probability density has some statistical properties, we can justify, a posteriori, the choice of a statistical model (like a Gaussian distribution) and test the validity of this choice using statistical tools like the χ^2 test. This procedure is fundamentally different than imposing a priori, a given statistical model (as for least squares for example).

A Study of Model Covariances in Amplitude Seismic Inversion

Wences Gouveia

Colorado School of Mines, Center for Wave Phenomena, Golden, Colorado, USA

1 Introduction

Seismic amplitude or travel time inversion methods are a major topic of geophysical research due to their potential for extracting detailed lithologic information about the subsurface. Several inversion methodologies are described in the literature. Although the procedures differ, it is acknowledged in all of them that the underlying model has poorly-resolved features not constrained by the data. To reduce the non-uniqueness of the inverse problem it is necessary to incorporate *a priori* information about the underlying model. The Bayesian approach for geophysical data inversion (Tarantola 1987) paves the way for the incorporation of such knowledge. In this methodology, the general solution of an inverse problem is defined as a probability density $\sigma(\mathbf{m})$ over the space of models $\mathbf{m}$, that is proportional to the product of two probability density functions. One, known as the likelihood function $L(\mathbf{m})$, measures the extent that the observed data are fit by model data. This function accounts for uncertainties in the data, i.e., features that were not taken into account in the forward modeling step. The other probability density function, $\rho(\mathbf{m})$, quantifies the *a priori* knowledge that is possibly available about the true underlying model problem. In this work I will assume that $\rho(\mathbf{m})$ and $L(\mathbf{m})$ are Gaussian probability distributions. In this situation the probability density $\sigma(\mathbf{m})$, also known as *a posteriori* probability density function is Gaussian as well and given by

$$\sigma(\mathbf{m}) \propto \exp\left[-\frac{1}{2}(g(\mathbf{m}) - \mathbf{d_{obs}})^T C_D^{-1}(g(\mathbf{m}) - \mathbf{d_{obs}}) \right.$$
$$\left. + (\mathbf{m} - \mathbf{m_0})^T C_M^{-1}(\mathbf{m} - \mathbf{m_0}) \right]. \tag{1}$$

Here, C_M and C_D are the model and data covariance matrices, respectively; $\mathbf{d_{obs}}$ is the observed data vector; $g(\mathbf{m})$ is the (linearized) modeling operator used in the synthetic data computation, and $\mathbf{m_0}$ is the mean or most likely prior model. The covariance matrix C_M of the probability density function $\rho(\mathbf{m})$ is a connection between the *a priori* information and the inverse problem. This information can be derived from regional geological considerations, well-logs, interpretative work and so on. To build covariance matrices from such sources is not trivial and is seldom attempted, at least in the published inversion literature. In view of this difficulty, simplifying assumptions are commonly made to construct model covariance matrices. Consequently, under the Bayesian framework, the significance of these matrices (and also of $\rho(\mathbf{m})$) is lost, and so is the confidence on the assessment of the uncertainties of the solution of the inverse problem.

The objective of this work is to illustrate with a simple synthetic example, the effects of using a model covariance matrix built using statistical considerations about the model one seeks, in the prior probability density $\rho(\mathbf{m})$. The specific problem studied here is the linearized (Born) waveform inversion, in which short wavelengths of the velocity field are estimated from seismic amplitudes. I assume known the kinematics of the true model, what makes the linearization a reasonable approximation to the problem. A quasi-Newton's method is used in the inversion process, which is based on the work of Jin et al. (1992). The amplitudes of the observed data (generated synthetically by ray tracing) are corrupted by a Gaussian time-correlated noise. Although the time-correlated noise is easier to cope with, as opposed for example to spatial-correlated noise, it suffices to illustrate the purposes of this paper.

I compare the result obtained with this approach with the one provided by the Tikhonov regularization method (Tikhonov and Arsenin 1977) based on model smoothness assumptions. As will be shown later, both results are roughly equivalent for the case considered here. However, as expected, the results given by the Bayesian methodology tend to preserve the roughness of the true solution at the expense of the stability of the inversion process. In the other hand, the regularized solution smoothes out features of the model that are not required to fit the data.

2 Regularization of Amplitude Inversion

2.1 Motivation

Consider the two-layer velocity model in Figure 1. The (acoustic) synthetic data generated for this model consists of shot gathers, in which a source is "exploded" at the surface of the Earth and the energy reflected from the geological interfaces are recorded by receivers located at the same depth as the source but at different offsets (source to receiver distance). Five of such shot gathers were generated using a ray-tracing procedure. After adding band-limited random noise the synthetic data is shown in Figure 2. The model is parameterized by rectangular cells, with thickness of 0.05km and width of 0.10 km. The dimension of the model space, i.e. the number of cells used in the parameterization is 125, and the dimension of the data space is 60000 (the length of the time window used in the misfit computations is 1 s, with a time sampling interval of 0.004 s. 240 seismic traces are used in the inversion process).

If one attempts to infer the velocity of the second layer of this model, given the velocity of the first layer, without acknowledging that the amplitudes are corrupted by noise, the result illustrated in Figure 3 is obtained. This "solution"

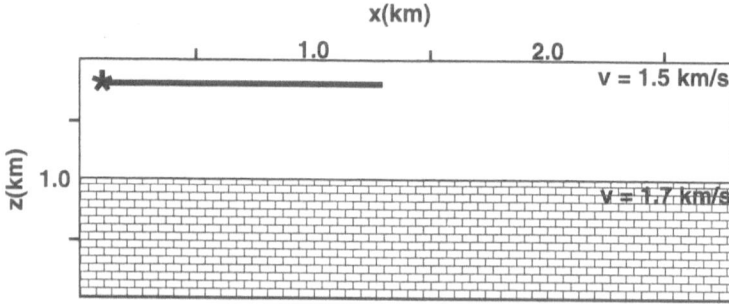

Fig. 1. Two-layer velocity model. The length of a shot gather is illustrated by the horizontal line in the figure.

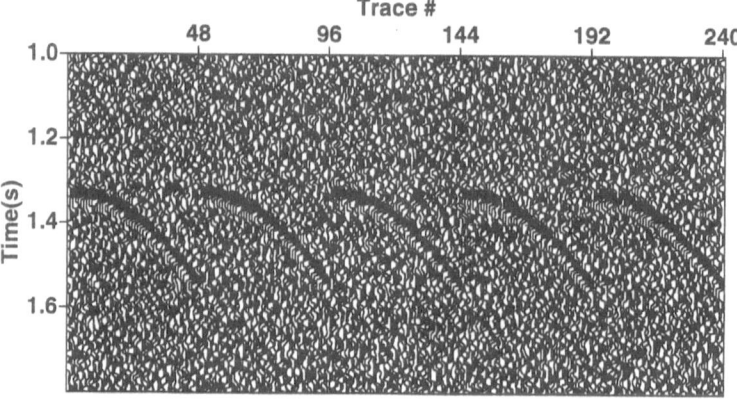

Fig. 2. Synthetic data with band-limited noise generated for the model shown in Figure 1. The signal to noise ratio is 2.

to the inverse problem contains errors up to 50% but it still fits the data to one standard deviation of the noise.

I applied Tikhonov regularization in the inversion algorithm of Jin et al. (1992), aiming at reducing its sensitivity with respect to perturbations (noise) in the data. I show the results later in this section, but first I briefly discuss the basics of the regularization approach.

2.2 Basics of regularization

Consider the linear system

$$A\mathbf{x} = \mathbf{b}, \tag{2}$$

where A is an operator of the forward problem that computes the data $\mathbf{b}$ for a given model $\mathbf{x}$. The solution for $\mathbf{x}$ in Equation (2) is said to be ill-conditioned if it is non-unique and/or if a small perturbation on the data $\mathbf{b}$ corresponds to a large perturbation in the model $\mathbf{x}$. The fundamental idea of Tikhonov's regularization method is to replace the operator A by a family of approximated operators, functions of the so-called regularization parameter α, such that the solution $x_\alpha{}^*$ for each one of those parameters is invertible, but in some sense tends to $\mathbf{x}$ as α goes to zero. The approximated solution $\mathbf{x}_\alpha{}^*$ can be defined as the minimizer of the quadratic functional:

$$\|A\mathbf{x} - \mathbf{b}\|^2 + \alpha\|R\mathbf{x}\|^2, \tag{3}$$

where R defines the correlation between different elements of model space according to some criterion, for example smoothness.

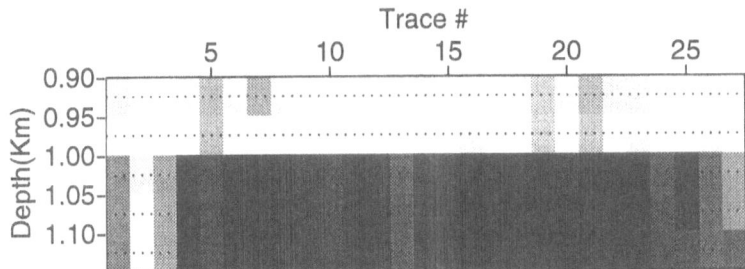

Fig. 3. Result of the inversion for the data in Figure 2. The velocity ranges from 1.2 Km/s (white) to 2.7 km/s (black).

The regularization parameter α controls the influence of the penalty term on the optimization problem described in Equation (3). If it is chosen too small, Equation (3) is close to the original ill-posed problem, and the regularization would be of no or little effect. If α is too large, the problem solved would have little connection with the original Equation (2). An "optimum" value for the regularization parameter can be chosen by a trial-and-error procedure: The inversion procedure is carried on for different values of α, providing several estimated models. The one associated with the largest value of this parameter that still fits the data to some extent (usually one standard deviation of the noise) is picked as the solution to the inverse problem.

2.3 Regularization of Amplitude Inversion

To incorporate Tikhonov's regularization in the amplitude-inversion algorithm one must add to the original objective function (a weighted least-squares error) of the problem a regularization term as described by the following expression:

$$\min_{\mathbf{m}} \left[(\mathbf{u_s} - G\mathbf{m})^T Q(\mathbf{u_s} - G\mathbf{m}) + \alpha\, \mathbf{m}^T R^T R\, \mathbf{m} \right]. \tag{4}$$

Here, G is the forward linear operator matrix, Q is a diagonal weighting matrix defined in Jin et al. (1992), $\mathbf{u_s}$ is the recorded wave field and $\mathbf{m}$ is the unknown model. As mentioned before the matrix R is constructed according to assumptions made about the underlying model.

2.4 An Example

Carrying on the inversion procedure for the data set shown in Figure 2 using the familiar tridiagonal second-order difference operator for the matrix R (Constable et al. 1992), I obtained the result illustrated in Figure 4. The regularization matrix R used here implies that the model is laterally smooth, that being the case for this problem. The inverse problem was solved for several values of the regularization parameter α. Figure 4 illustrates the model associated with the largest value of α that still fits the data up to one standard deviation of noise. The regularization was effective in reducing the sensitivity of the inversion procedure to the noise in the data, since the solution presents errors less than 5% in the velocity field. Compare this result with the model shown in Figure 3, in which regularization was not used.

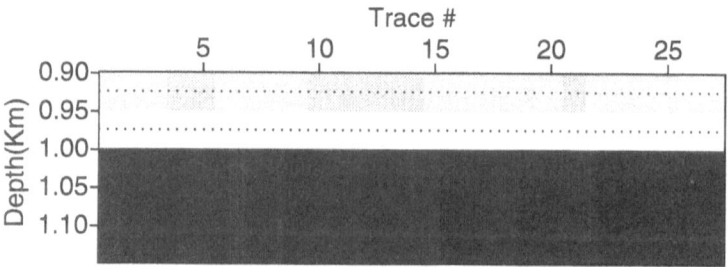

Fig. 4. Result of the inversion for in Figure 2 using regularization. The gray shading indicates the correct velocity.

3 Relating Bayesian Inversion and Regularization

In the inversion procedure proposed by Jin et al. (1992) the solution of the inversion problem was defined as a minimizer of a weighted least-squares norm defined in the model space, described in Equation (4). Under the Bayesian framework, this approach implies assumptions about the statistical nature of the noise in the observed data and on the correlation between model parameters. The purpose of this section is to describe what those assumptions are.

As mentioned before, for Gaussian statistics the *a posteriori* probability density function $\sigma(\mathbf{m})$ is given by Equation (1). Tarantola (1987) showed that this equation reduces to:

$$\sigma(\mathbf{m}) \propto \exp\left[(\mathbf{m} - \mathbf{m_{map}})^T C_{M'}(\mathbf{m} - \mathbf{m_{map}})\right], \tag{5}$$

where $\mathbf{m_{map}}$, the maximum *a posteriori* model, is the mean model of this distribution, and $C_{M'}$ is the *a posteriori* covariance matrix, given by:

$$C_{M'} = \left[G^H Q G + C_M^{-1}\right]^{-i}. \tag{6}$$

Here, $\mathbf{m_{map}}$ is found by minimization, via a quasi-Newton's optimization algorithm, of the exponent of Equation (1), written below for convenience:

$$\min_{\mathbf{m}} \left[(g(\mathbf{m}) - \mathbf{d_{obs}})^T C_D^{-1}(g(\mathbf{m}) - \mathbf{d_{obs}}) + (\mathbf{m} - \mathbf{m_0})^T C_M^{-1}(\mathbf{m} - \mathbf{m_0}) \right]. \quad (7)$$

The square-root of the main diagonal elements of the *a posteriori* model covariance matrix are the standard deviations of the solution of the inverse problem, representing the resolution provided by the data and the *a priori* information. Such standard deviations would be of little significance if this matrix is built without resorting to the statistics of the model parameters.

Notice that if m_0 is a null vector, Equations (7) and (4) are mathematically equivalent. The inverse of the diagonal weighting matrix Q^{-1} plays the role of a data covariance matrix C_D, and the product $\left[\alpha R^T R \right]^{-1}$ implements the model covariance matrix C_M. Therefore, the least-squares formulation of the inverse problem expressed in Equation (4) implies, in the Bayesian sense, the assumption of uncorrelated Gaussian noise in the data and that the model parameters are also described by a Gaussian probability density function. The model covariance matrix is defined by the regularization scheme.

Although mathematically equivalent, both approaches are fundamentally different. In the Bayesian methodology the *a priori* covariance matrix provides a mechanism to incorporate data-independent information (for instance well-logs) into the inversion process. Whereas in the regularization, we admit the fact that data *alone* can just resolve functionals (for example averages) of the subsurface parameters. If the regularization is a smoothing operator we are content with the smoothest model that still explains the data according to a certain criterion.

In the next section, I consider the situation where the second layer of the model illustrated in Figure 1 is inhomogeneous with a given correlation length, which will be used to build the model covariance matrix. The results will be compared with those obtained by the regularized inversion procedure.

4 Model covariance estimation

Consider the model illustrated in Figure 5. The velocity of the second layer was drawn from a Gaussian probability distribution of zero mean and of a given correlation length. I generated with a ray-tracing code five (acoustic) shot gathers for this model that, after addition of band-limited noise, are shown in Figure 6.

An estimation of the model covariance matrix of the velocity of the second layer can be given by the following expression (Priestley 1981):

$$C[i,j] = \frac{1}{N} \sum_{r=1}^{N} (v(r) - \overline{v})(v(r + j - i) - \overline{v}). \quad (8)$$

Here N is the number of grid points in the x direction, $v(r)$ is the velocity perturbation at position r and $\overline{v}$ is the mean velocity.

Again, the objective of the inversion is to determine the velocity profile of the second layer given the velocity of the first layer. In this example, I compare the

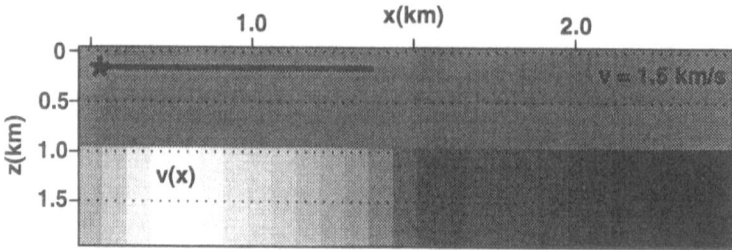

Fig. 5. Laterally inhomogeneous velocity model. The velocity of the second layer ranges from 1.1 km/s (white) to 2.2 km/s (black).

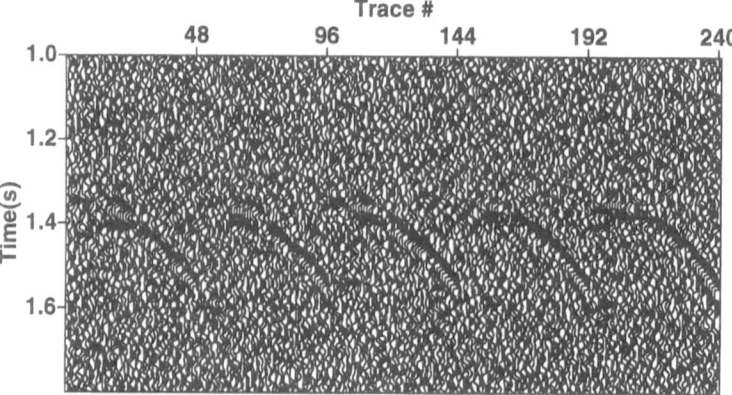

Fig. 6. Synthetic data with band-limited noise generated for the model shown in Figure 5. The signal to noise ratio is 2.

results obtained with the inversion of the model shown in Figure 5 at the target depth of 1 km in the following cases: 1) non-regularized inversion (See Figure 7); 2) regularized inversion using the matrix R as a second-order derivative operator (See Figure 8), and 3) Bayesian inversion with the covariance matrix as given by Equation (8) (See Figure 9). In Figure 9 the error bars are the standard deviations provided by the *a posteriori* covariance matrix given by Equation (6).

Figures 7, 8 and 9 show models that fit the data up to one standard deviation of the noise. The following important aspects are illustrated by these results: 1) The fact that the linearized amplitude inversion problem is ill-posed and some stabilization procedure is required to handle the noise in the data; 2) regularization provides a way to that end, by, as implemented here, smoothing out model features that are not required to fit the data, and 3) when model covariances matrices are derived from some source of information on model space the effects of the noise in the data are also reduced, to an extent that depends on the *a priori* model covariance. Uncertainties on the model $\mathbf{m_{map}}$ can be assessed as the error bars in Figure 9 illustrate.

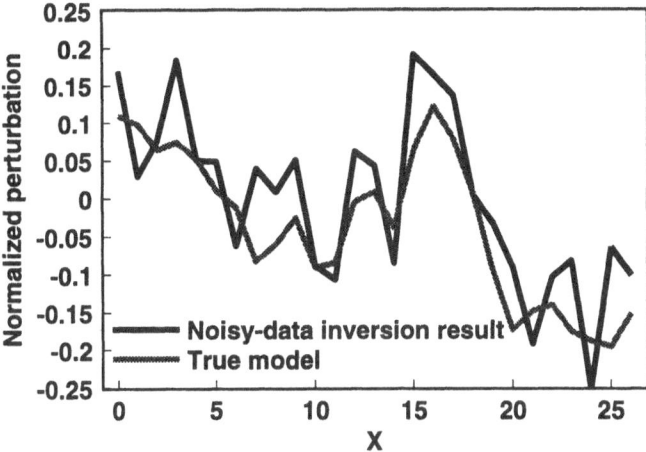

Fig. 7. Result of the non-regularized inversion. The vertical axis refers to a normalized quantity that should be added to the velocity of the first layer to obtain the velocity of the second layer.

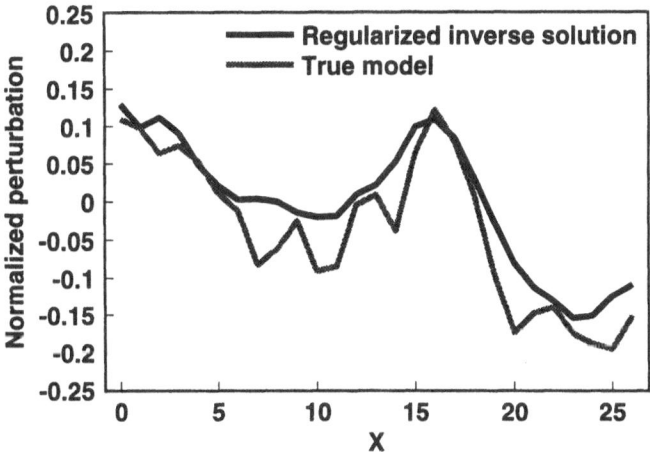

Fig. 8. Result of the regularized inversion.

5 Conclusions

Here, I presented a study on model covariances in the inversion of seismic amplitudes. For simple examples I compared inversion results when model covariances were built taking into account statistical considerations about the underlying model with Tikhonov regularized solutions. Although mathematically similar, both approaches are fundamentally different. In the Bayesian approach, we are able to incorporate data-independent information into the inverse process via covariance matrices or more general *a priori* probability distributions. In the

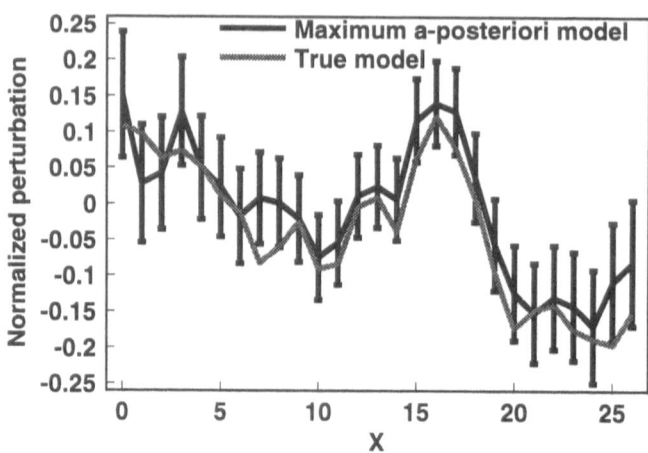

Fig. 9. Result of the Bayesian inversion.

Tikhonov regularization method we aim at estimating functionals of the subsurface parameters (in this work averages), that should be consistent with the data according to some criterion.

In the Bayesian case the lack of regularization parameter is attractive, however the construction of realistic priors (even Gaussians) is an open issue, and probably one of the most important research topics in this theory.

6 Acknowledgments

I thank Professor John Scales for his constant encouragement and advice. I am grateful to acknowledge the financial support of the Army Research Office.

References

1. Constable, S., Parker, R. and Constable, C., 1987. Occam's Inversion: A practical algorithm for generating smooth models from electromagnetic sounding data. Geophysics, 52, 289–300.
2. Jin, S., Mandariaga, R., Virieux, J. and Lambare, G., 1992. Two-dimensional asymptotic iterative elastic inversion. Geophys. J. Int., 108, 575–588.
3. Priestley, M.B., 1981. Spectral Analysis and Time Series, Chapter 9. Academic Press.
4. Tarantola, A., 1987. Inverse problem theory. Elsevier.
5. Tikhonov, A.N. and Arsenin, V.Y., 1977. Solution of ill-posed problems. Wiley, NY.

Resolution and error propagation analysis for tomographic data with correlated errors

Lars Nielsen and Bo Holm Jacobsen

Department of Earth Sciences, Geophysical Laboratory, Aarhus University, Denmark

Introduction

Tomographic data may under different circumstances be contaminated by error types which are not realistically described as uncorrelated Gaussian noise. We show how expected error correlation properties can be formulated in terms of covariance matrices for spatially irregular distributions of sources and receivers. The popular inverse estimate based upon simple Tikhonov regularization does not incorporate correlated errors. The proper stochastic inverse estimate that includes specified covariance matrices in the inverse operator is shown to produce less smearing and a more realistic analysis of error propagation.

In the following we address situations where datasets in seismic tomographic problems (e.g. McMechan, 1982; Spakman 1986; Zelt and Smith, 1992) are contaminated by errors from one or more sources. The statistics of these errors may not be trivial in the sense that they can be treated as "white noise".

Based on physical reasoning on the origins of the data errors one can specify a data error covariance matrix, which quantifies the correlation of the individual data errors. The specified covariance matrix is included in the linear inverse estimate (e.g. Jackson, 1979; Tarantola and Valette, 1982):

$$\underline{x}_{est} = (\underline{\underline{A}}^T \underline{\underline{C}}_e^{-1} \underline{\underline{A}} + \underline{\underline{C}}_x^{-1})^{-1} (\underline{\underline{A}}^T \underline{\underline{C}}_e^{-1} \underline{d}_{obs} + \underline{\underline{C}}_x^{-1} \underline{x}_0) \tag{1}$$

where $\underline{\underline{C}}_e$ is the data error covariance matrix, and $\underline{\underline{C}}_x$ is the covariance matrix of the a priori model parameters, $\underline{x}_0$. The popular estimate based upon simple Tikhonov regularization:

$$\underline{x}_{est} = (\underline{\underline{A}}^T \underline{\underline{A}} + \theta^2 \underline{\underline{I}})^{-1} \underline{\underline{A}}^T \underline{d}_{obs} \tag{2}$$

may be viewed as a special case of (1), where $\underline{\underline{C}}_e = \sigma_e^2 \underline{\underline{I}}$, $\underline{\underline{C}}_x = \sigma_x^2 \underline{\underline{I}}$ and $\underline{x}_0 = \underline{0}$ so that $\theta = \sigma_e / \sigma_x$. This study has $\underline{\underline{C}}_x = \sigma_x^2 \underline{\underline{I}}$ throughout in order to focus on the effects of the off-diagonal elements in $\underline{\underline{C}}_e$. Similarly, we have $\underline{x}_0 = \underline{0}$ in all the examples below.

In two examples we investigate how the estimate in (1) differs from the estimate in (2). First, we illustrate that (1) provides a more realistic treatment of error propagation. Second, we compare the qualities of the estimates given in (1) and (2) in the presence of statics-like correlated errors.

Sources of correlated errors and error covariance specification

Fig. 1 shows an example of seismic wide-angle data. By using tomographic inverse techniques such data may generally provide a relatively coarse and smooth picture of the seismic velocity distribution in the subsurface of the earth. Smaller inhomogenities (e.g. varying sedimentary cover close to the earths surface) that might be disregarded by the coarser model parameterization will produce 'geological noise', which is likely to have a spatially correlated character (statics-like errors).

Seismic wide-angle data may have a complicated and ringy character in some areas, which can lead to a misinterpretation of the traveltime-distance course of a given phase (e.g. cycle skip errors). Such errors are also likely to be spatially correlated, because the data themselves have a spatially correlated character (see Fig. 1).

Many tomographic algorithms use ray-tracing techniques in order to describe the propagation of the seismic wavefield. Due to shortcomings of ray-tracing methods certain phases (diffrated arrivals) can not be treated correctly. This is particularly critical in areas where large low-velocity anomalies exist (Wielandt, 1987). Such effects may lead to correlated errors.

The correlation properties of the total error budget in a given dataset (other sources than the three examples mentioned above may produce correlated errors) can be quantified by specifying a data error covariance matrix: Let r_i and R_i be the end points of ray number i. Then the data error covariance matrix element describing the errors $e_d(r_i,R_i)$ and $e_d(r_j,R_j)$ can reasonably be defined:

$$(\underline{\underline{C}})_{i,j} = \Phi_A(|r_i - r_j|) + \Phi_B(|R_i - R_j|) \tag{3}$$

where Φ_A and Φ_B are the covariance functions describing the spatially stationary error processes. This restriction to isotropic stationary processes is a sensible approximation in many cases, and as many covariance functions as needed may be added in order to describe the total covariance matrix.

For illustration purposes we have chosen the following function to describe the data error correlation properties in this study (Walden & Hosken, 1985):

$$\Phi(s) \ = \ \Phi_0 \exp(-s \,/\, L) \tag{4}$$

where L is the correlation length, and Φ_0 is the variance of the data error. Fig. 2 shows three examples of covariance functions given by (4) and their realization. The examples only differ in the choice of L. Note the non-white character of the realizations.

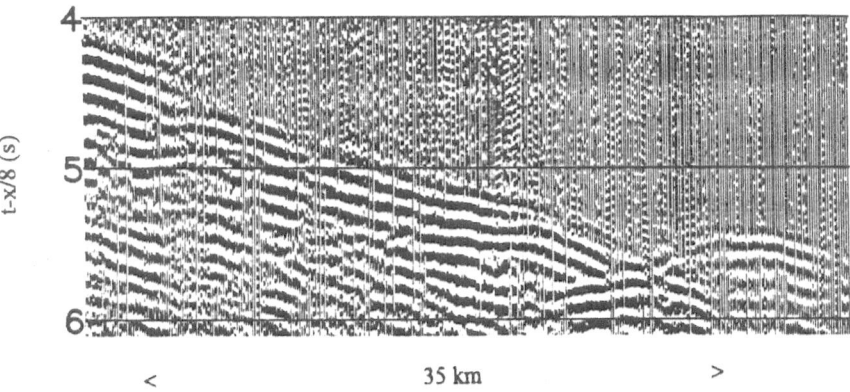

Fig. 1. A seismic wide-angle data section. Seismic phases are plotted in reduced traveltime as a function of distance. The section consists of c. 350 individual traces.

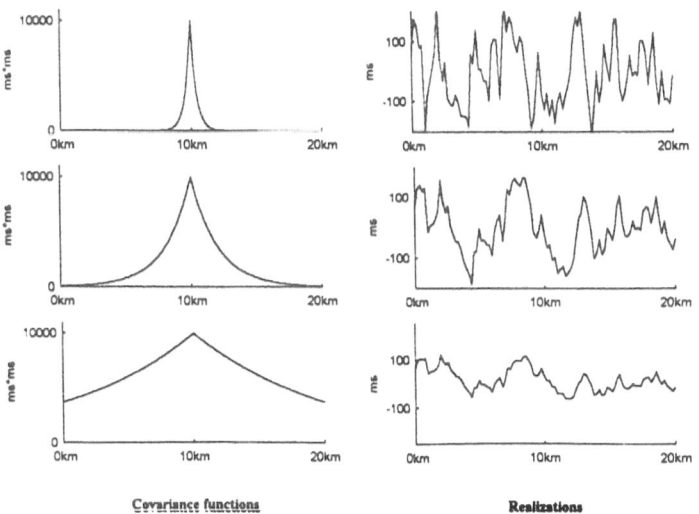

Covariance functions   Realizations

Fig.2. The covariance function given by (4) and its realization for three different choices of L: L = 0.4km (top), L = 2.0km (middle) and L = 10.0km (bottom).

Treatment of error propagation

Fig. 3a shows a geometry with 63 receivers (N=63) from which we want to estimate the nine model parameters in the slowness model below. This is a linear problem, which can be solved by using for instance (1) or (2). In the following we consider six datasets with two different noise components arising from: One uncorrelated error type representing simple 'white noise' (standard deviation = 1ms) and one spatially correlated error type representing receiver static-like errors (standard deviation = 10ms). Only the correlation length, L, of the correlated error type is different in the six datasets. We describe the correlated error component by using (4). In the example below we have σ_x = 10ms/km.

The ray densification from Fig. 3a to Fig. 3b will only add little new geometrical information, but the repeated sampling of the same cells will reduce the variance of the estimated model parameters to a degree which depends on L and N.

For L = 0 the data errors are independent random Gaussian numbers, and the posterior standard deviation, std_x, of the central model parameter drops off following the expected relation std_x $\propto$ N$^{-1/2}$ (see Fig. 3c). When L > 0 we see the same behavior for small N, i.e. receiver spacing larger than L, but for larger N where the receiver spacing becomes lower than L, the error reduction saturates as one would expect.

So, in this example the model uncertainty depends clearly on the degree of data error correlation. Moreover, if data error correlation is disregarded in the operator design, then the resulting Tikhonov inverse (2) is not optimal and the model uncertainty can be strongly underestimated (particularly for large N).

Proper account for statics-like error correlation

The source-receiver geometry in Fig. 4a consists of twelve rays, which sparsely cover the underlying 7 by 5 grid of cells. Two dense ray bunches travelling diagonally with respect to the underlying slowness model extend the geometry. This might be thought of as an extension of an old seismic explosion survey (e.g. Barton, 1984) with a modern airgun based wide-angle dataset (e.g. Meissner et al., 1992). Applying this geometry we want to recover the test pattern shown in Fig. 4b using (1) and (2). Each anomalous cell in Fig. 4b has a value of -10ms/km.

In the following we address the situation were the data are contaminated by statics-like spatially correlated errors. A static is an error component that is the same for rays sharing an end point. The standard deviation of the source and receiver statics are 100ms (equal to a travel time anomaly of a ray travelling a distance equal to one cell width through an anomaly), and the correlation length is set to be 10km. The correlation function given in (4) has been used to describe the error correlation. Additionally, uncorrelated errors with lower standard deviation (10ms) are added representing phase picking errors. Again we have $\sigma_x = 10$ms/km.

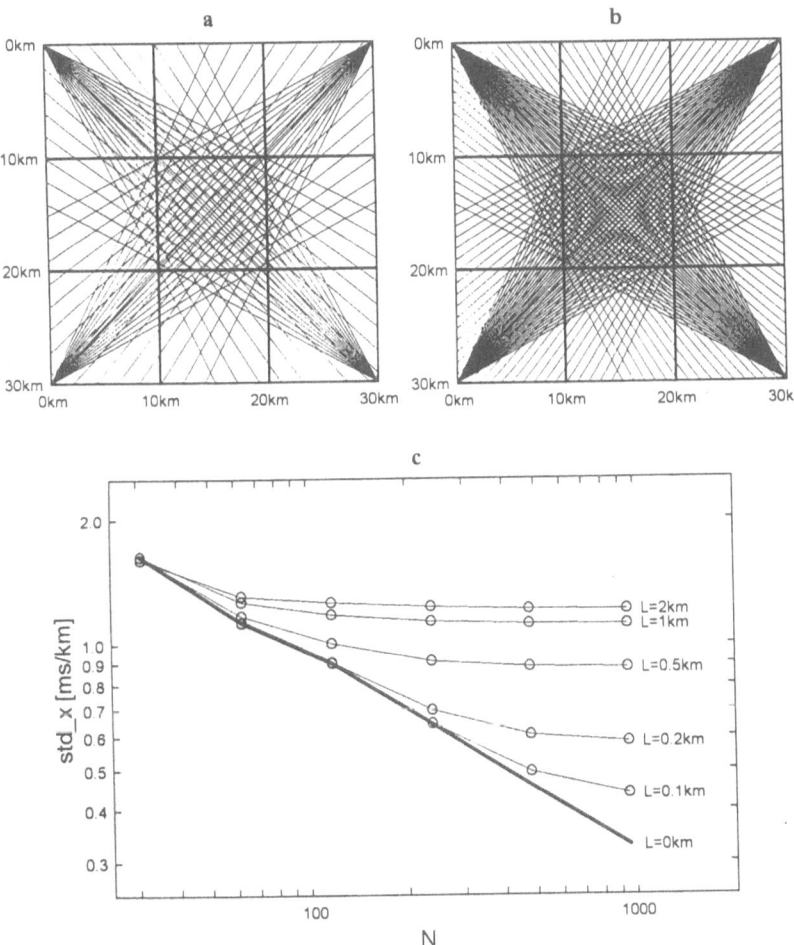

Fig. 3. **a** shows a 3 by 3 grid of cells overlain by a geometry consisting of four sources (ray starting points) marked by * and numerous receivers (ray end points) evenly distributed around the edge of the grid of cells. In **b** we see the same geometry as in **a**, but with more densely spaced receivers. In **c** std_x, the estimated standard deviation of the estimated central model parameter, is shown as a function of the number of rays, N.

Fig. 4c and 4d show the anomaly pattern recovered by the estimates given by (1) and (2). Clearly (1), where the correct error correlations are taken into account, offers the best estimate. By using (2), where only the correct variances of the data errors are known, it is only possible to recover a smaller part of the correct anomaly pattern, and more large, fictitious anomalies are introduced.

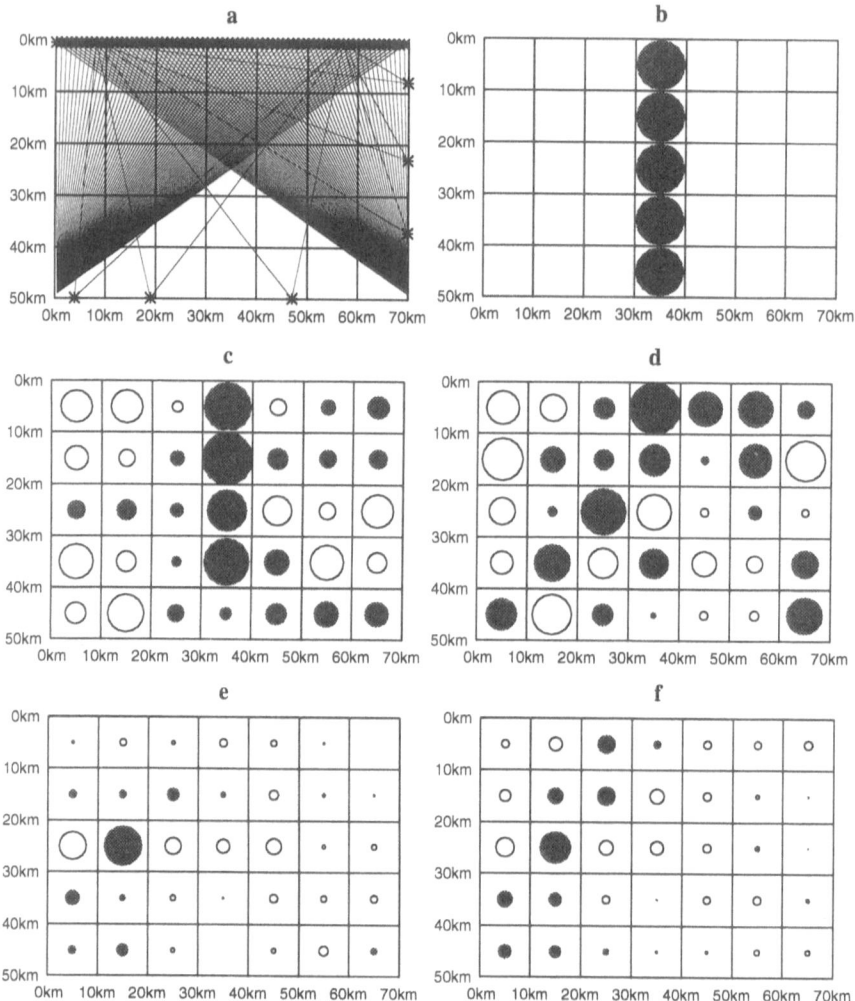

Fig. 4. The ray geometry shown in **a** is used to recover the slowness pattern shown in **b**. **c** shows the estimate recovered when using (1). The estimate shown in **d** is found by using (2). **e** shows the model resolution of the middle model parameter in the second column of the phantom achieved by (1). **f** shows the resolution of the same model parameter when using (2). Filled circles in **b**, **c** and **d** indicate negative slowness anomalies. A filled circle with diameter equal to one cell width corresponds to an anomaly of -10 ms/km. Open circles indicate positive slowness anomalies. In **e** and **f** a filled circle with diameter equal to one cell width corresponds to a value of 1, while open circles indicate negative values.

In Fig. 4e and 4f the model resolution for the model parameter in the middle of the second column of the pattern is shown for the two cases. The model resolution given when using (1) is good, whereas the model resolution is poorer when using (2). The estimate in (2) produces a pronounced smearing in the direction of the many rays. Smearing-effects are a well known phenomenon in seismic tomography (e.g. Hole, 1992). These effects seem to be reduced by taking the correlation into account. Reducing the damping in (2) would provide a better resolution in Fig. 4f, but a poorer inverse estimate in Fig. 4d.

Conclusion

Tomographic data are likely to be contaminated with correlated errors. We use an approach based on data error covariance specification in order to analyse the effect of such errors. This approach is valid for arbitrary irregular measurement geometries and therefore offers great flexibility.

We have shown that the estimates in (1) and (2) behave differently in the presence of correlated errors. Taking the correct error correlations into account we obtain a more realistic description of error propagation, and we produce better model resolution with less smearing.

Acknowledgements

The authors thank an anonymous reviewer for a thorough and valuable review of an earlier version of this paper.

References

Barton, P. J. and Matthews, D. H., 1984. Deep structure and geology of the North Sea region interpreted from a seismic refraction profile. Annales Geophysicae, 1984, 2, 6, 663-668.

Hole, J. A., 1992. Nonlinear High-Resolution Three-Dimensional Seismic Travel Time Tomography. Journal of Geophysical Research, vol. 97, No. B5, 6553-6562.

Jackson, D. D., 1979. The use of a priori data to resolve nonuniqueness in linear inversion. Geophysical Journal of the Royal Astronomical Society, 57, 137-157.

McMechan, G. A., 1983. Seismic tomography in boreholes. Geophysical Journal of the Royal Astronomical Society, 74, 601-612.

Meissner, R., Snyder, D., Balling, N., Staroste, E., (eds.), 1992. The BABEL Project. Commission of the European Communities. Directorate-General XII. Science, Research and Development. ISBN 2-87263-078-3.

Spakman, W., 1986. The upper mantle structure in the Central European-Mediterranean region, In: European Geotraverse (EGT) Project, the central segment, R. Freeman, Mueller, St. and Giese, P. (eds.), European Science Foundation, Strassbourg, 215-222.

Tarantola, A. and Valette, B., 1982. Generalized non linear inverse problems solved using the least squares criterion. Rev. Geophys. Space Phys., 20, 219-232.

Walden, A. T. and Hosken, J. W. J., 1985. An Investigation of the Spectral Properties of Primary Reflection Coefficients, Geophysical Prospecting 33, 400-435.

Wielandt, E., 1987. On the validity of the ray approximation for interpreting delay times. In: G. Nolet (ed.). Seismic Tomography, 85-98. D. Reidel Publishing Company.

Zelt, C. A. and Smith, R. B., 1992. Seismic traveltime inversion for 2-D crustal velocity structure. Geophys. J. Int., 108, 16-34.

Estimating Background Velocity from Reflection Seismic Data - A Feasibility Study

Lisbeth Engell-Sørensen[1], Tijmen-Jan Moser[1] and Jan Pajchel[2]

[1] University of Bergen, Institute of Solid Earth Physics, Bergen, Norway
[2] Norsk Hydro, Bergen, Norway

1 Introduction

The inverse problem of reflection seismics is to estimate the background velocity and the slowness perturbations along a reflector, which minimize the seismogram residual in a least-squares sense. This problem can be approximated by an iterative procedure of finding the perturbations along the reflector for fixed background velocity in one step (a linear problem) and optimizing the contrast of the image obtained by the slowness perturbations in another (a nonlinear problem). The first order Born approximation (Aki and Richards, 1980) was used to obtain the slowness perturbation. Since the contrasts in the image enhance with improved velocity in the medium we choose the lower entropy of the image contrast as a measure of the enhanced background velocity model. The approach was preposed by Ryzhikov et al. (1995) in a more restricted form than applied here. The inversion problem is hence to minimize the entropy of the image contrast in order to find the background velocity and the slowness perturbations simultaneously. The present work is restricted to a 2D smooth velocity model and to 2D surface acquisition. No *a priori* information concerning the velocity model, which would be possible to obtain from borehole data, is included in this work. Only *a priori* information in the form of regularization of the slowness perturbation is included.

Many have described the theory for the asymptotic linearized inversion problem (Beylkin and Burridge, 1990; Beydoun and Mendes, 1989; Bleistein, 1987; Miller et al., 1987, Beylkin, 1985). With our application of the first order Born approximation of the image of slowness perturbations we aim at obtaining a complete illumination of the scatterers in the target region (within the duration of the data signal, i.e. within the duration of the source signal in the case of synthetic ray data). In the linear inversion algorithm we hence sum over all the source-receiver pairs, whereas others, although also their formulas are given for all source-receiver combinations, aim at saving computer time and storage by restricting the target being illuminated by a specific source-receiver combination (Thierry and Lambaré, 1994). The work done yet for imaging diffractors and reflectors in the earth mainly assumes the smooth background is known, and solves for reflection coefficients or velocity perturbations (see e.g. Bleistein, 1987). However, it has been shown (for instance by Jin and Madariaga, 1992; Jin and Madariaga, 1993; Jin et al., 1992; Lambaré et al., 1992) that it is possible to determine the smoothly varying part of the velocity model by the use of global optimization. Preliminary results using genetic algorithms applied on the present object function in 3D using straight rays and a homogeneous background-velocity model has been obtained by Engell-Sørensen et al. (1994). Many have recently published analyses and improvements of the global optimization algorithms: simulated annealing (Press et al., 1992; Ingber, 1989; Ingber, 1993; Ingber and Rosen, 1992; Ruppeiner et al. 1991) and genetic algorithms (Grefenstette, 1990; Stoffa and Sen, 1991).

We define an optimum global optimization method, to be a method, which uses the minimum CPU-time in order to obtain a pre-defined accuracy of the final model estimate (i.e. a pre-defined minimum value of the object function or a pre-defined minimum difference between the new and the old object-function value in the optimization procedure) and which starting from any model representation in the model-parameter space always will converge towards the same solution (within some error). Since we cannot know from the final "best" estimates whether we have searched the model-parameter space globally we can afterwards choose to do a more "slow" search in the model parameter space in order to obtain a more stable global minimum or be more certain that it really is the global minimum we have found. As a first step towards finding the optimum global optimization method for solving the above inversion problem we have chosen one of the simulated annealing algorithms (Press et al., 1992) for a feasibility study. In order to ensure a global method we run the algorithm on parallel computer-nodes and require that almost all nodes converge towards the same minimum. The mean value and the standard deviation of the estimated solutions define the average model estimate and the error of the average model estimate, respectively. Here we use the estimate with the least minimum entropy as the "best" estimate. The main purpose of this work is to obtain the optimum number of parallel processors to be applied and analyse the cooling schedule in order to obtain a global method. In addition the work initiated and yet to be done in order to improve the object function and the global inversion method is discussed. The tests have been performed on an Intel Paragon super computer and a suite of fast work stations using the computer program NX (Paragon) and the public domain computer program Parallel Virtual Machine (PVM) (Paragon and work stations). Since it is crucial not to exceed the physical memory on each node of the Paragon computer (26 Mb) the target area used in this paper has as maximum possible size about 15 Mb (for 119 receivers x (97 x 161) grid points).

2 Method

The unknown squared slowness perturbation m as a function of virtual scatter point x^* (a vector in space) is given by Ryzhikov et al. (1995):

$$m(x^*) = \frac{\sum_\xi (\phi_{obs}, U_{x^*}(\xi))}{\sum_\xi (U_{x^*}(\xi), U_{x^*}(\xi)) + \alpha \sum_\xi (\phi_{obs}, \phi_{obs})}, \tag{1}$$

where ϕ_{obs} is any observed data, $\sum_\xi$ is over all source-receiver combinations for the chosen acquisition and scatter grid, α is a damping factor making the second term small compared to the first term in the denominator, and

$$(\phi_{obs}, U_{x^*}(\xi)) = \int_{t_1}^{t_2} \phi_{obs}(t) U_{x^*}(\xi, t) dt, \tag{2}$$

where t_1 and t_2 are, respectively, the beginning and end of the data signal. The Born signal caused by the virtual scatterer at x^* for the source-receiver combination ξ (source s, receiver r, all vectors in space) is given by (see Beydoun and Mendes, 1989)

$$U_{x^*}(\xi, t) = -G(r, s, t, \tau(r - x^*) + \tau(x^* - s)) * \ddot{f}(t)$$

$$= -A(r, x^*)A(x^*, s)\ddot{f}(t - \tau(r - x^*) - \tau(x^* - s)) \tag{3}$$

where $*$ denotes convolution, A and τ are, respectively, amplitude (geometrical spreading) and travel time for a ray traveling from the source to the scatterer or from the scatterer to the source, and G is the complete Green's function from source to receiver. In this work we are not restricted by any experimental source-time function, and define $\ddot{f}(t)$ to be a Berlage wavelet.

As object function we use the pseudo-statistical measure: "entropy" introduced to reflection seismics by De Vries and Berkout (1984):

$$H(m(x)) = -\int_\Omega p(m(x)) \ln p(m(x)) dx, \tag{4}$$

where p is "the probability density function", Ω is the analyzed spatial region of m, i.e. the target region, and we have ommitted the asterisk used above. The image contrast is introduced as p by Ryzhikov et al. (1995) by the expression

$$p(m(x)) = (\nabla m(x))^2 / \int_\Omega (\nabla m(x))^2 dx. \tag{5}$$

The complete functional H is then called the entropy of image contrast ("EnIC"). p can be interpreted as the probability density function since $\int_\Omega p(m(x)) dx = 1$. However, we do not have any other statistical requirements for p, hence we use the word "pseudo" above. H has maximum value in the case of uniform probability distribution and H has minimum (zero) value if p is different from zero for only one value of x. H increases if the variation in p decreases, i.e. if the variation in $(\nabla m(x))^2$ decreases. This means also that a target region with one single scatterer ($p = 1$) would give less H than a target region with a reflector represented by e.g. 97 scatterers ($p(m(x_i)), i = 1, ..., 97 = 1/97$).

The Born signals used for linear inversion in (1) are obtained by tracing wavefronts (rays) from the sources and receivers to the (virtual) scatterers in the target region. The ray tracing is carried out using the computer routine Green2D developed by Hanyga et al. (1995). The procedure used for obtaining the Born signals while taking account of the computer time saved by calculating a propagating wavefront from a point source to a grid of virtual scatterers and not vice versa, and by excluding the calculations of wavefronts from sources and receivers located in points of previously calculated sources and receivers has been described by Moser et al. (1995). In this work we have used simulated annealing on parallel processors, and obtained the "best" model estimate by comparing the final model estimates on every processor. A flow chart of one processor in the complete global optimization algorithm, where we have included one step of local optimization (see below), becomes:

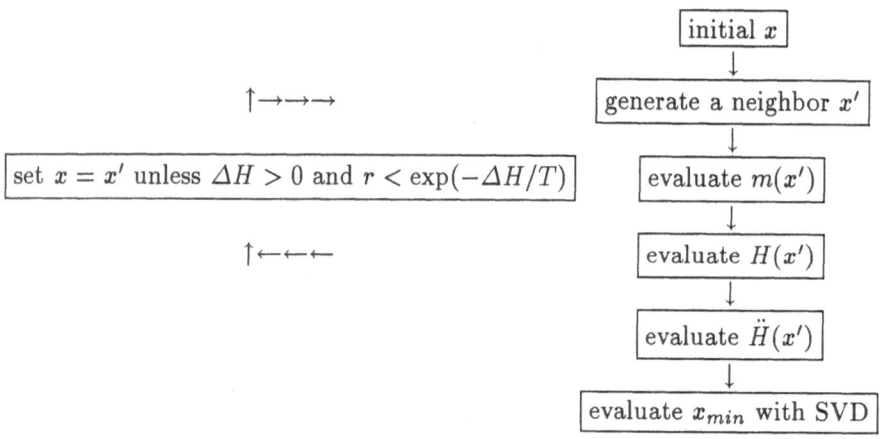

In this algorithm x and x' are, respectively, initial and trial representatives for the background velocity model, $\Delta H = H(x') - H(x)$, r is a random number between zero and one, and T is the cooling temperature. Two dots denotes the second order derivative with respect to the background-velocity model-parameters calculated numerically. In the algorithm applied here, the neighbour is generated by reflecting or contracting a simplex with N+1 corners, where N is the number of model parameters (N is three in this case) (see Press et al., 1992). The two boxes in the right bottom corner will be described later. In order to avoid m to be zero everywhere in the target area, we pre-define the model-parameter region. If a neighbour is generated outside the pre-defined model parameter region it is given very high H values. The main factor in our algorithm is the cooling temperature: if it is infinite, all new neighbours are accepted, if it is zero, only neighbours with less H are accepted. In this work we search for the optimum cooling schedule.

3 Results

3.1 Velocity Model and Data

The inversion problem is to obtain the three constants (a, b, and c) in the inhomogeneous 2D smooth background-velocity model

$$v(x, z) = ax + bz + c, \tag{6}$$

where v is expressed in ms^{-1}, x and z are in m, a and b are in s^{-1}, and c is in ms^{-1}. Figure 1 shows the analysed background velocity area, target region,

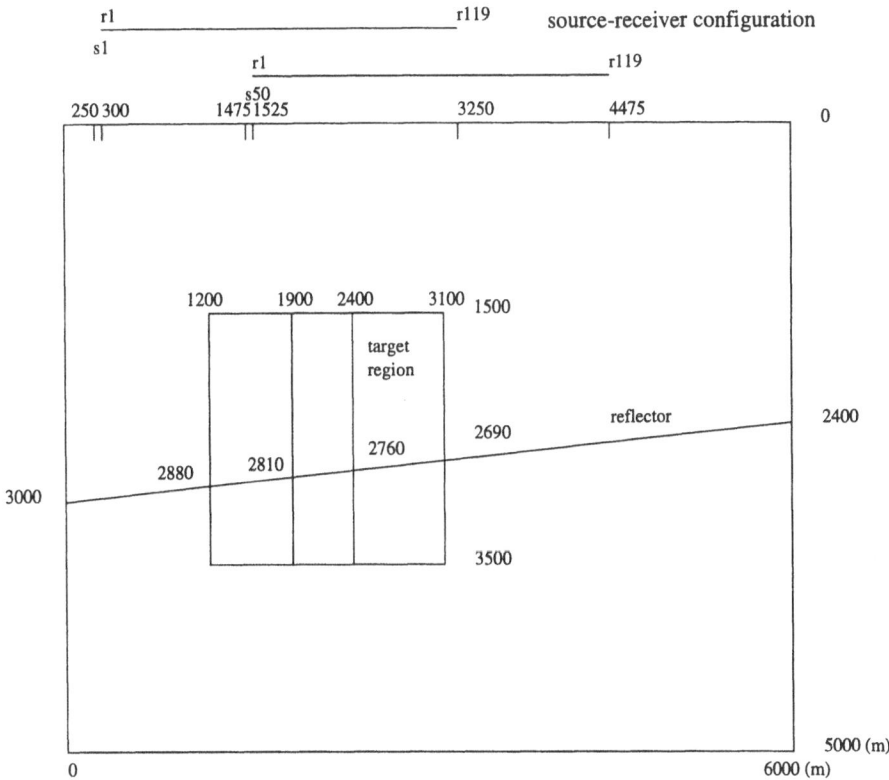

Fig. 1. Source-receiver configuration, target region, reflector, and region in which the background model is defined. s1 and s50 are first and last source positions, and r1 and r119 are first and last receiver positions along the streamers. The target region is the region of the virtual scatterers. All numbers are in m. The region to the left of the target region is the best enlightened target region by the data.

and source-receiver configuration used in all tests in this paper. The dimension

of the inhomogeneous 2D smooth background velocity model is 6000 and 5000
m in horizontal and vertical directions, respectively. The synthetic reflected ray
data (see Pajchel et al., 1988) are based on a velocity model with an interface
dipping from $(x, z) = (0$ m,3000 m$)$ to $(x, z) = (6000$ m,2400 m$)$ and a true velocity model above the reflector determined by a, b, and c values of, respectively,
0.01 s^{-1} , 0.133 s^{-1}, and 1800 ms^{-1}. The velocity below the reflector is 3000
ms^{-1}. The number of grid points in the target region in horizontal and vertical
directions are 97 and 161, respectively. The synthetic data simulate a real 2D
recording situation with a simultaneously moving "few-source/many-receiver"
(here: 50/119) acquisition system. In the following figures we have used the total number of entropy (object function) calculations from all tests in Table 1
in order to generate the gridded surface describing the entropy as a function
of two model parameters. The entropy as a function of two model parameters
are displayed for a selected "window" of the third model parameter. Isolines
of the surface are shown on the figures. Straight lines describe successive new
entropy values calculated for one node (Figure 2) or three nodes (Figure 3) in
the specific simulated annealing tests. The type of dashing of the line indicates
the temperature at the time of entropy calculation and the number one on the
figures indicate the first entropy calculation in one of the simulated annealing
procedures displayed in the specific window.

3.2 Optimum Number of Parallel Processes

Assuming optimal cooling schedule, Azencott (1992) has shown that equal accuracy of the global extremum can be obtained by simulated annealing on different number of parallel processors (nodes), as long as $n = N/m$, where n is
the number of object function calculations per node, N is the total number of
calculations of the object function, and m is the number of processors. Azencott
(1992) here assumed the same cooling rate on the nodes and not dependent upon
the number of nodes in use. We choose a cooling rate of 0.99 with a platform
of 4 or 7 (i.e. at every fourth or seventh calculation of entropy, the temperature
is decreased by a factor of 0.99) and an initial cooling temperature of 1. Since
we do not know if this cooling schedule is optimal we investigate here which
number of nodes achieve the smallest global minimum when the total CPU-time
is equal, i.e. with a fixed number of total calculations. Table 1 shows the "best"
EnIC after about equal spent CPU-time for a platform of 4 using 32, 16, and
8 nodes (test 1, 4, 7). For a platform of 7 the results are given below (test 2,
5, 8). From Table 1 it is seen that there is no significant difference between the
"best models" defined by the least value of EnIC for all number of nodes (i.e. 8,
16, and 32) for the individual cooling schedules. It is surprising, that although
the 8 nodes reach lower final temperatures than the 32 nodes, the "best" EnIC
are not less for 8 nodes than for 32 nodes. Figure 2 illustrates the results for the
cooling schedule with a platform of four calculations when 32 (test 1) nodes were
used. Hence also for our tests $n = N/m$. Azencott (1992) derived this relation
assuming an optimum cooling schedule, however the results in this section might
suggest that his conclusion remains true for non-optimal cooling schedules. The

Table 1. Number of Parallel Processes and Cooling Schedule. L: Test lable; CPU: Total CPU-hours spent; C: Number of times the temperature is multiplied by F; T: Starting temperature in every sequence of the cooling schedule (represented by one line); P: Length of platform with constant temperature; N: Number of nodes (processes). X-grad, Z-grad, and Const: "Best model"; EnIC: EnIC of "best model". G: Number of nodes having EnIC of "best model" below 10.36. The starting values for the three model parameters for all tests are randomly distributed within +/- 10 percent off the true model parameters.

							Best Model:				
L	CPU	C	T	P	N	F	X-Grad	Z-Grad	Const	EnIC	G
1	986	21	1.0	4	32	.99	-9.11939	-7.57125	0.947460	10.34715	17
4	990	42	1.0	4	16	.99	-5.41635	-8.18587	0.966651	10.34593	10
7	972	84	1.0	4	8	.99	8.29706	1.10822	- 0.079297	10.34649	6
2	1367	21	1.0	7	32	.99	9.94035	-8.86872	0.908421	10.34073	20
5	1360	42	1.0	7	16	.99	7.56558	-8.57463	0.893401	10.34192	10
8	1379	84	1.0	7	8	.99	9.94074	-9.72762	1.00356	10.34076	7
10	953	21	1.0	4	32	.72	9.99708	-8.80863	0.906802	10.34094	30
11	1401	21	1.0	7	32	.72	9.55244	-9.62777	0.968069	10.34067	29
3	3643	21	1.0	20	32	.99	9.89638	-9.07099	0.954947	10.34242	28
12	3612	21	1.0	20	32	.72	9.99864	-9.71084	0.980927	10.34019	31
14		28	1.0	4	8	.92					
		28	0.1	4	8	.92					
	953	28	0.01	4	8	.92	9.96608	-9.29445	0.940523	10.34046	7

object function tested here is well behaved and almost linear. Since the result makes the simulated annealing flexible it should be tested also for other more complex background velocity-models.

3.3 Cooling Schedule for Parallelized Simulated Annealing

In order to obtain a global method we compare the above "best" models with the "best" models obtained by choosing a faster cooling. Hence, we compare cooling rates of 0.99 and 0.72 (for 32 nodes) and 0.99 and 0.92 (for 8 nodes) for various platforms (Table 1) and determine which cooling schedule yields the least minimum after the same CPU-time spent. The new cooling rates (0.72, 0.92) are selected, in order to reach the temperature of 0.001 after the same CPU-time as in the test above. By comparing the tests in pairs (1 and 10, 2 and 11, 3 and 12, and 7 and 14) it is seen from Table 1 that the least global minimum is obtained with the fastest cooling for about equal spent time. In addition the number of nodes having minimum entropy below EnIC = 10.36 (G in Table 1) is larger for the faster cooling. In fact, almost all nodes give their minimum solution near the global minimum (defined in this way) for the fastest cooling. In addition the fastest cooling give convergence towards the global minimum on almost all nodes (not seen in the table). Figure 3 illustrates three nodes in the simulated annealing with the faster cooling. It is seen that the temperature

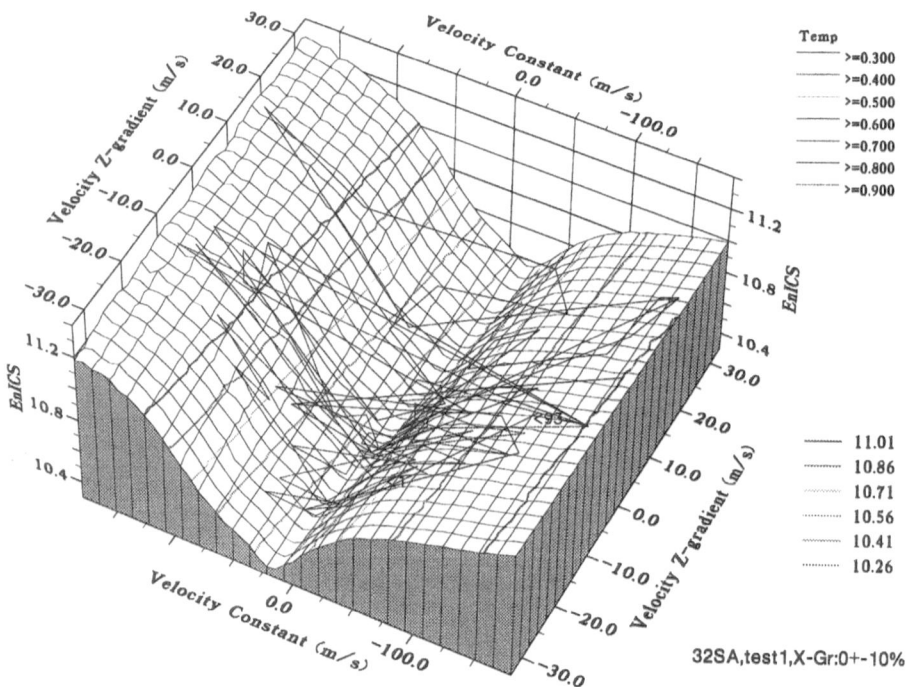

Fig. 2. Number of Parallel Processes. 32 nodes (test 1). Entropy as function of velocity constant and velocity z-gradient. The lable "S" after "EnIC" is used in order to show that the background velocity is assumed to be "smooth". X-gradient "window": 0 +/- 10 percent = 0 +/- 3 m/s. Origo refers to true model.The gridded surface is obtained using all tests. The track of straight lines shows successive new model parameters being tested on one node.

gets so low, that the advantage of using the simplex method for taking the next move becomes obvious: The simulated annealing takes steps only in the deep parts of the valley. The two tests 10 and 14 achieve approximately the same "best" minimum value although they have different cooling rates (0.72 and 0.92 for test 10 and 14, respectively). In a display of all simulated annealing tracks for these two tests it is seen that they both are well distributed in the model space. Hence, although 32 nodes (i.e. 32 starting points for the simulated annealing) seems to be more "safe" with respect to a global search than 8 nodes, the latter might be a sufficiently high number. We have seen that faster cooling than the initially chosen one speeds up simulated annealing and by choosing a fast cooling we have obtained a global method. The observations might be general for

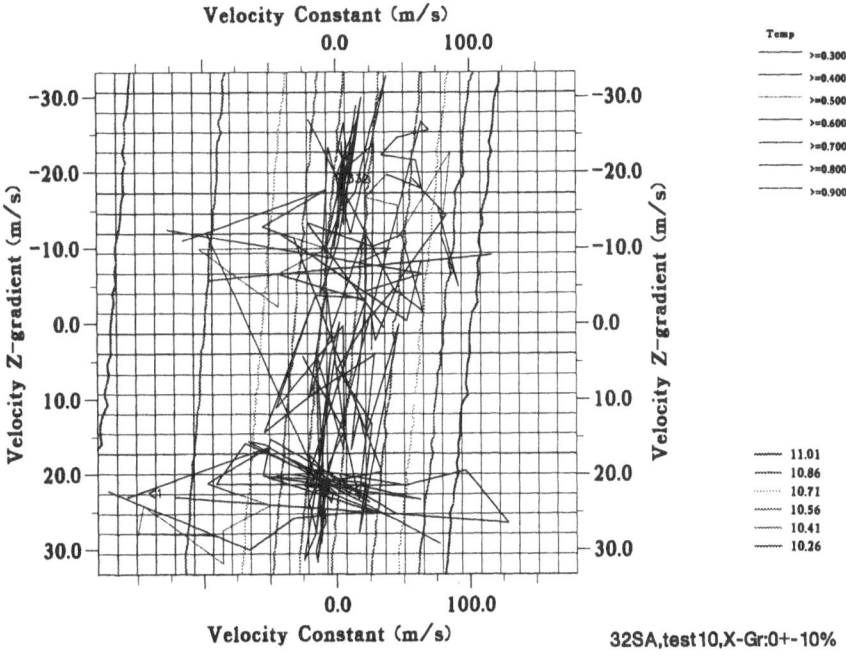

Fig. 3. Simulated annealing with faster cooling. 32 nodes (test 10). See explanation for Figure 2, Figure 3 is Figure 2 seen from above. The track of straight lines shows successively new model parameters being tested on three nodes.

the cases of few background-velocity model-parameters. However a block-model with many model parameters might have many local minima so by choosing a faster cooling one would not get a global solution. In Table 1 it is found that the tests with more calculations on each temperature platform systematically have lower "best" entropy values. This was expected since they contain the most calculations. Displaying all tracks in the various tests also shows that more platform calculations in addition gives a more distributed search, and are hence more stable.

3.4 Discussion of Global Minimum

The above two tests show that simulated annealing (for our velocity model) can be used to obtain the global minimum of the entropy of the image contrast (EnIC), and the faster cooling, the more accurate estimate of the global minimum. No other measure than EnIC has been used in the optimization. The global minimum is near the edge of the analysed region, and the gridded surface in Figure 3 shows the deepest valley-bottom near the edge of the analysed region. The explanation for this fact, is possibly the nature of EnIC: the enhanced focusing effect works well to obtain an approximate location of the reflector, however, the effect of less EnIC in the bottom of the valley reflects that the best EnIC is obtained by focusing the image to one point ($p=1$ in equation 4) in the final stage of the minization. A target grid placed more symmetrically in relation to the enlightening of the grid points than in the above tests (see Figure 1) does not overcome the problem.

Hence, it seems from the object function displays that the entropy of image contrast, as used here, is unsuitable for reflection seismic data with reflectors, i.e. when there are many diffractors. Only when the diffractors are few, or even better, there is only one diffractor, we might use the entropy as applied here. This conclusion was also made by De Vries and Berkhout (1984) for another use of the entropy function on reflection seismic data. They concluded that reflectors should be removed from the data. We prefer not to change the data, but to prepose a change to the method: Evaluate EnIC successively from left to right of the target region for scatterers on vertical lines (both in the 2D and 3D cases). Determine the final background velocity model (e.g. a block model) by a weighted average of the results obtained in each EnIC evaluation. The weighting depend on the ray density in each evaluation. Another possibility for global inversion to get the background velocity-model, however using much more CPU time, is to use the norm of the difference between observed and synthetic waveforms as object function. However, since the slowness perturbation is included in the synthetic seismogram, also this method might require sparsely distributed diffractors. Preliminary results have shown improved results when this object function is used alone.

4 Conclusion

In this work we have analysed inversion of reflection seismic data using simulated annealing for a simple velocity model with three model parameters. We have shown that we can obtain a specific accuracy of the global minima of EnIC using any number of parallel processors as long as the number of EnIC calculations is kept constant. In addition we have shown, that it is possible to obtain a global method using fast cooling and parallel processors. We still have to improve the speed of the cooling and still keep a global method, i.e. find the optimal cooling schedule by e.g. using adaptive cooling schedules (see Ruppeiner et al., 1991) or by applying Ingber's (1989) VFSA. We have preposed improvements to the tested object function and also suggested the residual as an alternative object function in order to avoid the false global minimum due to focusing. Also introduction of local optimization methods and SVD in the neighbourhood of the global minimum, where the object function is almost linear has to be tested (see flow-chart).

We have almost had constant velocity-x-gradient in the tests described. However varying the three model parameters within equal absolute velocity intervals shows also unresolvable dependency between the x-gradient and the z-gradient and between the x-gradient and the constant velocity, which is also the case, when travel time data alone is used. This problem of equivalent models will be an even bigger problem when more general velocity models (e.g. velocity block models) with several hundred model parameters are introduced. Since we cannot know the resolution of the model parameters without knowing an approximate analytical description of the object function, a procedure of first finding the global minimum approximately using simulated annealing and second improving the global minimum while assuming a quadratic form of the object function is suggested here. The last method will then in addition to providing the last step towards the global minimum provide the error estimates, the correlation between the individual model parameters, and information of which model parameters are unresolvable by the data (in the global minimum). The unresolved model parameters or the correlated model parameters can then *a priori* be fixed or added together, respectively, in further investigations as done by e.g. Böhm and Vesnaver (1995).

5 References

Aki, K., Richards, P., 1980: Quantitative Seismology, Theory and Methods, vol. 1 and 2. W. H. Freeman, San Francisco, 932 pp.

Azencott, R., 1992: Simulated Annealing: Parallelization Techniques, Azencott, R. (editor) John Wiley ,242 pp.

Beydoun, W.B. and Mendes, M., 1989: Elastic ray-Born l_2-migration/inversion, Geophysical Journal **97**, 151-160.

Beylkin, G., 1985: Imaging of discontinuities in the inverse scattering problem by inversion of a causal generalized Radon transform. J. Math. Phys. **26**, 99-108.

Beylkin, G., Burridge, R., 1990: Linearized inverse scattering problems in acoustics and elasticity. Wave Motion **12**, 15-52.

Bleistein, N., 1987: On the imaging of reflectors in the Earth. Geophysics **52**, 931-942.

Böhm, G., Vesnaver, A. L., 1995: 3D Tomographic Inversion of Traveltimes in Irregular Grids. EAEG, Glasgow.

De Vries, D. and Berkhout, A.J., 1984: Velocity analysis based on minimum entropy, Geophysics **49**, 2132-2142.

Engell-Sørensen, L., Ryzhikov, G., Biryulina, M., 1994: Global optimization in 3D reflection seismics. 6th International Symposium on Seismic Reflection Probing of the Continents and their Margins, Budapest, Hungary, 12-17 September 1994.

Grefenstette, J. J., 1990: GENESIS. A computer package.

Hanyga, A,, Thierry, P., Lambaré, G., Lucio, P. S., 1995: 2D and 3D asymptotic Green's functions for linear inversion. SPIE Proc. Vol. 2571, 40th SPIE Annual Meeting, San Diego, 9-14 July 1995.

Ingber, L., 1989: Very fast simulated re-annealing. Math. Comput. Modelling 12, 967-973.

Ingber, L., 1993: Simulated Annealing: Practice versus Theory. Math. Comput. Modelling 18, 29–57.

Ingber, L., Rosen, B., 1992: Genetic Algorithms and very fast simulated reannealing: A comparison. Math. Comput. Modelling 16, 87-100.

Jin, S., Madariaga, R., 1992: Velocity determination by optimizing the coherency of spatial short wavelength components. Geophysics, in preparation.

Jin, S., Madariaga, R., 1993: Background velocity inversion with a genetic algorithm. Geophys. Res. Lett. 20, 93-96.

Jin, S., Madariaga, R., Virieux, J., Lambaré, G., 1992: Two-dimensional asymptotic iterative elastic inversion. Geophys. J. Int. 108, 575-588.

Lambaré, G., Virieux, J., Madariaga, R., Jin, S., 1992: Iterative asymptotic inversion in the acoustic approximation. Geophysics 57, 1138-1154.

Miller, D., Oristaglio, M., Beylkin, G., 1987: A new slant on seismic imaging: Migration and integral geometry. Geophysics 52, 943-964.

Moser, T. J., Engell-Sørensen, L., Pajchel, J., 1995: Linear inversion as a first step in entropy of image contrast optimization, 3-D Asymptotic Seismic Imaging. Contract JOU2-93-0321, Second periodic report.

Pajchel, J., Helle, H. B., Froyland, L. A., 1988: A 2-D raytrace modelling software package: User's Manual. Internal Report, Norsk Hydro Research Centre.

Press, W. H., Teukolsky, S. A., Vetterling, W. T., Flannery, B. P., 1992: Numerical Recipes in C: the art of scientific computing. Cambridge University Press.

Ruppeiner, G., Pedersen, J. M., Salamon, P., 1991: Ensemble approach to simulated annealing. J. Phys. I 1, 455-470.

Ryzhikov, G. A., Biryulina, M., Hanyga, A., 1995: 3D nonlinear inversion by entropy of image optimization. Nonlinear Processes in Geophysics 2, 228-240.

Stoffa, P. L., Sen, M. K., 1991: Nonlinear multiparameter optimization using genetic algorithms: Inversion of plane-wave seismograms. Geophysics 56, 1794-1810.

Thierry, P., Lambaré, G., 1994: Ray + Born modelization: a strategy for fast computing, 3-D Asymptotic Seismic Imaging. Contract JOU2-93-0321, First periodic report.

Acknowledgements

This research was done in the framework of the Joule project "3D Asymptotic Seismic Imaging" and funded by The European Commission. This work has received support from The Research Council of Norway (Programme for Supercomputing) through a grant of computing time. We thank A. Hanyga and V. Sorin (Institute of Solid Earth Physics) for fruitful discussions and good suggestions.

Phase Migration: An Approach to Inversion of Wide-angle Seismic Data

Konstantin S. Osypov[1,2], Zhengsheng Yao[2], and Roland G. Roberts[2]

[1] Department of Geophysics, Uppsala University, Sweden
[2] On leave to Center for Wave Phenomena, Colorado School of Mines, USA

1 Introduction

Traditionally, the analysis of wide-angle seismic data includes only the inversion of travel-times of a few main arrivals, e.g. Pg, Pn, PmP, Sg, Sn, SmS (e.g. Zelt and Smith 1992). Thus only a tiny proportion of the information in the seismic data is actually used by this approach. Moreover, the assessment of the arrival-time of the various arrivals can be difficult and very subjective. In principle, the use of the complete waveform information (e.g. Tarantola 1987) can avoid these problems, but such analysis is much more complicated, is often less robust, poorly converges for long wavelength velocity anomalies and can be prohibitively computationally expensive. Faster techniques operating on the waveform data but making suitable simplifying assumptions are therefore often used. For example, a back projection technique of diffraction tomography using the Born approximation, may be regarded as a migration summing covariances between scattered and observed wave-fields (Ryzhikov and Troyan 1992). For the acoustic case this approach is similar to depth prestack Kirchoff migration which is known to be an effective method for imaging complex structures from reflection data (e.g. Claerbout 1971).

Especially in lithospheric near normal incidence reflection studies, background velocities are often poorly determined. With wide-angle data, travel time tomography can be used to estimate the low-frequency velocity distribution. However, the application of migration to wide-angle data is complicated by several problems. One of them is that poor coverage (low fold, possible spatial aliasing) makes the inverse problem underdetermined which may result in artifacts and poor resolution in the migrated image. In other words, the assumption that estimates for different medium points are independent is not valid, in contrast to the case in normal reflection seismics. To overcome this problem, adaptive migration approaches using wave-field correlations have been developed (Timoshin et al. 1989).

The second problem with wide-angle migration is the existence of converted and refracted waves which may corrupt an image. Diving waves at near-critical distances almost coincide with crustal reflections and therefore do not cause severe artifacts while the migration. Diving waves at larger offsets often can be filtered out before the migration.

One more problem is that amplitude distortions during signal propagation and registration can not be fully explained if the usual types of models are used. For example, elastic scattering (directionality, phase conversion) should be taken into account in true amplitude migration of wide-angle data, but this is difficult because elastic scattered patterns are very complicated (e.g. Li and Hudson 1995).

Thus for wide-angle data a migration method is required which is robust regarding effects such as phase conversion, rapid spatial variations in amplitude, and poor spatial coverage.

2   The Phase Migration Algorithm

The wave-form information contained in a seismic section can be decomposed into two parts: Instantaneous amplitude and phase. As discussed above, amplitude information can be difficult to use and for this reason much seismic processing utilizes automatic gain control (AGC) so that the significance of the amplitude information is downweighted. Thus conventional migration of data including the use of AGC employs mainly the phase information, as amplitudes are equalized.

Following this line of thought, we choose to base our migration algorithm on phase information explicitly. Note however, that temporally local variations in amplitude are utilized when calculating phase-lags. The kinematics of diffracted and reflected waves together with phase correlation are used for the identification of coherent arrivals, and the mapping of these into the spatial domain.

To extract the phase information we first perform complex demodulation using a modified Butterworth filter (Osypov 1994). The central and cut-off frequencies for complex demodulation are chosen by reference to the expected mean frequency and length of a source wavelet, assumed to be sufficiently well represented by the first arrivals.

The phase part of the cross-correlation function between two complex demodulated traces for a certain time lag and over a suitable short time window may be approximated by simple phase lag between two complex samples, one from each record taken with the corresponding time lag. Furthermore, we can estimate the total sum of phase lags for all cross-correlations along a travel-time curve within a certain aperture. Lindfors and Roberts (1996) showed that this approach can better identify arrivals than e.g. semblance analysis or beamforming. For many cases, especially when several signals interfere, it is preferable to sum phase lags only for a certain correlation radius less than the migration aperture.

As a measure of phase coherence we use a cosine function of the phase deviation over the chosen aperture. This measure differs from conventional coherence but has a similar meaning and varies from 0 to 1. Phase coherence is 1 for zero phase misfit. How the confidence level is defined will be discussed later.

The algorithm of phase migration is based on analyses of phase correlation along the isochrone for a certain medium point. This may a diffraction point or an element of a reflector. For the real wide-angle data we are mostly revealing reflectors. We assume that each reflector can be constructed from a number of diffraction points - the same idea of Huygens interference as used in Kirchoff migration. It is important to point out that while migrating a reflector, only part of the data along the diffraction isochrone fits the reflection curve (see Fig.1). Thus, as in Kirchoff migration, the result of the migration is primarily influenced only by the part of data in the vicinity of the points defined by normal moveout (NMO) or dip moveout (DMO) equations (see e.g. Yilmaz 1994). In this region phase coherence is high. In Kirchoff migration the data outside this area sums destructively, tending to cancel. Similarly, in phase migration, because the phases are randomly distributed the phase coherence will tend to average to zero.

The size of the area where the diffraction and reflection isochrones coincide corresponds essentially to the Fresnel zone. Therefore it is logical to limit the integration aperture by reference to the size of Fresnel zones for dipping reflectors (Timoshin et al. 1989):

$$\ell_m = \frac{4}{cos^2 i}\sqrt{mz\lambda\delta cos\varphi cos(i-\varphi)}$$

where m - an order of Fresnel zone (we deal with m=1), i - angle of incidence, φ - dip angle, λ - wavelength, δ - number of periods in the wavelet.

Note that according to this formula the size of the Fresnel zone implicitly depends on the frequency band-width rather than on the main frequency of a wavelet. The Fresnel zone estimated from this formula is different from the one commonly used in reflection seismic studies (e.g Yilmaz 1994) and is generally larger. It is natural to allow such considerations of resolution to determine the spatial density of the grid which is used for migration.

To increase the contrast of the migrated image we use a search for the best correlation in a running window in the vicinity of the predefined range of reflector dips. This adaptive search proves effective even in the presence of interfering phases.

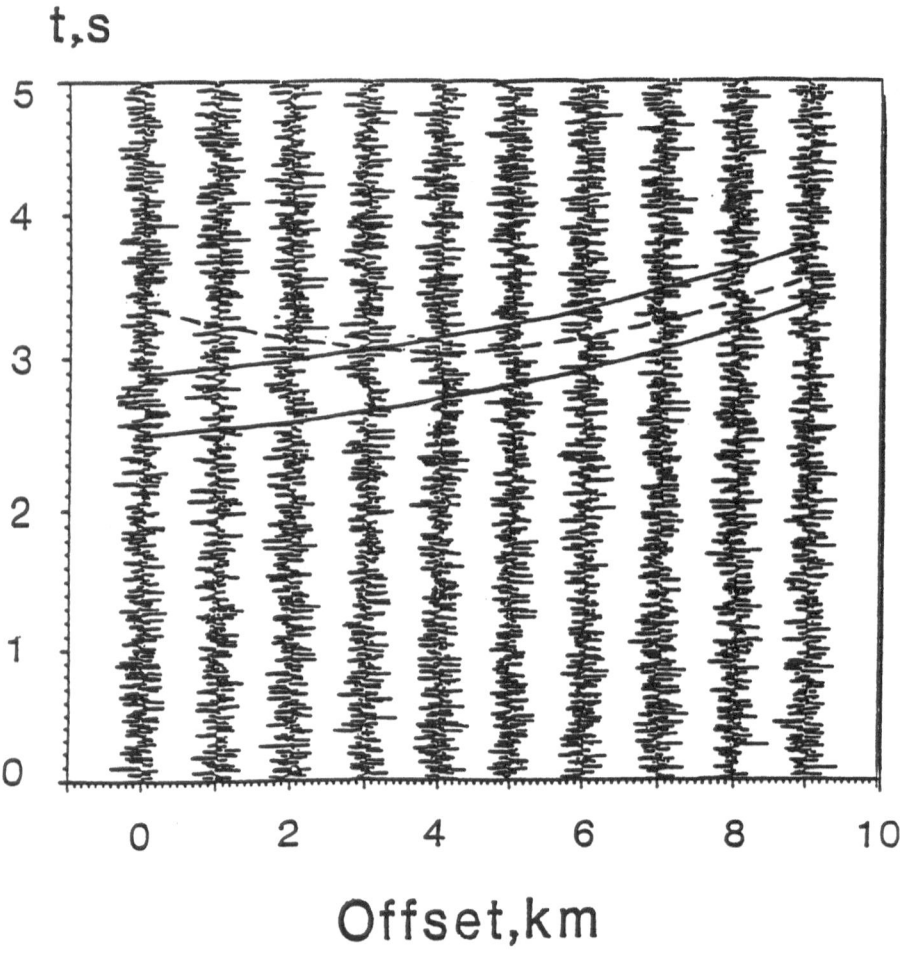

Figure 1: Synthetic shot gather for a pre-critical reflection from a horizontal boundary. Travel-times were calculated for a background velocity of 4 km/s, and then theoretical reflection coefficients were convolved with the wavelet (first derivative of Gaussian function with duration 0.4 s). White gaussian noise with SNR=1 was added. The shot is on the surface, 10 stations are evenly spaced on the surface from 0 to 10 km offset. The solid lines delineate the reflected wave, and the dashed line is the travel-time curve of diffraction from a medium point at depth 5 km and offset 5 km.

The result is that after scanning a grid of medium points, we map the maximum phase correlation on the isochrone onto a medium point position. Implicitly, the image obtained corresponds to a map of scattering from the Earth's discontinuities. Therefore, the phase migration may be regarded as an inversion procedure.

The relative speed of the phase migration depends primarily upon the density of the spatial grid used. For wide-angle data this grid can be rather coarse, and therefore computational costs are much less than for the prestack Kirchoff migration of wide-angle data without averaging.

3   Simulations

To test the properties of the method we produced synthetic seismograms for different models and using different techniques (ray, Born, finite-differences). The data shown in Fig.1 were generated under the ray approximation and simulate pre-critical reflection from a horizontal boundary in the presence of Gaussian uncorrelated noise with SNR=1.

Fig.2 shows two images produced from the data in Fig.1, by depth prestack Kirchoff migration and by phase migration respectively. The dashed line shows the position of the true boundary. Only part of the reflector is illuminated and therefore can be imaged. Both images are similar but phase migration produces an image closer to the true model with fewer artifacts. However, the main advantage of phase migration in comparison with Kirchoff migration is the possibility of analyzing array correlation patterns almost independently of amplitude fluctuations and scaling. On the other hand, a stronger reflector produces higher phase correlation for a certain noise level.

4   Application to BABEL Data

To illustrate the application of the phase migration to real wide-angle data we present here an example of analysis of BABEL data. The BABEL experiment was conducted in 1989 and provided good quality reflection and refraction marine data. For details of the experiment we refer to BABEL Working Group (1993). Our aim was to refine the crustal structure along BABEL line 2 previously modelled using travel-time inversion of arrivals picked by visual inspection (Riahi and Lund 1994). We used the records from station 201 out to an offset of 100 km. The distance between shots used was about 190 m. Pre-processing included frequency filtering 6-12 Hz, predictive deconvolution with lengths of auto-correlation and gap equal to 1000 ms and 125 ms respectively, and AGC within the time window 1500 ms.

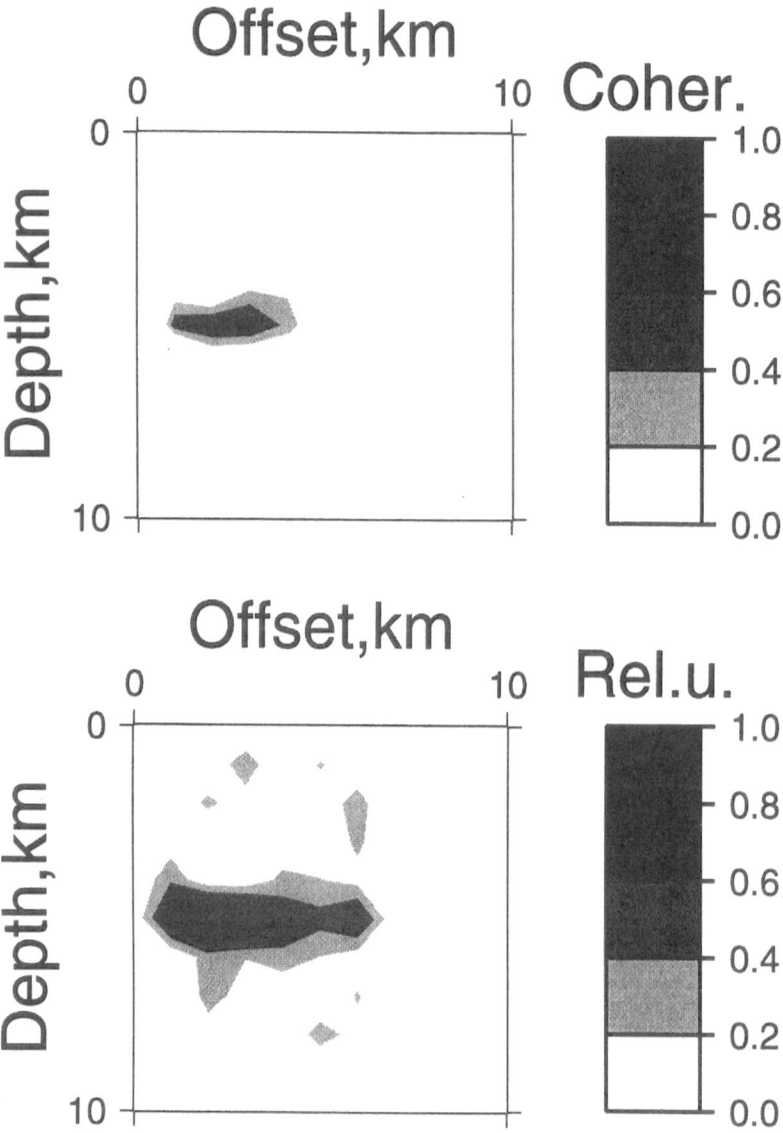

Figure 2: top - Phase migrated image produced from the data on Fig.1; bottom - Depth Kirchoff migrated image from the same data. An aperture of 7 km was used in both cases.

Fig.3 presents the part of data used for migration. Note that we used only the data in the area between Pg and Sg shown by solid lines. Fig.4 shows the results of phase migration (top) and corresponding normal-incidence reflection section (bottom) with reflective boundaries deduced from previous interpretation of the wide-angle data added (Riahi and Lund 1994). Confidence levels of phase coherence were assessed empirically by analysis of phase coherencies after shifting traces in a random manner. This test showed that for our case coherencies more than 0.2 may be considered significant.

As one can see, the crustal structure is very complicated and appears to consist of several reflecting boundaries. Note that only a part of the medium is well illuminated due to the experiment geometry. One of the most prominent features of the phase migrated section is the Moho revealed at about 40-45 km depth at offsets 20-30 km, 50 km depth at offsets 50-70 km and 47 km depth at offset about 80 km. Comparing the top and the bottom of Fig.4 we can see that this result basically agrees with previous analyses based on the data used here, other wide-angle data and the streamer data (Riahi and Lund 1994). Other features can also be correlated with previous analyses, notably the shape of the strongly reflecting region in the depth range from 30 to 40 km. The high coherence at about 60 km depth and offsets 30-40 km probably represents an upper mantle discontinuity.

5   Conclusions

The phase migration provides an interpretation method for wide-angle data which is an alternative to more traditional methods such as modelling or inversion of arrival time information picked e.g. by visual inspection, or classical migration. The phase migration extracts detailed kinematic information from even one wide-angle shot gather by estimation of phase coherency along travel-time curves corresponding to a grid of medium points. Our results indicate that the new method can produce results in some ways superior to those from the other methods, and it is simple and not particularly computationally demanding.

Our first application of phase migration to wide-angle BABEL data revealed some features of crustal structure similar to the streamer section but not before interpreted from wide angle data. Also some features of upper mantle structure were imaged for the first time.

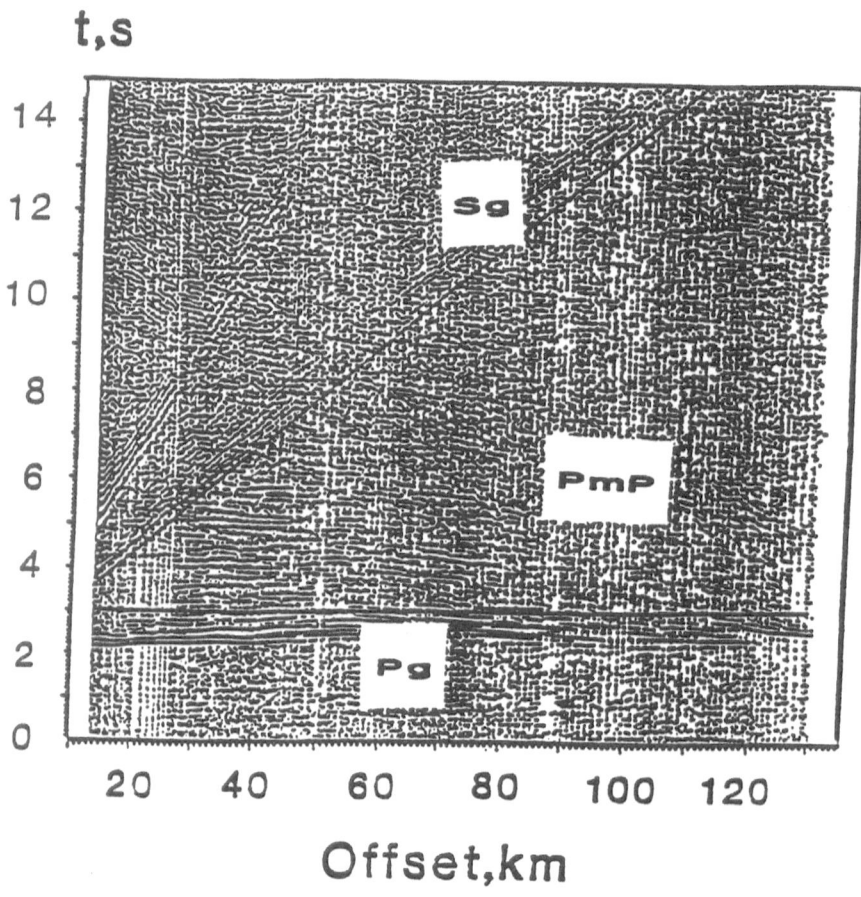

Figure 3: BABEL wide-angle seismic section recorded at station 201. Vertical component. Band-passed filter 0-10 Hz. Reduction velocity 6 km/s.

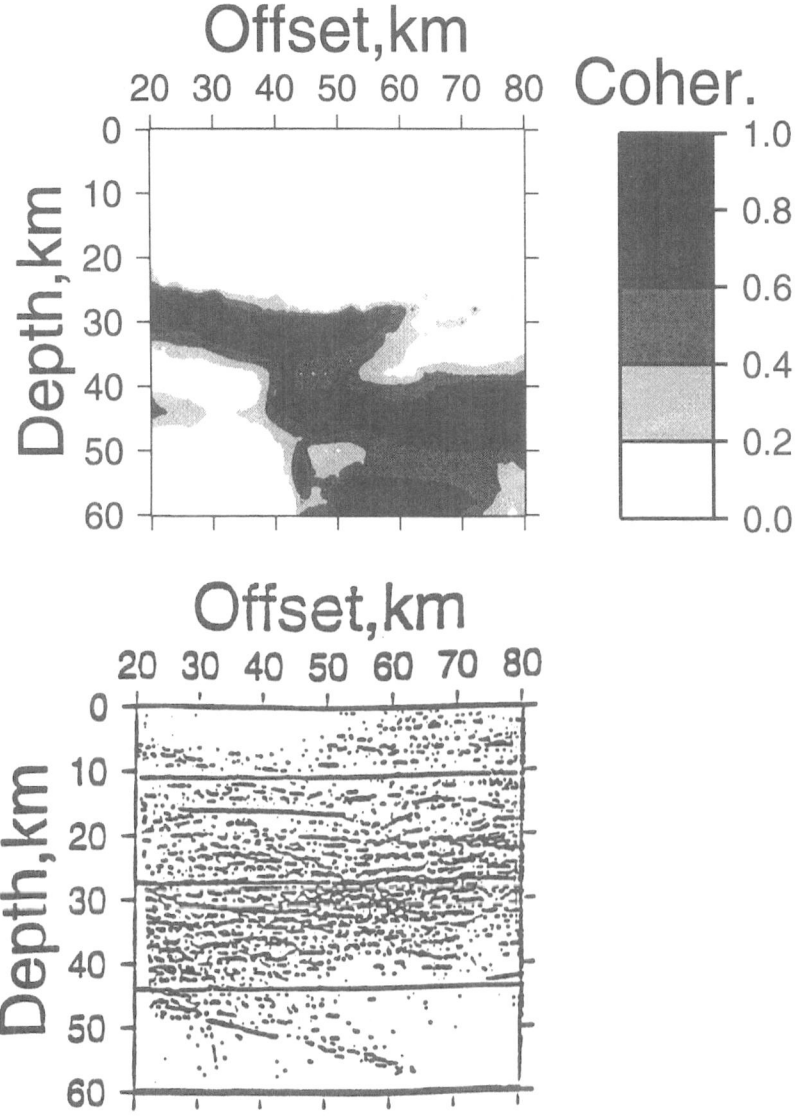

Figure 4: top - Phase migrated image produced from the data on Fig.3; bottom - Migrated normal-incidence reflection data of BABEL line 2 prepared by BIRPS, Cambridge University after application of a coherency filter. Black thick lines show reflective boundaries deduced from the wide-angle data by Riahi and Lund (1994). An aperture of 5 km and grid cells 2x2 km were used for the migration. A gradient velocity model with surface P velocity 5.8 km/s and gradient 0.052 1/s. was used for the migration

6 Acknowledgments

We are greatful to Laust Pedersen and Gennady Ryzhikov for fruitful discussions. We also appreciate Ali Riahi for providing us with the processed BABEL data and anonymous reviewer for useful comments. The work was supported by the Swedish National Research Council and Uppsala University.

7 References

BABEL Working Group: Integrated seismic studies of the Baltic Shield using data in the Gulf of Bothnia region. Geophys. J. Int. **112** (1993) 305–324

Claerbout, J.F.: Toward a unified theory of reflector mapping, Geophysics **36** (1971) 467– 481

Li, X., Hudson, J.A.: Elastic scattered waves from a continuous and heterogeneous layer. Geophys. J.Int. **121** (1995) 82–102

Lindfors, A. and Roberts, R.: Array processing via phase regression. In: E.S. Husebye and A.M.Dainty (eds.), Proceedings of "Nato Advanced Study Institute on Monitoring a comprehensive test ban treaty", Alvor- Algarve- Portugal, 23 Jan - 2 Feb,1995", NATO ASI Series E, **303** (1996) 629–644.

Osypov, K.S.: Algorithms and programs for statistical analysis of non-stationary geophysical processes. In: COSPAR Colloquia Series **5** (1994) 703–706

Riahi, M.A. and Lund, C.-E.: Two dimensional modelling and interpretation of seismic wide-angle data from the western Gulf of Bothnia. Tectonophysics **239** (1994) 149–164

Ryzhikov, G. and Troyan, V.: 3D diffraction tomography. Part 2: Reconstruction algorithm with statistical regularization, in Expanded Abstracts, Russian-Norwegian Oil Exploration Workshop II, Voss, May 5-7, (1992) 1–14

Tarantola, A.: Inverse problem theory. Elsevier (1987) Amsterdam

Timoshin, U.V., Birdus, S.A. and Merchij, V.V.: Seismic holography of complex structures. Nedra (1989) 255 p. (In Russian).

Yilmaz, O.: Seismic data processing. SEG (1994) 526 p.

Zelt, C.A., Smith, R.B.: Seismic traveltime inversion for 2-D crustal velocity. Geophys.J.Int. **108** (1992) 16-34

Location of Seismic Events - A Quest for Accuracy

Wojciech Dębski

Institute of Geophysics Polish Academy of Sciences, Warsaw, Poland

1 Introduction

One of the most frequent inverse problems in observational seismology is the location of seismic events such as earthquakes, nuclear explosions, mining tremors, etc. By "location" we understand determining the place (called the hypocenter) where an event has occurred and the time of its appearance (the origin time). This goal is achieved by the analysis of information contained in seismograms (registered ground motions caused by passing seismic waves). The most fundamental approach to the problem is attributed to Geiger (1910) and now used in a form adapted to modern computational facilities (see e.g. Buland 1976). In this method we use arrival times (time onsets) of different seismic phases (waves propagating along different paths) recorded at a number of seismograph stations. The next step involves looking for the source place and the origin time for which observed and theoretically predicted arrival times are as close as possible. This is done by searching for the minimum for the sum of squares of time residuals (differences between observed and modeled time onsets). The point for which the minimum occurs is the sought hypocenter location and the origin time of the event[1]

One very important thing in the seismic location problem is the question of accuracy of any hypocenter position found. An answer to this question requires a precise analysis of all possible sources of errors. One of them is the discrepancy between the real and the model velocity distributions. It leads to the modeling, or so-called theoretical, errors consisting of statistical and systematic parts. Statistical modeling errors are due to the neglecting of small scale space variations of velocity distribution and appear whenever we are unable to model wave propagation in a medium as complicated as a real one. Systematic modeling errors are generated by the erroneous velocity values used in calculat-

---

[1] The strict mathematical formulation of the seismic location problem as well as descriptions of other methods can be found in e.g. Aki (1985), Lee and Stuart (1981), Bullen and Bolt (1985) or Gibowicz and Kijko (1994).

ing theoretical time onsets,[2] and appear as a result of a lack of knowledge of the real medium's structure. The influence of the statistical modeling errors, when they are only modeling errors, with known statistics on location accuracy can be easily studied in the framework of the inversion schema (see Tarantola 1987). The situation is much more complicated when systematic errors also appear. However, even then, if the statistical part can be extracted from total modeling errors independently of the systematic one, then both types of errors can be consistently included in the location schema. This paper is devoted to the analysis of systematic location errors in such a special case.

The problem of error generation by erroneous values of velocities is reformulated in the language of the general inverse theory in the next section. The last section provides a simple numerical illustration of the problem.

2 Theory

Let the vector $\mathbf{m} = (x, y, z, t)$ be the vector of spatial and time coordinates of the point which is supposed to be the hypocenter of an event. Let also $\mathbf{d}_{\text{obs}} = (t_1^o, t_2^o, \dots t_N^o)$ be the vector of time readings for the observed phases and $\mathbf{m}_v = (v_1, v_2, \dots v_k)$ be the vector of velocities used in a model, further called the parameters. It has been shown by Tarantola (1987) that the a posteriori probability density σ of the event location at the point $\mathbf{m}$, being the most general solution of the location (or generally inverse) problem is given by the following formula:

$$\sigma(\mathbf{m}; \mathbf{m}_v, \mathbf{d}_{\text{obs}}) = f(\mathbf{m}) \times L(\mathbf{m}; \mathbf{m}_v, \mathbf{d}_{\text{obs}}) \tag{1}$$

where $f(\mathbf{m})$ describes a priori information and $L(\cdot)$ is a likelihood function measuring to what extent theoretically predicted data fit the observations. Let the observational errors defined as differences between true values of phase time onsets $\mathbf{d}$ and measured ones $\mathbf{d}_{\text{obs}}$ be described by statistic Θ_{obs}. Let also Ψ_{th} be a probability of theoretical prediction of $\mathbf{d}$ for a given source location $\mathbf{m}$ and parameters $\mathbf{m}_v$. Then, the likelihood function $L(\cdot)$ reads:

$$L(\mathbf{m}; \mathbf{m}_v, \mathbf{d}_{\text{obs}}) = \int \Theta_{\text{obs}}(\mathbf{d} - \mathbf{d}_{\text{obs}}) \times \Psi_{\text{th}}(\mathbf{d}; \, \mathbf{g}(\mathbf{m}, \mathbf{m}_v)) \, d\mathbf{d} \tag{2}$$

where $\mathbf{g}(\mathbf{m}, \mathbf{m}_v)$ describes the modeled time onsets (solution of the forward problem) for given $\mathbf{m}$ and $\mathbf{m}_v$.

This is the basic formula allowing the full analysis of errors of the location results as e.g. calculation of a covariance matrix, error ellipses, dependences of the hypocenter location on the parameters $\mathbf{m}_v$ and so on. To use it efficiently, a knowledge of Ψ_{th} is required[3]. The general form of Ψ_{th} depends on the velocity model underlying the assumptions and should take into account all errors caused by inaccuracy in forward modeling. Some simplification of (2) is possible if a statistical part with known statistic Θ_r independent of $\mathbf{m}$ and $\mathbf{m}_v$, can

---

[2] The other source of systematic errors is our inability to reliably describe ongoing physical processes and is beyond the scope of this paper.

[3] We do not consider here experimental errors assuming Θ_{obs} to be known.

be separated from the overall modeling errors. If additionally systematic errors are caused only by an erroneous estimation of the values of velocities then the likelihood function reads

$$L(\mathbf{m}; \mathbf{m}_v, \mathbf{d}_{obs}) = \int \theta_{obs}(\mathbf{d} - \mathbf{d}_{obs}) \times \Theta_r(\mathbf{d} - \mathbf{g}(\mathbf{m}, \mathbf{m}_v) - \mathbf{h}(\mathbf{m}; \mathbf{m}_v, \mathbf{m}_v^{true})) \, d\mathbf{d}$$

(3)

where

$$\mathbf{h}(\mathbf{m}; \mathbf{m}_v, \mathbf{m}_v^{true}) = \mathbf{g}(\mathbf{m}, \mathbf{m}_v^{true}) - \mathbf{g}(\mathbf{m}, \mathbf{m}_v) \ .$$

Let us stress, that this formula describes the likelihood function in a case when the erroneous parameter values are used in the forward modeling function $\mathbf{g}()$. The last term on the right side, $\mathbf{h}()$ takes into account (compensates) the presence of the systematic bias of the theoretical predictions caused by wrong values of used velocities. If it was known, the found hypocenter location would be not affected by the systematic modeling bias, but reality is different. The simplest possible way of dealing with this term is just to drop it, which means that systematic modeling errors are included into the likelihood function. The resulting solution of the location problem can thus be seriously biased. We proceed this way in a numerical experiment described in the following section.

3 Numerical experiment

To illustrate the above theoretical considerations we have performed a simple numerical experiment. Synthetic data for three stations were generated for a given (known) source position. Our "real" medium was assumed to be two dimensional, homogeneous space characterized by a constant wave velocity equal to 6km/s. The origin time was set to zero. The question was how the results of inversion of these data for the source location depend on the velocity used to solve the forward problem. For all considered values of the velocity we have disregarded the $\mathbf{h}$ term in (3), so the source locations found were subjected to the presence of a systematic modeling bias. Moreover, in the inversion procedure no a priori information was used, so $f(\mathbf{m})$ in (2) was assumed to be constant. The observational and modeling statistical errors were assumed to lead to the Cauchy distribution for the a posteriori probability distribution. We have considered two variants of source–station configurations: a symmetric one with the source placed in the center of an equilateral triangle formed by the stations (case A) and the second one (case B), when the stations are almost collinear (see Fig. 1). Table 1 summarizes the values of the source and station coordinates as well as the "data" used for location, in both variants. The results represented by the maximum likelihood solution (MLL)[4] and the one-dimensional marginal probability distributions are shown in Figs. 2 and 3 respectively.

Figure 2 shows the maximum likelihood solutions for x and y source coordinates as functions of the velocity. In the case A (upper row) the hypocenter position does not depend on the velocity at all. Simultaneously, the errors of

---

[4] The point for which the maximum of the a posteriori probability distribution occurs.

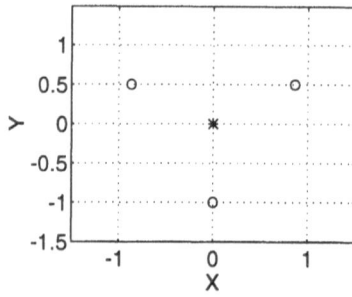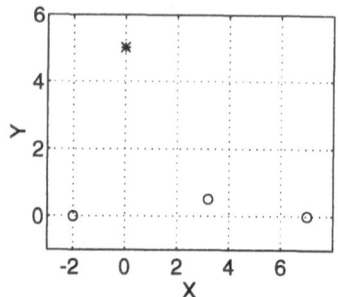

Fig. 1. The layout of the source (star) and stations (open circle) configuration for both A (left) and B (right) variants of the experiment.

Table 1. Source and station coordinates and synthetic time onsets

	Configuration A			Configuration B		
	x[km]	y[km]	t[s]	x[km]	y[km]	t[s]
source	0.0	0.0	0.0	0.0	5.0	0.0
station 1	0.866	0.5	0.167	3.2	0.0	0.920
station 2	-0.866	0.5	0.167	-2.0	0.32	0.897
station 3	0.00	-1.0	0.167	7.0	0.0	1.437

the estimated hypocenter position have increased, which is manifested as the spread-out of the a posteriori distribution for velocities other than the true one, as is shown in Fig. 3.

In the case B, a big shift of the maximum likelihood solution from the vicinity of the true source location to the spurious one for velocity much smaller than the true one (see Fig. 2, lower row) is observed.

Figure 3 shows the one-dimensional a posteriori marginal probability distributions for both considered cases. Contrary to the MLL solution discussed above, these distributions also provide information about overall location errors (Tarantola 1987). As can be seen, wrong velocity values cause the spread of the a posteriori distribution, which means a decrease of the location accuracy. For the case A (upper row), this is the only result of the presence of systematic modeling errors. In such a situation a reasonable location error estimation can be carried out in a standard way, e.g. by means of a covariance matrix. In the case B (lower row) the essential change of the distribution shape is observed for velocities much smaller than the true one, the highest and secondary picks change their position. The final errors are then dominated by systematic ones that can be estimated by means of the distance between the highest and the secondary pick. The standard error analysis now has no sense. However, for velocity close to the true one the standard methods can be applied, and in spite of overall errors consisting of a statistical and a systematic part they may give a rough estimate of the location errors.

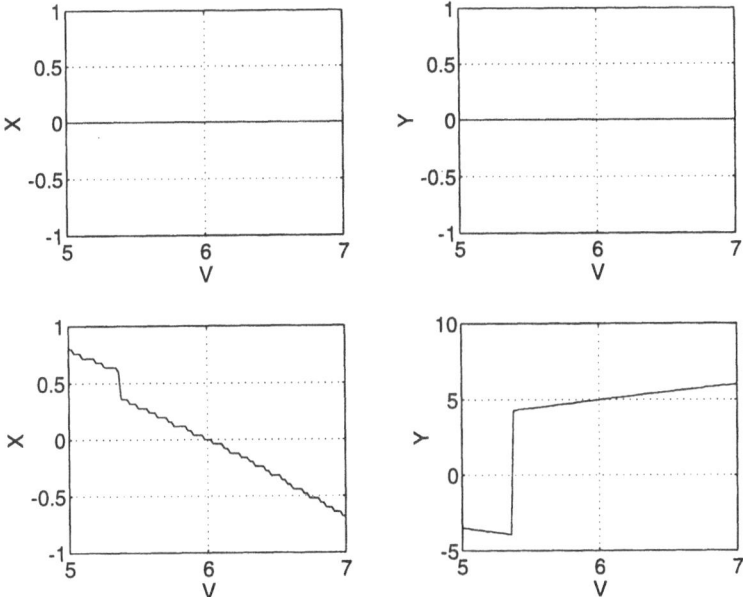

Fig. 2. The dependence of the maximum likelihood solution for x (left column) and y (right column) source coordinates for two different station configurations (see text) A (upper row) and B (lower row).

4 Conclusion

Apart from variables estimated from data, fixed parameters (e.g. seismic velocity distribution) necessary for theoretical modeling of wave propagation are also involved in the seismic location problems. It is a general feature of all inverse tasks that in spite of "variables" to be estimated from a given data set, the parameters needed for theoretical modeling with values known from somewhere else are present in the problem. The (possibly) erroneous values of these fixed parameters introduce additional errors to the solution of the inverse problem. They sometimes manifest themselves as a shift of the maximum likelihood solution and sometimes lead to a decrease in the accuracy of inversion. If the parameters deviate little from their "true" values the maximum likelihood solution shift is masked by the spread of the a posteriori probability distribution. In this case the standard methods of error estimation based on the a posteriori distribution give reasonable results. On the other hand, for large deviations, the systematic errors become dominating and lead to a completely wrong maximum likelihood solution. Then, error estimation which does not take into account the presence of systematic bias is not reliable.

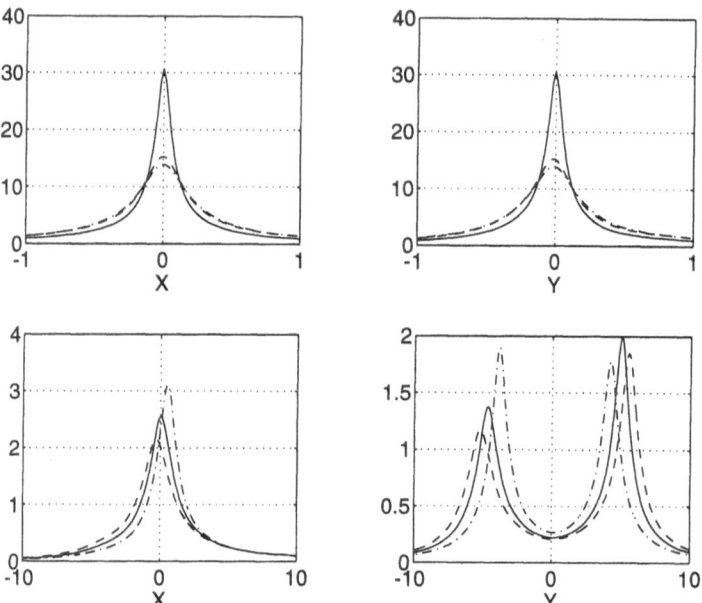

Fig. 3. One-dimensional marginal probability distributions for the x (left column) and y (right column) source coordinates calculated for velocities smaller (5.3 km/s) (dash-dotted), equal (solid line) and greater than the true one (6.5 km/s) (dashed line) for case A and B (upper and lower row respectively)

References

Aki, K. and Richards, P.G., 1985. Quantitative seismology. Freeman and Co, San Francisco.

Bullen, K.E., and Bolt, B.A., 1985. An Introduction to the Theory of Seismology. Cambridge University Press.

Buland R., 1976. The mechanics of locating earthquakes, Bul. Seis. Am., 66, 173–187

Gibowicz S.J. and Kijko, A., 1994. An Introduction to Mining Seismology. Academic Press, San Diego.

Geiger, L., 1910. Herdbestimmung bei Erdbeben aus den Ankunftzeiten. K. Gessel. Wiss. Goett., 4, 331–349

Lee, W.H.K. and Stewart, S.W., 1981. Principles and Application of Microearthquake Networks. Academic Press, New York.

Tarantola, A., 1987. Inverse Problem Theory: Methods for Data Fitting and Model Parameter Estimation. Elsevier, Amsterdam.

Inverse Estimation of Parameters in Petroleum Reaction Networks

Søren B. Nielsen[1] Tanja Barth[2] and Louis C. Dahl[2]

[1] Department of Earth Sciences, Geophysical Laboratory, Aarhus University, Denmark
[2] Department of Chemistry, University of Bergen, Bergen , Norway

1 Introduction

Petroleum phases such as oil and gas are generated by thermal degradation of kerogen, which is the remains of dead organisms preserved in sedimentary rocks. Generally, the temperature range for oil generation in sedimentary basins is 120-150°C and for gas generation 150-200°C (Quigley and Mackenzie, 1988), which corresponds approximately to depths of burial from 2 to 5 km for oil generation and from 3 to 6 km for gas generation for the likely ranges of temperature gradients in different types of sedimentary basins.

As an aid in petroleum exploration it is desirable to be able to simulate the history of petroleum generation during basin evolution. For example, it is important that trap formation preceds petroleum migration and that the volume of petroleum generated in the source rock exceeds the petroleum retained in the source rock and the possibly very large loss during migration to the trap.

The basic requirements for the simulation are the source rock temperature history and knowledge about the processes of petroleum generation and migration. The temperature history can be reconstructed with some certainty from the burial history and palaeothermal indicators and present day temperatures obtained from boreholes (Tissot et al., 1987; Nielsen, 1995).

The chemical complexity of kerogen and the petroleum phases that are generated precludes quantitative chemical description of the process by traditional means. Classical chemical theory is not suited for prediction of the temperature range of petroleum generation from a particular source rock. However, petroleum generation can be experimentally investigated by artificial maturation under controlled conditions. The results of such experiments are used to determine bulk parameters of petroleum generation models (Ungerer, 1989; Barth and Nielsen, 1993). This paper discusses how this can be done and how accurate the results are.

Hydrous pyrolysis (Lewan et al., 1979; Lewan, 1985; Barth et al., 1989) is a technique of artificial maturation which produces petroleum products in close resemblance to natural petroleum. Pulverized source rock samples are kept at constant temperature in a sealed pressure vessel for typically 72 hours. Due to the requirement that a liquid water phase be present, the maximum possible experimental temperature is 365°C. The temperature at which incipient reaction activity takes place defines the lower experimental temperature limit and is usually around 220°C for an experiment duration of 72 hours. At the end of the experiment the residual concentration of the reactant (kerogen) is measured along with the concentrations of generated products such as bitumen (liquid petroleum), asphaltenes, gasses (e.g. CH_4 and CO_2), and the inert residue 'coke', which is formed during the reactions. A suite of such experiments performed at e.g. 220, 250, 260, 280, 300, 320, 340, and 360°C provide an important part of the data needed to estimate the parameters of petroleum generation models.

Open system pyrolysis (Tissot and Espitalié, 1975; Jüntgen and Klein, 1975; Burnham et al., 1987) is a technique for investigating the process of kerogen degradation only. A source rock sample is heated at a constant rate of e.g. 25°C min^{-1} and the rate of gas generation is recorded. The temperature of maximum kerogen degradation rate (maximum gas generation rate), T_{max}, is of special significance as it shows a heating rate dependence which can be described by first order reaction kinetics. T_{max} generally ranges from 380 to 500°C for laboratory heating rates between 0.1 and 50°C min^{-1}.

A number of studies show that petroleum generation models based on first order reaction kinetics can describe both the results of laboratory pyrolysis experiments at elevated temperatures and borehole observations of petroleum generation in sedimentary basins (Ungerer and Pelet, 1987; Quigley and Mackenzie, 1988). The equations governing the temporal evolution of a reaction kinetic network read

$$\frac{\partial}{\partial t}y = By$$
$$y(t=t_0) = y_0 \tag{1}$$

$y = (y_1, y_2,.., y_n)^t$ is a vector of reactant and product concentrations with initial concentrations defined by y_0, and B is a matrix of reaction rates and stochiometric factors. In the case of first order reactions with Arrhenius reaction rates the reaction rate k_i governing the decay of the ith reactant is independent of concentration and reads

$$k_i = A_i exp(-\frac{E_i}{RT}) \tag{2}$$

A_i is the frequency factor and E_i is the activation energy. Both are assumed to be constant and independent of temperature for each reaction. R is the gas constant and T is the absolute temperature. The reaction parameters A_i and E_i govern the temperature dependence of the ith reaction and their accurate determination is a prerequisite for accurate simulation of the petroleum generation history in sedimentary basins.

2 Inversion

2.1 The Forward Model

The forward model is given by equations (1) and (2). As an example consider a simple reaction network in which kerogen (y_1) degrades to bitumen (y_2) and gas (y_3) under the formation of coke (y_4). In turn bitumen degrades to gas under the formation of coke. Gas and coke are the stable end products of the system. The system of equations reads

$$\frac{\delta}{\delta t}\begin{pmatrix} y_1 \\ y_2 \\ y_3 \\ y_4 \end{pmatrix} = \left\{ \begin{bmatrix} -k_1 & 0 & 0 & 0 \\ a_{21}k_1 & -k_2 & 0 & 0 \\ a_{31}k_1 & a_{32}k_2 & 0 & 0 \\ a_{41}k_1 & a_{42}k_2 & 0 & 0 \end{bmatrix} \right\} \begin{pmatrix} y_1 \\ y_2 \\ y_3 \\ y_4 \end{pmatrix} \tag{3}$$

with the initial condition $y(t=t_0) = (y_{01}, 0, 0, 0)^t$ which says that initially neither bitumen and gas nor coke is present.

The stochiometric factors a_{ij} are constrained by the relationships

$$0 \leq a_{ij} \leq 1$$

$$\sum_i a_{ij} = 1 \tag{4}$$

which are incorporated in the inversion procedure and used to reduce the number of unknown stochiometric factors by one for each column.

Figure 1 shows the theoretical response of the system of equation (3) to simulated maturation experiments of duration 72 hours. The equations are integrated by the Rosenbrock method (Press et al., 1992) for integration of stiff equations. At each temperature the theoretical concentration of each component is calculated. Initially kerogen degrades into bitumen, gas, and coke. At higher temperatures bitumen degrades into gas and coke.

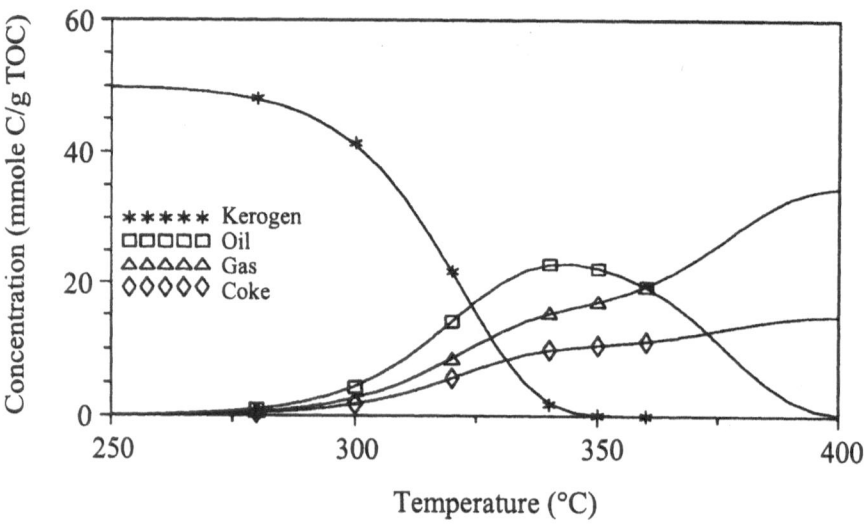

Figure 1. The response of the system of equations (3) to isothermal experiments of duration 72 hours. Symbols at temperatures 280, 300, 320, 340, 350, and 360°C indicate a typical suite of hyrous pyrolysis data available.

2.2 Inverse Method

The inversion procedure applied is outlined by Tarantola and Valette (1982). Since application of this procedure to kinetic problems is well established (Nielsen and Barth, 1991) we concentrate here on the parameter transformations applied in order to incorporate constraints on the stochiometric factors.

The constraints of equation (4) are incorporated by repeated application of the fact that $\cos^2\theta + \sin^2\theta = 1$. For example, consider the first column in the matrix of equation (3) and define the following transformation of $\{\theta_2, \theta_3\} \in (0, \frac{\pi}{2}) \times (0, \frac{\pi}{2})$:

$$
\begin{aligned}
a_{21} &= \sin^2\theta_2 \\
a_{31} &= \sin^2\theta_3 \cos^2\theta_2 \\
a_{41} &= \cos^2\theta_3 \cos^2\theta_2
\end{aligned}
\tag{5}
$$

By this transformation the constraints are satisfied and the number of variables is reduced by one. The variables θ_i are related to the stochimetric factors by the relations

$$\theta_2 = \sin^{-1}(\sqrt{a_{21}})$$

$$\theta_3 = \sin^{-1}(\sqrt{a_{31}}/\cos\theta_2)$$

(6)

An additional transformation defined by

$$\theta_i = \tfrac{1}{2}(tg^{-1}(x_i) + \tfrac{\pi}{2})$$

(7)

relates the bounded variables θ_i to the unbounded variables x_i on the interval $(-\infty, +\infty)$. The relations defined by equations (5), (6), and (7) define a transformation from n-1 unbounded variables on to the space of n stochiometric factors satisfying the constraints of equation (4).

3 Examples

3.1 Artificial Example

In order to illustrate the constraints imposed by the hydrous pyrolysis data of Figure 1 together with a borehole derived T_{max} of 128°C associated with a heating rate of 1°C Ma^{-1}, the process of determining the unknown model parameters was Monte Carlo simulated. A priori model parameter values are selected at random from the a priori distribution of model parameters and the artificial data are perturbed according to their standard deviations. The optimum model is then determined iteratively according to the algorithm of Tarantola and Valette (1982). Figure 2 shows the resulting probability density functions (p.d.f.) of the a priori model parameter values and the optimum model parameter values based on 300 random simulations.

The reaction parameters E and A of kerogen degradation (Figure 2A) are well determined as evidenced by the narrow posterior p.d.f.'s as compared to the much wider a priori p.d.f.'s. This is because both laboratory and sedimentary basin data of widely differing thermal regimes and time scales of reaction constrain this process. For comparison the reaction parameters of bitumen degradation (Figure 2B) are not well determined as evidenced by the similarity of the prior and posterior p.d.f.'s. This is because only a few data points at very close temperatures

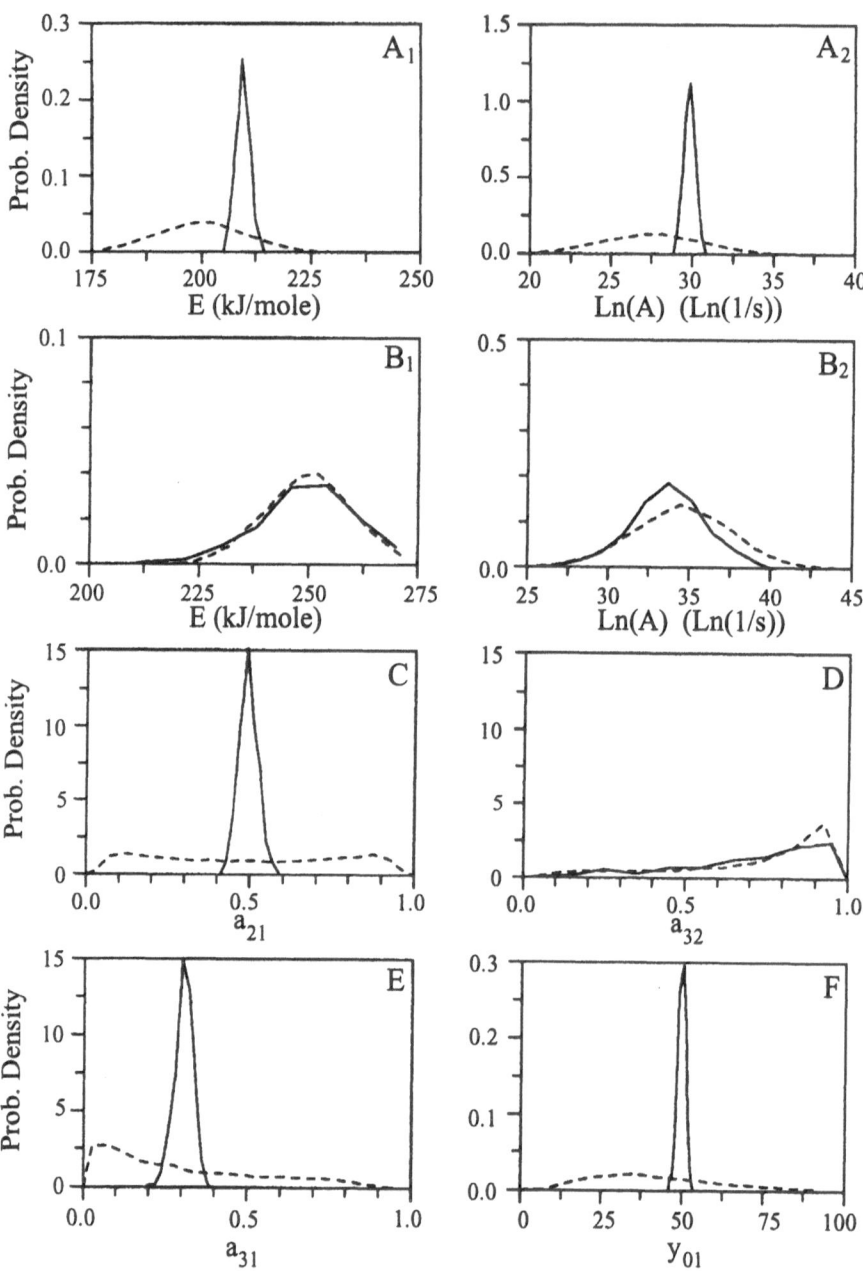

Figure 2. A priori (dashed) and a posteriori (solid) probability density functions of the variable parameters of the network of equation (3). A_1 and A_2: kerogen degradation; B_1 and B_2: bitumen degradation; C, D and E: stochiometric factors a_{21}, a_{32} and a_{31}; E: initial concentration of reactant.

(340-360°C) associated with very similar reaction rates sample bitumen degradation. The stochiometric factors a_{21}, a_{31}, and a_{41} controlling the production of bitumen, gas, and coke from kerogen are all well determined (Figure 2C and 2E). However, the factors a_{32} and a_{42} controlling the production of gas and coke from bitumen are not well determined (Figure 2D). The inital concentration of the reactant (Figure 2F) is well determined.

Removing the borehole derived T_{max} value from the inversion has severe consequences for the accurate determination of the kinetic parameters of kerogen degradation. Their values are now generally determined by the a priori distributions. The resolution of the remaining parameters is similar to the above case. In particular, the well determined stochiometric factors a_{21} and a_{31} show only slightly reduced resolution in spite of the increased uncertainty in E and A.

3.2 Real Example

Figure 3 shows 10 interpretations of a data set obtained from hydrous pyrolysis of a lacustrine shale. The experimental temperatures are 260, 300, 310, 320, 330, 340, and 360°C. Kerogen degradation is modelled by 3 parallel and independent reactions with relative weights of 0.85, 0.10 and

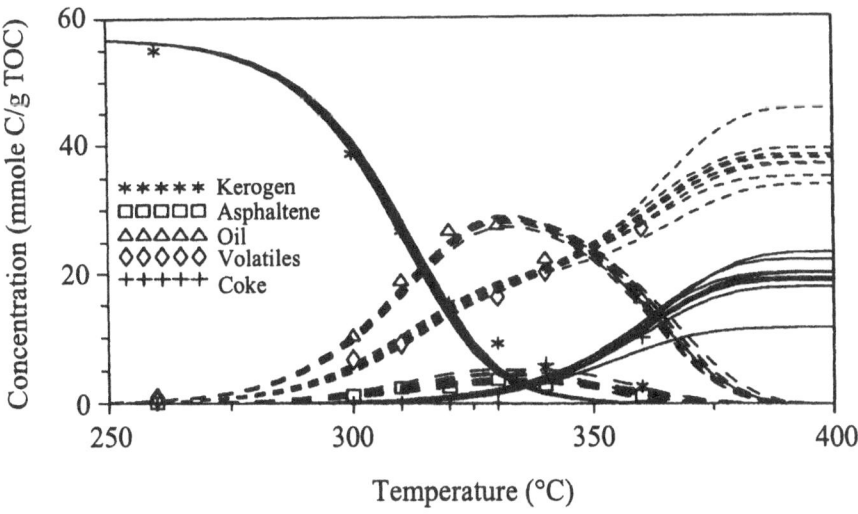

Figure 3. 10 interpretations of hydrous pyrolysis data of a lacustrine shale. The variability in the interpretaions is induced by data uncertainties and variable a priori parameter values.

0.05, respectively. For each of the kerogen degradation reactions four stohiometric factors control the distribution of the breakdown products between the phases asphaltene, oil, volatiles, and coke. The results of Monte Carlo simulating the process of parameter determination are shown in Figure 4. Figure 4A shows that the stochiometric factors a_{41}(kerogen → asphaltene), a_{51}(kerogen → oil), and a_{61}(kerogen → volatiles) associated with the first kerogen component of weight 0.85 are all well constrained by the data; their posteriori p.d.f.'s are narrow as compared to the a priori p.d.f.'s. Contrary to this Figures 4B and 4C show that the posterior p.d.f's of the stochiometric factors associated with degradation of the second and third kerogen components of weights 0.10 and 0.05, respectively, are almost entirely determined by the respective a priori p.d.f.'s and hence are not appreciably influenced by the data. Figure 4D shows the results for the factors a_{54} (asphaltene → oil), a_{64} (asphaltene → volatiles), and a_{65} (oil → volatiles). It is apparent that the degradation of asphaltene is not well constrained, while the degradation of oil is somewhat better constrained.

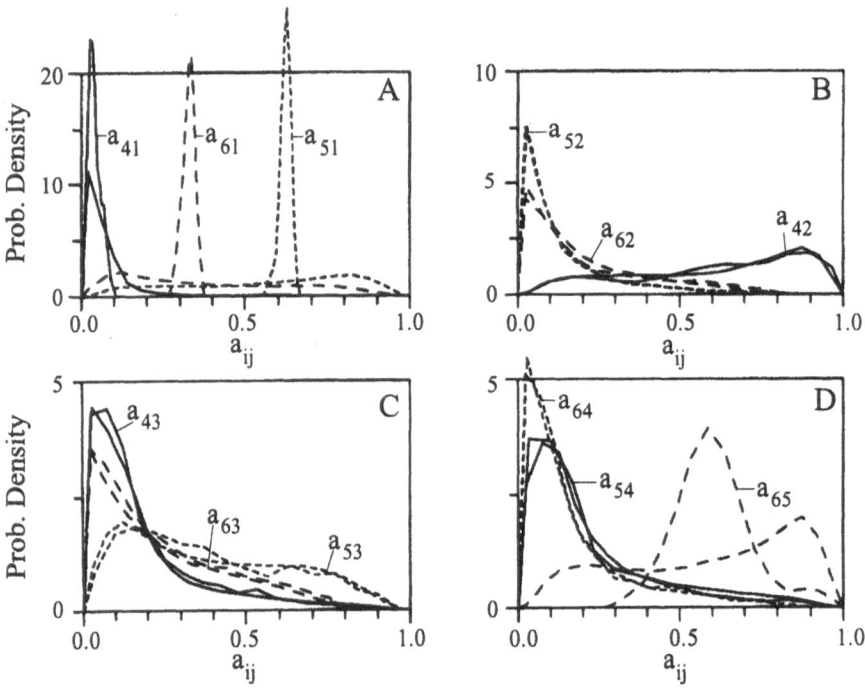

Figure 4. A priori and a posteriori p.d.f.'s of the stochiometric parameters of the reaction network used in Figure 3. Associated a priori and a posteriori probability density functions have the same annotation. See text for details.

The very variable degree of resolution of the different stochiometric factors can be simply explained by the mass flow associated with each column in the system matrix since the mass flow is proportional to the signal amplitude. Degradation of kerogen component 1 (column 1; a_{41}, a_{51}, a_{61}, a_{71}) involves 85% of the mass in the system. The stochiometric factors are all well resolved. Degradation of kerogen component 2 (column 2; a_{42}, a_{52}, a_{62}, a_{72}) involves 10% of the mass in the system and degradation of kerogen component 3 (column 3; a_{43}, a_{53}, a_{63}, a_{73}) involves 5% of the mass in the system. Neither of the stochiometric factors are well resolved. Degradation of asphaltene (column 4; a_{54}, a_{64}, a_{74}) involves 12% of the mass in the system and the stochiometric factors are not well resolved. Degradation of oil (column 5; a_{65}, a_{75}) involves 57% of the mass in the system and the stochiometric factors are relatively well resolved.

4 Conclusions

This paper has demonstrated how Monte Carlo simulation of the process of inverse parameter determination can give useful insight into how well the available data constrain the unknown parameters of petroleum reaction networks.

The results emphasize that data associated with widely differing thermal regimes and therefore widely differing reaction rates are a prerequisite of accurate estimation of the kinetic parameters E and A of petroleum reaction networks. The best case is provided by the combination of laboratory and borehole data involving a range of reaction rates of 12 to 13 orders of magnitude. When borehole data are not available, as is generally the case for the degradation of heavy petroleum products to lighter ones, the non-uniqueness in the kinetic parameter values must be reduced by including such data a priori. This has the effect of selecting inverse solutions which yield petroleum products in the right temperature range at sedimentary basin heating rates.

The stochiometric parameters of the reaction network determine the flow of mass in the system and are hence determined by the relative amounts of the different phases measured. The accuracy with which the stochiometric parameters can be determined is related to the bulk mass flow which they control. Columns which distribute a large fraction of the mass have relatively well determined stochiometric factors, and vice versa. Furthermore, accurate determination of the stochiometric factors is relatively independent of accurate determination of the kinetic parameters. This means that a suite of accurate hydrous pyrolysis experiments can constrain the values of the leading stochiometric factors, even though accurate values of E and A for the network reactions cannot be determined from these data alone.

This result is important to the simulation of petroleum generation in sedimentary basins as an aid in exploration. Because E and A of reaction newtworks are notoriously difficult to determine accurately, and because of the uncertainty in determining the source rock temperature history, the only safe conclusions which can be drawn from petroleum generation modelling are either that no significant generation has taken place, that some generation has taken place, or that most of the petroleum potential of the source rock has been released (Nielsen, 1996). In the case of petroleum generation, hydrous pyrolysis experiments can determine the relative amounts of petroleum phases generated.

This result must be tempered with the fact that a maximum experimental temperature of 365°C does not always allow for sufficiently detailed observation of the secondary degradation processes, leaving a considerable uncertainty in the model for high temperature processes.

References

Barth, T., Borgund, A.E., and Hopland, A.L., 1989. Generation of organic compounds by hydrous pyrolysis of Kimmeridge oil shale - Bulk results and activation energy calculations. Org. Geocehm., 14, 69 - 76.

Bart, T. and Nielsen, S.B., 1993. Estimating kinetic parameters for generation of petroleum and single compounds from hydrous pyrolysis of source rocks. Energy and Fuels, 7, 100 - 110.

Burnham, A.K., Braun, R.L., Gregg, H.R., and Samoun, A.M., 1987. Comparison of methods for measuring kerogen pyrolysis rates and fitting kerogen kinetic parameters. Energy and Fuels, 1, 452 - 458.

Jüntgen, H. and Klein, J., 1975. Entstehung von Erdgas aus kohligen Sedimenten. Erdöl und Kohle, 2, 65 - 73.

Lewan, M.D., Winthers, J.C., and McDonald, J.H., 1979. Generation of oil-like pyrolysates from organic rich shales. Science, 203, 897 - 899.

Lewan, M.D., 1985. Evaluation of petroleum generation by hydrous pyrolysis experimentation. Phil. Trans. R. Soc. Lond., A 315, 123 - 134.

Nielsen, S.B. and Barth, T., 1991. An aplpication of least squares inverse analysis in kinetic interpretation of hydrous pyrolysis experiments. Mathematical Geology, 23, 565 - 582.

Nielsen, 1995. Inverse estimation of the palaeothermal structure of sedimentary basins. Geophys. J. Int., accepted for publication.

Nielsen, 1996. Sensitivity analysis in thermal and maturity modelling. Marine and Petrol. Geol., 13, 415 - 425.

Press, W.H., Teukolsky, S.A., Vetterling, W.T., and Flannery, B.P., 1992. Numerical Recipes in C. Cambridge University Press, 994 pp.

Quigley, T.M. and Mackenzie, A.S. , 1988. The temperature of oil and gas formation in the sub-surface. Nature, 333, 549 - 552.

Tarantola, A. and Valette, B., 1982. Generalized nonlinear inverse problems solved using the least squares criterion. Rev. Geophys. Space. Phys., 20, p. 219 - 232.

Tissot, B.P. and Espitalié, J., 1975. L'evolution de la materiere organique des sediments: aplpications d'une simulation mathematique. Rev. Inst. Fr. Petr., 30, 743 - 777.

Tissot, B.P., Pelet, R., and Ungerer, P., 1987. Thermal histoy of sedimmentary basins, matyration indices, and kinetics of oil and gas generation. Am. Assoc. Petro. Geol. Bull., 71, 1445 - 1466.

Ungerer, P. and Pelet, R., 1987. Extrapolation of kinetics of oil and gas formation from laboratory experiments to sedimentary basins. Nature, 327, 52 - 54.

Ungerer, P., 1989. State of the art of research in kinetic modelling of oil formation and expulsionl. Org. Geochem., 16, 1 - 25.

Inversion of a Simple Petroleum Expulsion Model

Jens Martin Hvid[1] *and Søren B. Nielsen*[2]

[1] Department of Geology and Geotechnical Engineering, The Technical University of Denmark, Lyngby, Denmark
[2] Laboratory of Geophysics, Department of Earth Sciences, University of Aarhus, Denmark

1. Introduction

This contribution is a synthetic resolution study of a numerical model for calculation of petroleum generation and expulsion. Numerical models for estimation of the maturity evolution of source rocks and subsequent formation and expulsion of petroleum are used as tools in the search for petroleum prospects. One main problem in this context is to find reliable model parameter values and to quantify the uncertainties in the results. The problem is addressed by Monte Carlo simulation in order to highlight how uncertain data affect parameter estimation and model outcome. Inversion methods have been applied to maturity related aspects of basin modelling by Lerche and coworkers. (eg. Lerche, 1988a,b; Nakayama and Cao, 1993; Cao and Lerche, 1994). Modelling of petroleum generation by use of several forward calculations has been carried out (Cao and Lerche, 1990; Hermanrud et al., 1990; Irwing et al., 1993). These studies indicate that joint consideration of as many different data types as possible with complimentary information is important in order to reduce the variability in the calculational results.

2. The Forward Model

The reaction kinetic network for petroleum generation is of the serial type $\partial x/\partial t = Ax+Bx$ used by Espitaliér et al. (1988). The vector x contains reactants and products while A and B are matrices of reaction rates of primary and secondary cracking respectively. This stiff system of first order differential equations is integrated by means of the Rosenbrock integration method (Press et al., 1992). The petroleum expulsion out of the source rock is controlled by the build up of petroleum saturation inside the source rock as suggested by Ungerer et al. (1988). The modelled maximum petroleum saturation S_{max} for expulsion to occur depends on source rock compaction, volume expansion as a result of kerogen degradation, and cracking of heavy oil to less dense fractions. What cannot be retained is expelled. This purely volumetric concept holds no matter which mechanisms control the process of petroleum expulsion. The following seven parameters critical to petroleum generation and expulsion are chosen for inversion: 1) initial weight percent of organic carbon TOC_0, 2) initial hydrogen index

HI_0 (petroleum potential per TOC_0), 3) petroleum saturation threshold S_{max} for expulsion to occur, 4) mean activation energy for kerogen degradation E_a^P, 5) mean activation energy for cracking of petroleum E_a^S, 6) matrix density ρ_m, and 7) petroleum density represented by the density of oil ρ_{C14+}. The modelled data are 1) TOC, 2) Rock Eval S_1 (petroleum in source rock) 3) Rock Eval S_2, (petroleum yield by artificial maturation of source rock), 4) bulk rock density ρ_{bulk}, 5) transformation ratio TR (ratio of kerogen converted to petroleum), 6) expelled oil Oil_E, 7) gas/oil ratio of expelled petroleum GOR, and 8) petroleum expulsion efficiency PEE (ratio of petroleum expelled to petroleum generated). As an example of the parametric formulation of the model the equation used to calculate the amount of oil expelled is found as oil generated minus the oil retained in the source rock

$$Oil_E = TOC_0\, HI_0\, TR(T(t), E_a^P, E_a^S) - \phi S_{max}\, (\rho_{C14+}/\rho_{bulk}) \times 10^3 \qquad (2.1)$$

where the porosity ϕ is given by $(\rho_m-\rho_{bulk})/(\rho_m-\rho_w)$ and ρ_w is the pore water density. Transformation ratio TR is calculated by the kinetic network using the temperature history $T(t)$. The non-linearity of the model is introduced by the time-depth dependance on T, TOC_0, ρ_{bulk}, ρ_{C14+}, and ρ_w whereas the parameters HI_0, E_a^P, E_a^S, and S_{max} are independent variables which maintain their values throughout the model calculations. The data observable at present day are assigned tentative uncertainties (table 2.1). Petroleum expulsion models of the volumetric type have previously been used in forward modelling (Forbes et al., 1991; Braun and Burnham, 1992; Skervøy and Sylta, 1993; Rudkiewicz and Behar, 1994).

Table 2.1. Variable parameters p_j and modelled data $g_i(p)$ used in this study. Predefined a priori parameter intervals are given together with standard deviations σ_i assigned to synthetic data.

i,j	p_j		A priori parameter range	$g_i(p)$		Data std. dev. σ_i
1	TOC_0	(%)	0-6	TOC	(%)	0.2
2	HI_0	(mg HC/g rock)	100-500	S_1	(mg HC/g rock)	0.2
3	S_{max}	(fraction)	10-50	S_2	(mg HC/g rock)	0.5
4	E_a^P	(kJ/mole)	215-240	ρ_{bulk}	(kg/m³)	20
5	E_a^S	(kJ/mole)	215-240	TR	(fraction)	
6	ρ_m	(kg/m³)	2600-2800	Oil_E	(mg HC/g rock)	
7	ρ_{C14+}	(kg/m³)	600-1100	GOR	(fraction)	
8				PEE	(fraction)	

3. Method of Inverse Parameter Estimation

The ability of some data to constrain model parameters can be investigated by statistical Monte Carlo simulation in which model inversion plays a key role (Press et al., 1992). The aim is to achieve knowledge of the a posteriori distribution for each parameter by simulating the case of having many sets of repeated data measurements available. Modelled synthetic data sets $d^s = g(p^{mean})$ generated from mean values of the a priori parameter ranges (table 2.1) are added random gaussian noise **e** to simulate data uncertainties. Inversion is then used to find sets of a posteriori parameters $\mathbf{p}^{post}$ associated with each set of synthetic data $\mathbf{d}^s + \mathbf{e}$. The objective function is formulated in the least square sense. The criteria for accepting an inverse solution **p** is given by

$$(1/N) \sum_i ((g_i(\mathbf{p})-(d_i^s+e_i)/\sigma_i)^2 < \varepsilon \qquad (3.1)$$

where i = 1 to N counts modelled data and σ_i is the standard deviation of the i'th datum. A value of $\varepsilon = 1$ has been used. Searching for minima of the objective function can be done with or without the use of derivatives $\partial g(\mathbf{p})/\partial \mathbf{p}$. During this study both types of methods were tested. Minimization of the nonlinear, multidimensional vector function in eq. 3.1 was obtained by means of the Down Hill Simplex method (Press et al., 1992) which does not require derivatives. The use of a simplex method is a fairly easy way of solving the nonlinear problem with a limited number of variable parameters. In this case the simplex method was found to be a good alternative to the gradient method. The advantage of the gradient method is that it offers information on the a posteriori parameter covariance matrix. The main principles of the inversion procedure used here are to sample the model space in broad regions around likely areas of the best solutions and then use the sampled points as starting points to advance a solution by means of the simplex method. A priori parameters are sampled uniformly by use of the latin hypercube sampling method (McWilliams, 1987). This method ensures a good coverage of the model space. A number of 500 a priori samples of the model space proved large enough to gain insight in the shape of a posteriori parameter density distributions and the variability of the outcome space.

4. Results of Synthetic Model Experiments

The presentation of simulation results is limited to a few graphic examples. The results of an experiment are the range and shape of each constrained parameter distribution which is estimated by a histogram. A posteriori parameter distributions resulting from eq. 3.1 with the conditions of table 2.1 are shown together with the uniformly sampled a priori parameters (fig. 4.1). The resolution of the different parameters is not equally successful. Model parameters for calculation of petroleum generation such as TOC_0, HI_0, and E_a^P are generally better constrained by the data than parameters sensitive to petroleum expulsion. Well constrained parameters are those which have an appreciable influence on data. The level of noise in the synthetic data determines the possible control on the resulting parameter estimates. Figure 4.2 shows the sensitivity of model outcome as function of depth to a priori and a posteriori parameters respectively. The scatter in the model results due to imprecise data are represented by 20% and 60% confidence intervals. Error bars indicate the uncertainties given to synthetic data. The modelled depth interval covers the full range of petroleum generation and expulsion. Figure 4.2 clearly demonstrates the capability of the chosen synthetic data to constrain

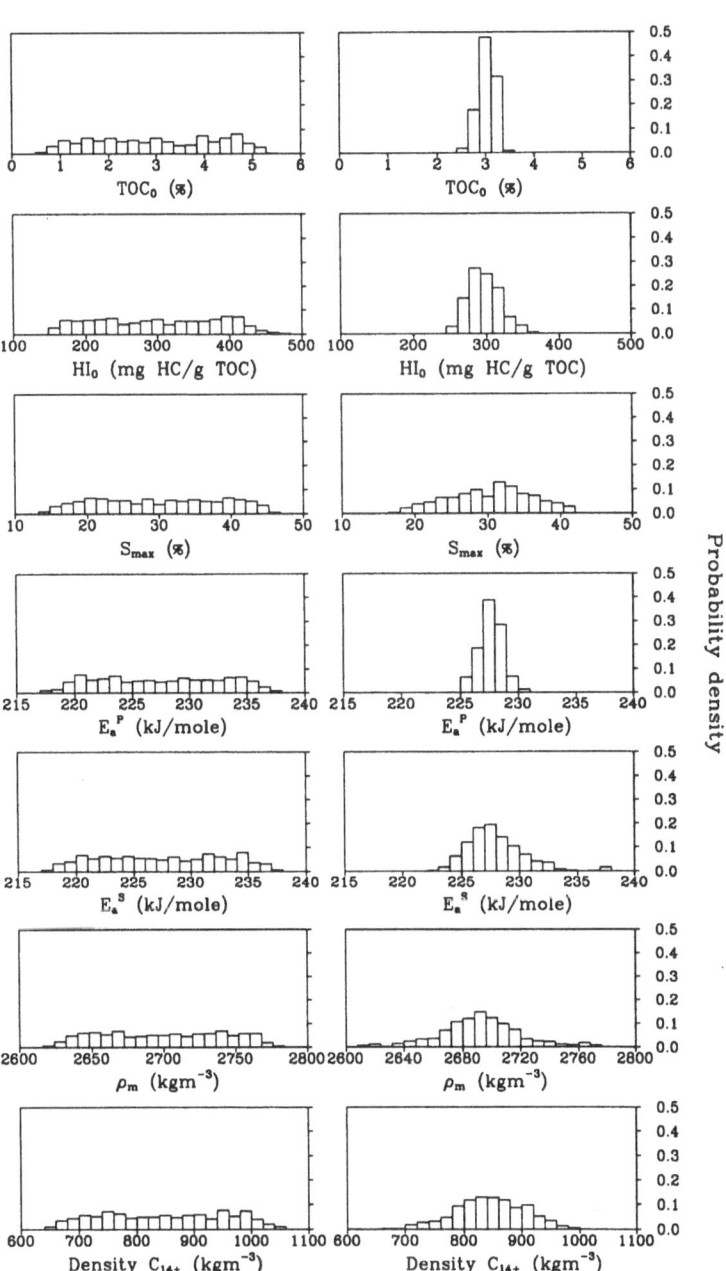

Fig. 4.1. Probability density distributions representing model parameters before (left column) and after inversion (right column) respectively. Petroleum expulsion parameters are relatively badly constrained compared to parameters for petroleum generation.

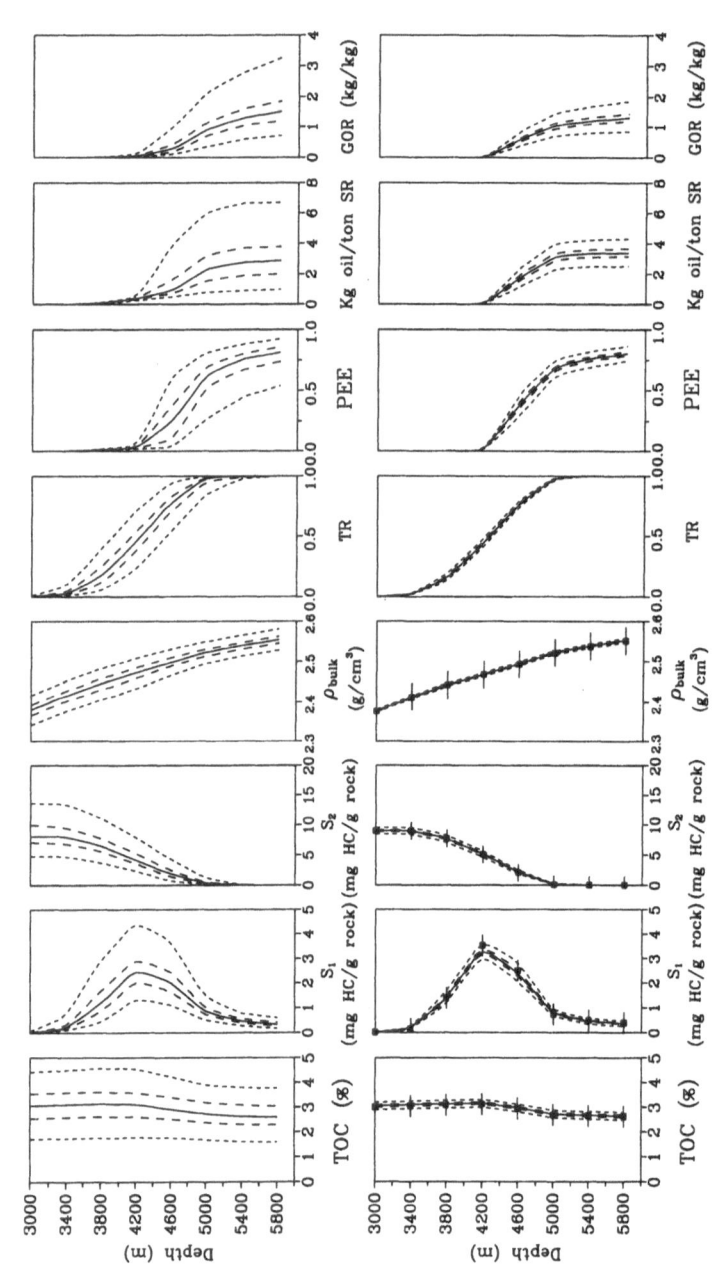

Fig. 4.2. Model outcome g(p) as function of depth. Upper and lower graphs are produced by a priori and a posteriori parameter values, respectively. The error bars indicate uncertainties assigned to synthetic borehole data. Modelled uncertainties are illustrated by confidence limits of 20% (long dashed) and 60 % (short dashed).

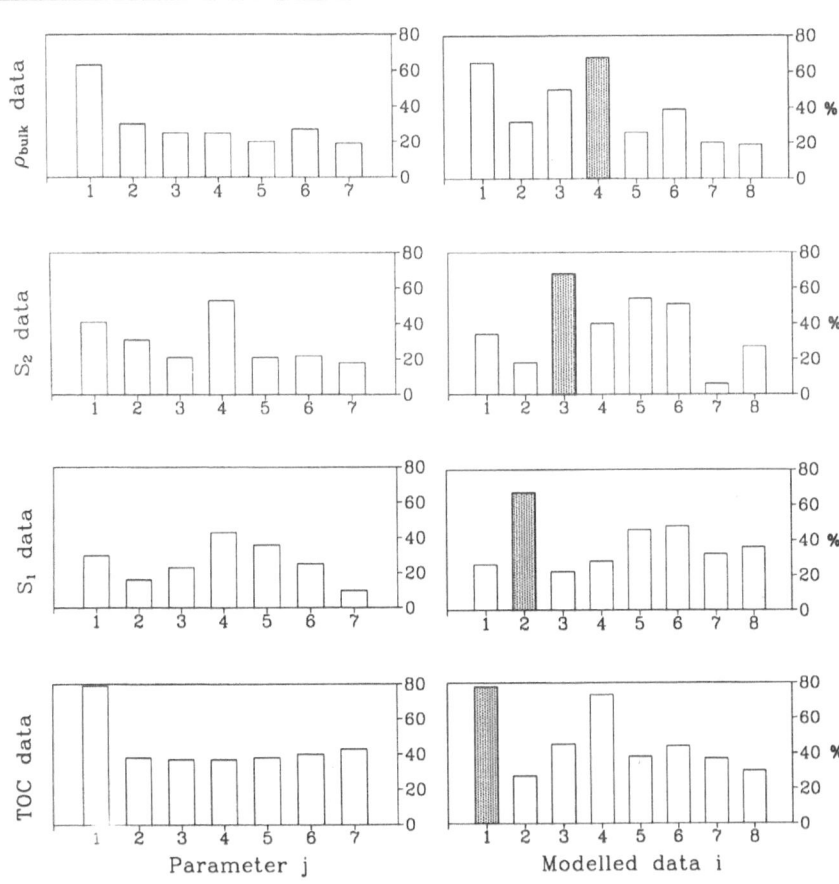

Fig. 4.3. Improvements in % of model uncertainties gained by inversion using one type of data at a time. Improvements are given as the relative differences in standard deviations of model parameters p_j and in the corresponding modelled data $g_i(\mathbf{p})$ before and after inversion respectively (See table 2.1 for index numbering). The sizes of the shaded bars are relative measures of the uncertainties given to TOC, S_1, S_2, and ρ_{bulk} respectively.

modelling results which are not easily observed, such as transformation ratio and petroleum expulsion efficiency.

We will now study the consequences of having only one of the data types available for inversion at a time. Uncertainties of the individual data types are still those of table 2.1. Figure 4.3 summarizes the results of four experiments of the type shown in fig. 4.1 and fig. 4.2. In each experiment the parameters are only constrained by one type of data (i.e. TOC, S_1, S_2, or ρ_{bulk}). Figure 4.3 shows the percentage improvement in standard deviations after inversion compared to standard deviations modelled by use of a priori information. Reduction of uncertainty by inversion with single data types is given for parameters as well as for model outcome. The modelled data of fig. 4.3 are taken from the depth of 4.6 km where no values of the modelled data approach zero. TOC data as well as bulk density are efficient in constraining TOC_0. Likewise, the uncertain synthetic data constrain their corresponding modelled data (shaded bars) better than 70% relative to data modelled by a priori parameters. However, the degree of constraint depends on the data uncertainties (Table 2.1). The best estimates of the mean activation energy of kerogen degradation are obtained by use of Rock Eval data. The use of S_1 data is the best way of improving knowledge of petroleum expulsion efficiency. Rock Eval S_2 data are excellent for improvement of transformation ratio and the kerogen activation energy, whereas little improvement is gained in the gas/oil ratio. The uncertainty in quantities of oil expelled is almost equally sensitive to the four synthetic data types. The use of non-gaussian a priori parameter distributions does not change the results after inversion significantly but it reduces the time of calculation. This indicates that data control is most important.

5. Conclusion

A posteriori parameter distributions of a simple model of petroleum generation and expulsion have been estimated by means of Monte Carlo simulation. Generally, the most successfully constrained parameters are those of petroleum generation while the more weakly constrained parameters are those of petroleum expulsion. The importance of individual data types in parameters estimation as well as the resulting influence on the model outcome is clarified. This kind of information is valuable in the selection of data and data uncertainties needed for inverse estimation of model parameters. Other types observable data should be analyses in this manner to investigate their ability to improve the confidence in model results.

References

Braun, R., and Burnham, A. K., 1992. PMOD: a flexible model of oil and gas generation, cracking and expulsion. In: Advances in Organic Geochemistry 1991. (Edited by Eckardt et al.), Org. Geochem. Vol. 19 Nos 1-3, p. 161-172.

Cao, S. and Lerche, I., 1990. Basin modelling: applications of sensitivity analysis. J. Pet. Sci. Eng., 4, p. 83-104.

Cao S. and Lerche I., 1994. Thermal history reconstruction by inversion of down-hole S2 pyrolysis data: synthetic studies and case histories. Applied Geochem. Vol. 9, pp. 93.117, 1994.

Espitaliér, J., Ungerer, J., Irwing, H., Marquis, F., 1988. Primary cracking of kerogen. Experimentation and modelling C_1, C_2-C_5, C_6-C_{15} and C_{15+} classes of hydrocarbons. In: Mattaveli,

L., Novelli, L. (eds) Advances in Organic Geochemistry 1987, Org. Geochem., 13, p. 893-899.

Forbes, P. L., Ungerer, P., Kuhfuss, A. B., Riis, F., and Eggen, S., 1991. Compositional modelling of petroleum generation and expulsion. Trial application to a local mass balance in the Smørbukk Sør Field, Haltenbanken Area, Norway. Am. Assoc. Pet. Geol., Bull., Vol. 75, No. 5, p. 873-893.

Hermanrud, C., Eggen, S., Jacobsen, T., Carlsen., E.M. and Pallesen, S., 1990. On the accuracy of modelling hydrocarbon generation and migration: the Egersund basin oilfind, Norway. Org. Geochem., Vol. 16, Nos 1-3, p. 389-399.

Irwing, H., Hermanrud, C., Carlsen , E. M., Vollset, J. and Nordvall, I., 1993. Basin Modelling of hydrocarbon charge in the Egersund Basin, Norwegian North Sea: pre- and post-drilling assesment. In: A. G. Doré et al. (Editors), Basin modelling: Advances and Applications. Norwegian Petroleum Society (NPF), Special Publication 3. Elsevier, Amsterdam, p. 539-548.

Lerche, I., 1988a. Inversion of multiple thermal indicators: quantitative methods of determining paleoheat flux and geological parameters, I. The theoretical development for paleoheat flux. Math. Geol., 20, p. 1-36.

Lerche, I., 1988b. Inversion of multiple thermal indicators: quantitative methods of determining paleoheat flux and geological parameters, I. The theoretical development for chemical, physical, and geological parameters. Math. Geol., 20, p.73-96.

McWilliams, T. P., 1987. Sensitivity analysis of geological computer models: a formal procedure based on Latin Hypercube sampling. Math. Geol., 19, p. 81-90.

Nakayama, K. and Cao, S., 1993. An inverse method for determining kinetic parameters for kerogens from routine pyrolysis data, and application to the Peace River arch area, Alberta, Canada. In: A. G. Doré et al. (Editors), Basin modelling: Advances and Applications. Norwegian Petroleum Society (NPF), Special Publication 3. Elsevier, Amsterdam, p. 251-263.

Press, W. H., Teukolsky, S. A., Vetterling, W. T., and Flannery, B. P., 1992. Numerical recipes in C. The art of scientific computing. (2. ed.), Cambridge Univ. Press, Cambridge, 994 pp.

Rudkiewicz, J. L. and Behar, F, 1994. Influence of kerogen type and TOC content on multiphase primary migration. Org. Geochem. Vol 21, No. 2, p. 121-133.

Skjervøy, A. and Sylta, Ø., 1993. Modelling of expulsion and secondary migration along the southwestern margin og the Horda Platform. In: A. G. Doré et al. (Editors), Basin modelling: Advances and Applications. Norwegian Petroleum Society (NPF), Special Publication 3. Elsevier, Amsterdam, p. 499-537.

Ungerer, P., Espitaliér, J., Behar, F., Eggen, S., 1988. Modélisation des mathématique des interaction entre craquage et migration lors de la formation du pétrole et du gaz. C. R. Acad. Sci. Paris, Sér II, p. 927-934.

Inversion of Transient Hydraulic Head Variation Caused by a Controlled Point Source

Dorte Dam, Niels B. Christensen and Kurt I. Sørensen

Department of Earth Sciences, University of Aarhus, Denmark

Introduction

An increasing need for surveillance of the quality of our ground water resources requires a comprehensive hydraulic modeling. A thorough knowledge of the hydraulic parameters is a necessity for reliable hydraulic modeling. Pumping and slug tests have usually been the main source of information about the hydraulic conductivity and the specific storage. Pumping tests result in conductivities and storage coefficients averaged over large volumes, and slug tests are often seriously affected by the well screens and filter packs.

A new method for in situ determination of the hydraulic conductivity and the specific storage has been developed at the University of Aarhus. By making use of a hollow auger drill stem it is possible to inject water directly into the formation and measure the rise in hydraulic head at a short distance from the injection point. Other procedures for in situ determination of hydraulic parameters have been developed (Fejes and Jósa, 1990).

The measured data (the transient hydraulic head variations) are interpreted using an inversion algorithm based on a least squares iterative formalism. To make the problem more linear the inversion is carried out in the log parameter space. The model response applied is the transient head variations due to a point source in a homogenous isotropic fullspace, where the source is considered as a step source. Since there are only three model parameters to be estimated and typically more than one hundred data, the inverse problem is strongly overdetermined.

The forward problem

The solution to the 3-D differential equation of flow in a homogeneous isotropic aquifer with appropriate initial and boundary condition is given by Carslaw and Jaeger (1959).

$$\phi(t,r) = \frac{Q}{4\pi K r}\,\text{erfc}\left(r\sqrt{\frac{S}{4\,Kt}}\right) + H = \frac{Q}{4\pi K r}\,\text{erfc}\left(\sqrt{\frac{u}{t}}\right) + H \qquad (1)$$

ϕ : change in potential head (m)
Q : water injection rate (m^3 sx^{-1})
K : hydraulic conductivity (m s^{-1})
S : specific storage (m^{-1})
r : radial distance from the injection point (m)
t : time since injection start (s)
H : initial head (m)
u : characteristic time constant, u=Sr^2/4K (s)

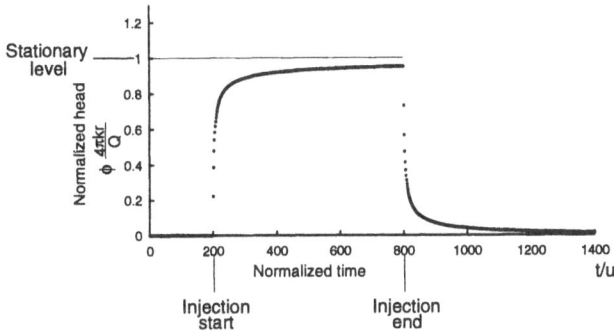

Fig. 1. The theoretical curve, when K = 3·10^{-3} m/s, S = 3·10^{-3} m^{-1}, r = 1 m and Q = 3·10^{-4} m^3/s, which gives a characteristic time constant, u = Sr^2/4K = 1/4 seconds. The injection is turned on after 200 time constants and switched off after another 600 time constants.

Inversion

The task is to estimate the hydraulic conductivity and the specific storage from a time series of typically more than 300-700 observations. The least squares inverse formalism applied is presented below (Tarantola and Valette, 1982; Jacobsen, 1993; Nielsen, 1992):

$$m_{n+1}^{est} = \left(G_n^T C_d^{-1} G_n\right)^{-1} G_n^T C_d^{-1}\left(d - g(m_n^{est})\right) + m_n^{est} \qquad (2)$$

d   : measured data values (dimension N)
C$_d$: measured data covariances (dimension N x N)
m$_{n+1}^{est}$: improved parameter estimate (dimension M)
m$_n^{est}$: previous parameter estimate (dimension M)
G$_n$: the Jacobian matrix at m = m$_n$ (dimension N x M)
g($\cdot$)   : non-linear data-parameter relationship: ϕ = g(m)

To make the problem more linear the variables K, and S are changed to log$_e$(K) and log$_e$(S) (eq. 3). Misjudgment of the initial head will introduce a bias on all data and result in inconsistency with the model. To overcome this problem the initial head was included in the model parameter space, and as a secondary profit this also minimizes the data processing before inversion. As it will appear later the initial head is well determined, and it is not very correlated neither to S nor to K.

$$
G_n = \begin{bmatrix} \dfrac{\partial\phi_1}{\partial lnk} & \dfrac{\partial\phi_1}{\partial lns} & \dfrac{\partial\phi_1}{\partial H} \\[2mm] \cdot & \cdot & \cdot \\ \cdot & \cdot & \cdot \\ \cdot & \cdot & \cdot \\ \dfrac{\partial\phi_n}{\partial lnk} & \dfrac{\partial\phi_n}{\partial lns} & \dfrac{\partial\phi_n}{\partial H} \end{bmatrix} = \begin{bmatrix} \dfrac{\partial\phi_1}{\partial lnk} & \dfrac{\partial\phi_1}{\partial lns} & 1 \\[2mm] \cdot & \cdot & \cdot \\ \cdot & \cdot & \cdot \\ \cdot & \cdot & \cdot \\ \dfrac{\partial\phi_n}{\partial lnk} & \dfrac{\partial\phi_n}{\partial lns} & 1 \end{bmatrix} \tag{3}
$$

When the effect of the non-linearity is not too severe, the inversion will converge to the same solution, independent of initial guess. When non-linearity is significant, secondary minima or maxima may exist. To get an impression of the error surface iterations can be started with a variety of initial guesses (Menke, 1989; Tarantola and Valette, 1982; Nielsen, 1992). In the case of a parameter space consisting of only two parameters, $log_e(S)$ and $log_e(K)$ the error surface can be directly mapped (Charles et al., 1989). This analysis gives no indication of ambiguity concerning the least squares solution.

For the purely overdetermined linear problem the matrix of posterior parameter covariances C_m can be obtained by:

$$
C_m = \left(G^T C_d^{-1} G\right)^{-1} \tag{4}
$$

For simplicity the data errors are assumed not to be correlated i.e. the data covariance matrix is diagonal and the posterior parameter covarians matrix, C_m can then be described by:

$$
C_m = \left(G^T \sigma_d^{-2} I\, G\right)^{-1} \tag{5}
$$

where σ_d is the data standard deviation. If g(m) is only weakly non-linear the posterior covariance matrix, C_m for the overdetermined non-linear problem is well approximated through eq. 4 respectively eq. 5, (Jacobsen, 1992).

Analysis of the inverse problem

The standard deviation of the various model parameters as a function of model parameters K and S has been examined (fig. 2). The length of the time series, the sample density, the injection rate, and the standard deviation of data are fixed at values relevant to the present design of the tool.

The surfaces in figure 2 present the analysis of the response from various choices of S and K. The length of the time series is 2 x 120 + 10 seconds, i.e. 10 seconds before the source is turned on, 120 seconds after the source is turned on, and another 120 seconds after the source is switched off. The source is considered as a step source with a constant injection of $3 \cdot 10^{-4}$ m^3/s, the radial distance from the source is 1 m and the absolute standard deviation on data is 1.5 mm.

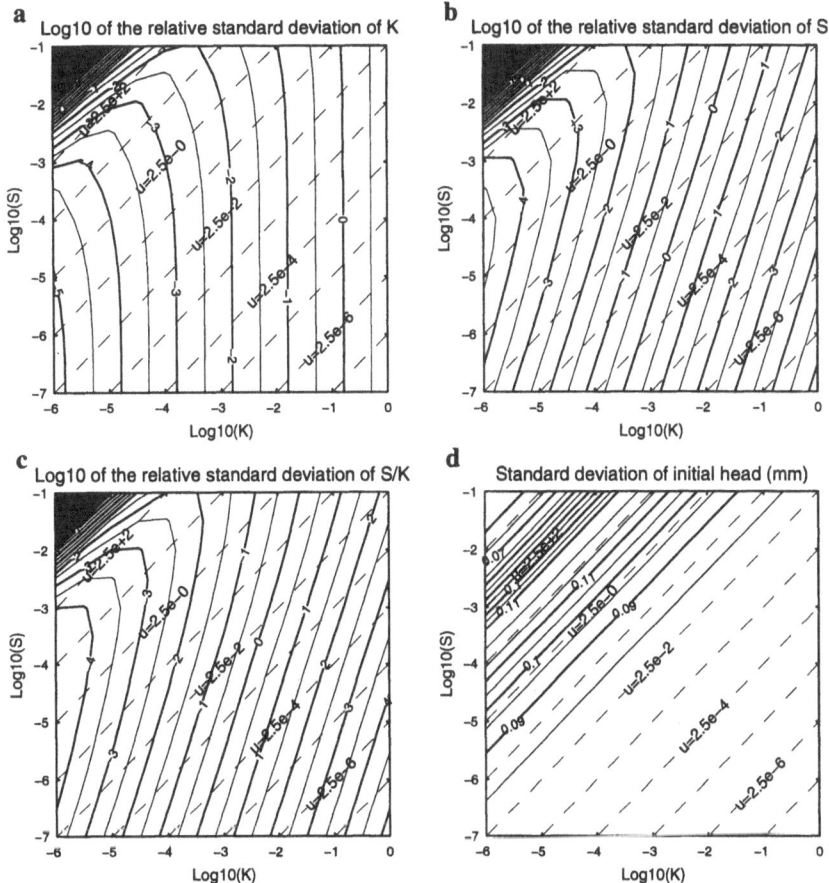

Fig. 2. Standard deviation of the model parameters as a function of model parameters K (hydraulic conductivity) and S (specific storage). The length of the time series is fixed at 2x120 + 10 seconds with a sample density of 3 samples per second, the injection is $3 \cdot 10^{-4}$ m^3/s and the standard deviation of data is 1.5 mm. The dashed lines are lines of equal characteristic time constants, u. **a** the relative standard deviation of K, **b** the relative standard deviation of S, **c** the relative standard deviation of S/K, and **d** the standard deviation of the initial head.

The corresponding analysis of the response from a time series of 120 seconds, solely for the increasing part of the response, from 10 seconds before turn-on and 120 seconds after, or from the decreasing part of the response after the source is turned off, differ mainly at the general level of the surfaces increased by a factor of $\sqrt{2}$ compared to the analysis of the response from the time series of 2 x 120 seconds.

The level of the standard deviation surfaces depends partly on the standard deviation of data, σ_d, partly on the injection, Q. The relative standard deviations are inversely proportional to Q and directly proportional to σ_d.

For characteristic time constants, u (eq. 1), less than approximately 2.5 seconds, the standard deviation of K depends solely on K (fig. 2a), and the surface of the relative standard deviation of K slopes with a gradient of approximately 1. This shows that the data noise is dominating. The hydraulic head is inverse proportional to K, when stationary conditions are achieved (cf. eq. 1: $\phi(\infty,r) = Q/4\pi K r$). Therefore, assuming absolute noise on data, an increase in K-values causes an increase in the relative standard deviation of the model parameters.

The standard deviation of S and S/K depends on K and S, independent of characteristic time constant. The standard deviation of S and S/K decreases with increasing S-values. The surface of the relative standard deviation of S slopes with a gradient larger than 1. Besides the above-mentioned effect of the data noise, the information about S also diminishes because the stationary level is reached faster when the K-value increases (i.e. less time and fever samples in the early stage).

When the K-value is small and the S-value large, neither K, S nor S/K are determined. When the time constant, u, exceeds approximately 1000 seconds, K, S, and S/K are completely undetermined. Large S-values result in a smooth response curve, and small K-values find expression in a high amplitude. This means that the stationary level is never reached.

The direction of the ravine in the surfaces of relative standard deviation of K, S, and S/K corresponds to a characteristic time constant of approximately 0.25 seconds or a length of the time series of 1000 time constants (fig. 2a-c). This implies that the location of the ravine is dependent on the length of the time series. The longer time series - the longer towards the upper-left corner the ravine is located.

Experimental Design

For a given demand on the maximum standard deviation of a model parameter it is possible by means of figure 3a-d to find the required length of the time series and the required number of samples per time constant.

It is not surprising to notice that the standard deviation of K decreases by approximately $\sqrt{N}$, where N is the number of samples, as the sample density increases (corresponding to decreasing time interval between samples), or as the length of the time series increases (fig. 3a).

Likewise, the standard deviation of S and S/K diminishes as the sample density increases, but the decrease in the standard deviation of S and S/K is not as pronounced for increasing length of time series as in the case of the standard deviation of K (fig. 3b and 3c). When the time series is longer than 200 time constants ($\log_{10}(t/u) \sim 2.3$), the contours are approximately vertical. This implies that the certainty of the determination of the model parameters, S and S/K is not improved notably by further sampling in time. As we would expect, the information about the S- or S/K-parameter lies at the early times, just after the source is turned on, or just after it is turned off. Comparing figure 3b and 3c it is seen that, for short timeseries, S alone is better determined than the fraction S/K.

At all times, except for time series shorter than the time between samples, the initial head is determined at least with the certainty of the data, which is 1.5 mm (fig. 3d). The standard deviation of the initial head also decreases as the sample density and time series length increases. An increase in time series length results in more measurements of the last part of the response (cf. fig. 1 t/u =1200-1400) and hence more measurements of the model parameter we wish to determine.

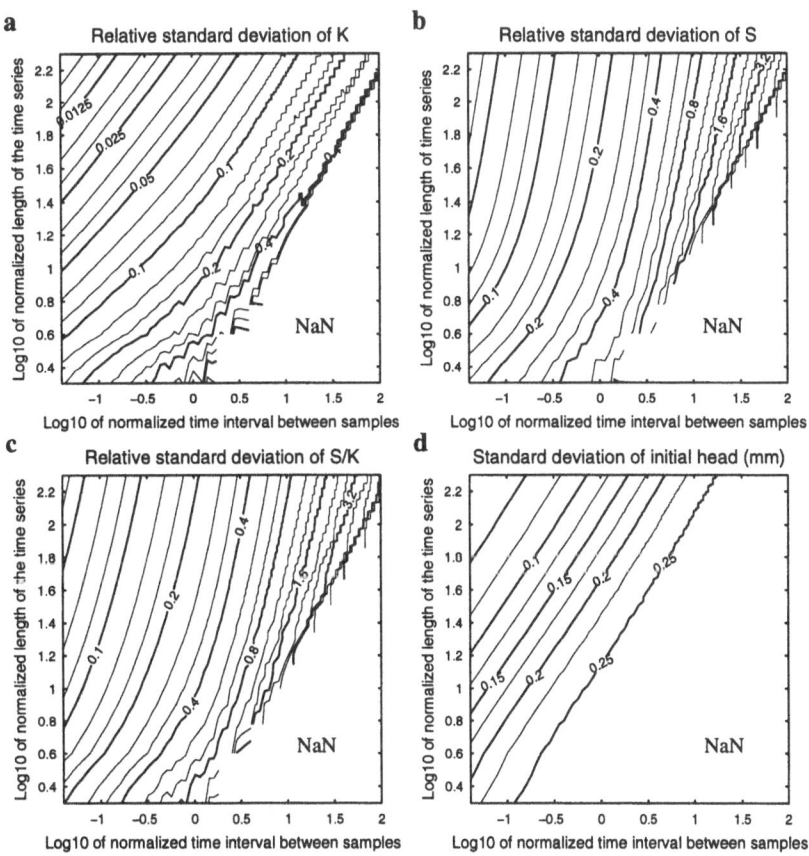

Fig. 3. Standard deviation of the model parameter as a function of the time interval between samples and the length of the time series. The x-axis is in time constants of $\log_{10}(\Delta t/u)$ and the y-axis is in time constants of $\log_{10}(t/u)$, where $u = Sr^2/4K$, (cf. eq. 1). **a** is the relative standard deviation of K, **b** is the relative standard deviation of S, **c** is the relative standard deviation of S/K, and **d** is the standard deviation of the initial head.

Field Example

To evaluate practically the implementation of the method, preliminary field measurements have been carried out in Stjær gravel pit, East Jutland, Denmark. This location was chosen because of the expected homogeneity of the formation. Prior to the hydraulic log an electrical log and a gammalog were made (fig. 4e-f), using the Ellog Auger Drilling Method (Sørensen, 1989; Auken et al., 1994). Hydraulic measurements were carried out at six different depths in the phreatic aquifer. At each depth four injections were recorded, and for each injection the rise in potential head was measured at three distances from the injection source for both rising and decreasing potential corresponding to source-on and source-off. A separate inversion of the three distances and a joint inversion has been made on each hydraulic measurement, and the estimated hydraulic conductivities are shown in figure 4.

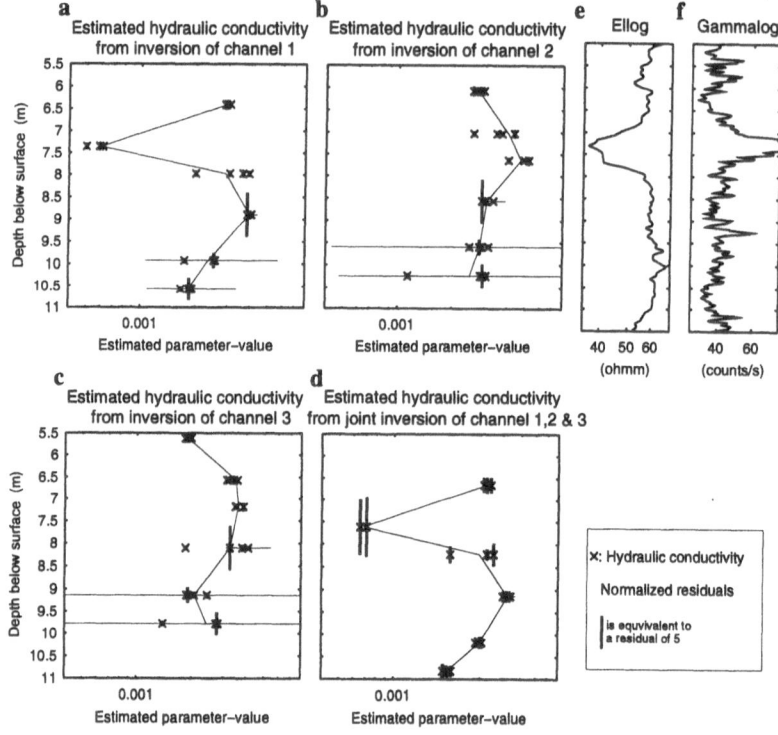

Fig. 4. a Estimated hydraulic conductivity from the inversion of channel 1, **b** channel 2, and **c** channel 3. The depth reference point is the midpoint of the configuration. **d** Estimated hydraulic conductivity from the joint inversion of channel 1, 2, and 3. The depth reference point is the depth of the source. Error bars are plotted as horizontal bars. The vertical bars are not error bars in depth but the normalized residuals (eq. 6). The residual scale is seen in the lower right corner. **e** Ellog and **f** Gammalog the depth scale is equal to the depth scale of the conductivity plots.

The ellog and gammalog show minor variations in the geology, which is also reflected in the injection log. A clayey layer is seen around 7.5 meters below the surface. At the same depth the hydraulic conductivity estimated from the inversion of the smallest spacing decreases. The central and largest spacing, however, show a minor increase in the conductivity, probably due to the large spacing compared to the thickness of the layer. The inconsistency between the three channels is reflected in the joint inversion as an increased normalized residual at the depth considered, i.e. at source depth of approximately 7.5 meters.

The variations between the estimated parameters at each depth are small, but not always within the estimated standard deviations (error bars) of the parameter estimates. This small variation indicates a high reproducibility.

The standard deviation of the model parameter estimates and the normalized residuals are generally small. However the normalized residual (eq. 6) of the joint inversion is large, around 3 in most cases. This must be attributed to the inconsistencies between the actual formation and the assumed model: the homogeneous and isotropic full space.

$$Res = \sqrt{\frac{1}{N} \sum_{i=1}^{N} \frac{\left(d_i - g\left(m_i\right)\right)^2}{\sigma_{d_i}^2}} \qquad (6)$$

A result of the joint inversion of the three channels at source depth 8.3 meters is shown in figure 5. The systematic deviation between the channels is probably (as mentioned above) caused by anisotropy, local inhomogeneities or local unknown conditions along the drill stem. The use of more spacings, both individually and joint, permits an estimation of the homogeneity of the formation and more spacings is equivalent to repeated experiments.

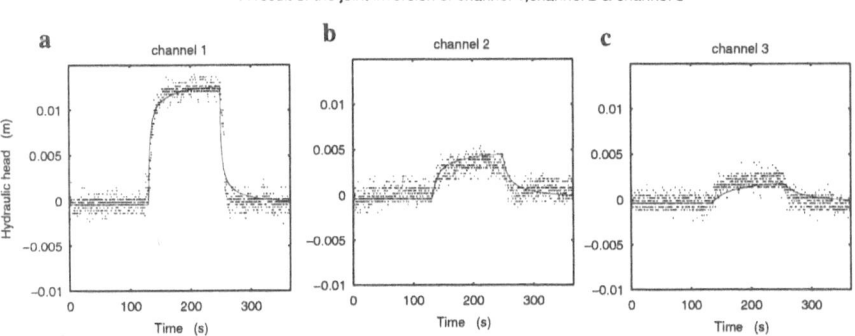

Fig. 5. An example of the result of the inversion on field data. The depth of the source is 8.3 meters below the surface. The dots are the measured data, the line is the response of the final model after inversion, **a** r = 0.5 m, **b** r = 1.16 m, **c** r = 2.085 m.

The source which is assumed to be a step source may very well not be so. A gradually increasing source, would result in a smoother rise in the pressure data.

This would either cause an over estimation of the specific storage or it would introduce an inconsistency between the assumed model response and the actual response.

If instead a linear increasing source is used, it is possible to model the smooth rise in data (fig. 6). This means that another parameter, the run-on time, i.e. the time it takes for the source to reach a constant flow rate, must either be implemented in the model parameter space - in which case it must be expected to be tightly coupled to the specific storage - or accurately specified from additional independent measurements of the flow rate.

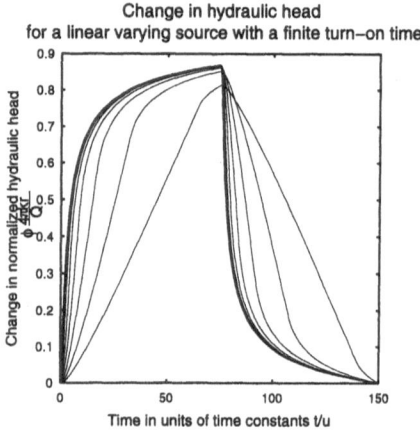

Fig. 6. Theoretical curves for a linear increasing source with finite turn-on times at: 2, 4, 8, 16, 32 and 64 time constants (eq. 8).

This implies that it is most essential to know the source function. In case of a varying source the resulting changes in the hydraulic head can be calculated employing the principle of superposition:

$$\phi(t,r) = \sum_{i=1}^{N} (Q_i - Q_{i-1}) \cdot \phi_0(t - t_i, r) \tag{7}$$

In case of linear variation in Q the difference $Q' = Q_i - Q_{i-1}$ is constant for constant time steps $\Delta t = t - t_i$. Q' can then be isolated outside the summation and in the boundary where Δt approaches zero the sum converge towards the integral (fig 6):

$$\phi(t,r) = Q' \int_{t-\Delta}^{t} \phi_0(t - t', r) dt \tag{8}$$

Conclusion and future work

This method for in situ determination of the hydraulic parameters is an attractive alternative to the slug test and a good supplement to the pumping test. As appears from the analysis, it is possible to determine the hydraulic conductivity, but more difficult to determine the specific storage. Analysis of the inverse problem stresses the importance of dense measurements and, if possible, repeated experiments. Dense measurements together with strict control of the injection at early stages are most essential if the specific storage is needed. Analysis of the standard deviation of the model parameters, as a function of the model parameters K (hydraulic conductivity) and S (specific storage), show that the parameter standard deviation, in addition to the data noise and the injection, Q, mainly depends on the K-value. To get a satisfying determination of both K and S it is necessary to sample with a density of ten samples per characteristic time constant, whereas a time series length of only ten time constants should be sufficient. It is more efficient to make repeated measurements than just measure for a longer period. Due to inconsistency in data at an early stage, however, it is necessary to measure beyond ten time constants, and it is of primary importance to know the source function. The minimum length of the time series should be 30 time constants corresponding to approximately 80 % of the stationary level. It is also essential to measure before the injection is started to get information about the initial level.

Before further measurements are carried out the transducer system need to be improved. The assumption of the uncorrelated data is of course a presumption which should also be investigated in the future. Owing to air- or filter-resistance data might be correlated, although this correlation is expected to be small. Furthermore data processing, by means of a robust averaging, must take place in the field and finally a better control of the source is desirable.

More field measurements are planned in the near future, taking into account the experimental design resulting from this analysis.

Acknowledgement

We would like to thank Bo Holm Jacobsen for valuable advice, and we would also like to thank Kent Sørensen for the work he has done on the development of the equipment and the collection of the data from Stjær.

References

Auken, E., Christensen, N. B., Sørensen, K. I., Effersø, F.,1994, Large scale hydrogeological investigation in the Beder area. Proceedings of the Symposium on the Application of Geophysics to Engineering and Environmental Problems, Boston, March 1994, pp. 615-628

Carslaw, H. S., Jaeger, J. C., 1959. Conduction of heat in solids, Oxford University Press., pp. 195

Charlez, Ph., Herail, R., Despax, D., 1989, Interpretation of Hydraulic Fracture Parameters by Inversion of Pressure Curves. Int. J. Rock Mech. Min. Sci. & Geomech. Abstr. Vol. 26, No. 6 pp. 549-553

Fejes, I. and Jósa, E., 1990, The engineering geophysical sounding method: principles, instrumentation, and computerized interpretation, in Ward, Stanley H., Ed., Geotechnical and environmental geophysics, 2: Soc. Expl. Geophys., pp. 321-331.

Jacobsen, B. H., 1992, Nonlinear model construction and error propagation analyses: When does linearization work? Proceedings of the Interdisciplinary Inversion Workshop 1, pp. 63-68

Jacobsen, B. H., 1993, Practical methods of a priori covariance specification for geophysical inversion. Proceedings of the interdisciplinary Inversion Workshop 2, pp. 1-10

Menke, W., 1989, Geophysical Data Analysis: Discrete Inverse Theory. Academic Press. pp. 285

Nielsen, S. B., 1992, Least squares inversion applied to thermal and hydrocarbon modelling. Proceedings of the Interdisciplinary Inversion Workshop 1, pp. 35-44

Sørensen, K. I., 1989, A method for measuring the electrical formations resistivity while auger drilling, First Break, Vol. 7, No. 10, pp. 403-407.

Sørensen, K. I., 1994, The Ellog Auger Drilling Method. Proceedings of the Symposium on the Application of Geophysics to Engineering and Environmental problems, Boston, March 1994, pp. 985-994

Tarantola, A. and Valette, B., 1982, Generalized nonlinear inverse problems solved using the least squares criterion. Reviews of Geophysics and Space Physics. vol. 20. No. 2. pp. 219-232

Deconvolution of Geophysical Multioffset Resistivity Profile Data

Ingelise Møller, Niels Bøie Christensen and Bo Holm Jacobsen

Department of Earth Sciences, Aarhus University, Aarhus, Denmark

Introduction

Near-surface structures in the earth can be mapped by geoelectrical methods by measuring the earth response to galvanic current. A measurement is carried out by transmitting galvanic current through the ground and measuring the potential difference between two potential electrodes. Increased distance between the electrodes results in increased penetration depth. A 2-dimensional data coverage is obtained by combining sounding and profiling.

During the last decade computerized data acquisition systems for collecting profile data have been developed resulting in at least two different approaches, the Continuous Vertical Electrical Sounding (CVES) (Overmeeren 1988, Dahlin 1993) and the Pulled Array Continuous Vertical Electrical Sounding (PA-CVES) (Sørensen 1995). The PA-CVES system is capable of producing 10–15 km profile data per day, which results in about 100,000 data values.

Large amounts of data require rapid interpretation procedures. These procedures fall into two groups: one-pass inversion algorithms and the iterative algorithms. Oldenburg and Ellis (1991) introduce approximate inverse mapping inversion, which uses an approximate inverse and an accurate forward modelling in the iterative scheme. Li and Oldenburg (1992) present 2D and 3D one-pass inversion algorithms for DC resistivity problems. The 3D algorithm is expanded into an iterative algorithm in Li and Oldenburg (1994). Loke and Barker (1996) use the quasi-Newton approach for an iterative 2D DC resistivity algorithm.

We present a fast one-pass inversion algorithm for 2D data following the same approach as the one presented by Li and Oldenburg (1992) for 3D data i.e. deconvolution using the Frechet kernel for the homogeneous halfspace solved in the Fourier domain. Further we use correlated covariance matrices as regularization.

Born Approximation and Inverse Mapping

The electric potential expressed on integral form (Snyder, 1976) is linearized by applying the Born approximation. The data, $d(x, s)$, the apparent resistivity rather than electrical potential, are the measurements at position x on the profile for a specific measurement geometry, s. In the Born approximation (cf. Jacobsen 1996) $d(x,s)$ is described as

$$d(x,s) \simeq d_{ref}(0,s) + \int_0^\infty dz' \int_{-\infty}^\infty dx'\, \Phi(x'-x,z',s)\, \delta\rho(x',z') \qquad (1)$$

where d_{ref} is the data predicted from the reference model, ρ_{ref}, which in our case is the homogeneous halfspace, $\delta\rho(x,z) = \rho(x,z) - \rho_{ref}$ is the anomalous model parameter, and $\Phi(x,z,s)$ is the 2D Frechet kernel for the homogeneous halfspace.

When we have a translational invariant reference model (the homogeneous halfspace or another 1D model), the Frechet kernel becomes translational invariant; therefore the inner integral of (1) is a cross correlation. The 2D Frechet kernel, as displayed in fig. 1, is given analytically and derived from the 3D kernel (Li and Oldenburg 1992) by integration in the y-direction.

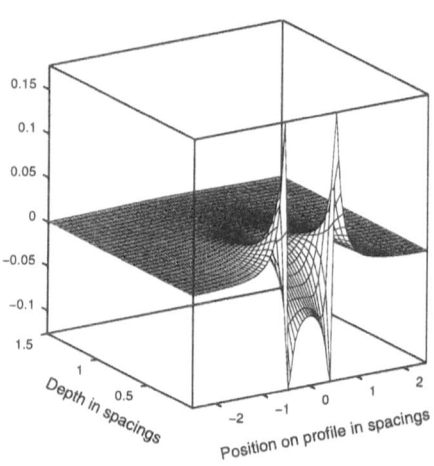

Fig. 1. The 2D Frechet kernel for apparent resistivity for pole-pole array. The current electrode is located in $(-1/2\,a, 0, 0)$ and the potential electrode in $(1/2\,a, 0, 0)$.

The anomalous model parameter, $\delta\rho(x,z)$, is approximated with averaged resistivity functions, $\delta\rho(x,m)$, in M layers. Due to the discretization in layers of the model the 2D Frechet kernel, $\Phi(x,z,s)$, must be integrated over each layer in the z-direction. This vertically integrated Frechet kernel $\overline{\Phi}(x,z,s)$ is inserted into (1). When a Fourier transform is performed, and a data anomaly, $\widetilde{\delta d}(k,s)$ is defined, we obtain

$$\widetilde{\delta d}(k,s) = \tilde{d}(k,s) - \tilde{d}_{ref}(k,s) \simeq \sum_{m=1}^M \widetilde{\overline{\Phi}}^*(k,m,s)\, \widetilde{\delta\rho}(k,m) \qquad (2)$$

Equation (2) describes for each wave number, k, a set of S linear equations with M unknowns. Thereby the large 2D spatial problem is decoupled into many small 1D problems in the wave number domain. As an example, take a profile of 1000 measurements in 10 configurations, then a 10,000-by-10,000 problem is to be solved in the space domain. Transformed into the wave number domain it becomes 1000 10-by-10 problems.

For each k we have a minimum variance estimate (e.g. Jackson 1979) of $\tilde{\delta}\rho(k, m)$

$$\tilde{x}_{\text{min.var}} = \left(\mathbf{A}^{T}\mathbf{C}_{e}^{-1}\mathbf{A} + \mathbf{C}_{x}^{-1}\right)^{-1}\mathbf{A}^{T}\mathbf{C}_{e}^{-1}\mathbf{y} \tag{3}$$

where $A_{sm} = \widetilde{\overline{\Phi}^{*}}(k, m, s)$, and $y_{s} = \widetilde{\delta d}(k, s)$. The minimum variance estimate in the space domain is obtained by inverse Fourier transforms. The model parameters and measurement used in (3) are logarithmically transformed resistivities and apparent resistivities.

The determination and use of covariance matrices will be illustrated and discussed in the following.

Determination of Model Covariance Matrices

Data errors coming from the measuring process are assumed to be uncorrelated and gaussianly distributed with a standard deviation of 0.05 (i.e. $\sim \pm 5\%$). Then the data error covariance matrix, $\mathbf{C}_{e}$ is diagonal.

Equation (1) describes an ill-posed problem. To get an estimate of $\tilde{\delta}\rho(k, m)$ in (2) we need to add information. This can be done in terms of a priori data (Jackson 1979) where the elements in the a priori data vector are zero, as we assume the a priori model to be the reference model. The uncertainty of the prior model is assumed to be correlated in a way that can be described as a power-law fractal process (Jacobsen 1993). The corresponding covariance functions for fractal exponent 1 and for fractal exponent 2 are

$$\Psi_{1}(s) = \Psi_{0}\,K_{0}(s/L) \tag{4}$$
$$\Psi_{2}(s) = \Psi_{0}\,\exp(s/L) \tag{5}$$

where Ψ_{0} is the variance, K_{0} the modified Bessel function, s a spatial distance and L the correlation length.

Fig. 2a shows a 2D model. Linearized forward responses of the model are calculated using (2) for 10 Wenner configurations with spacings logarithmically distributed between 2 and 48 m (fig. 2b). Using linearized forward responses the influence of the model covariance matrix on the model estimate can be inspected without interference of non-linearites.

A 5% white gaussian noise is superimposed on the predicted data, and we obtain a minimum variance estimate from (3) using the diagonal $\mathbf{C}_{e}$ and either (4) or (5) as $\mathbf{C}_{x}$. The model estimate using (5) as $\mathbf{C}_{x}$ is displayed in fig. 2c and using (4) as $\mathbf{C}_{x}$ in fig. 2d.

The variance of $\mathbf{C}_{x}$ in fig. 2c and fig. 2d is 0.25 in logarithmic units which is found to be realistic from simulations of the model. Using (4) as $\mathbf{C}_{x}$ allows vertical changes to take place over shorter distance and must be preferred over (5) based on this limited analysis. The correlation length, L, is assumed to be

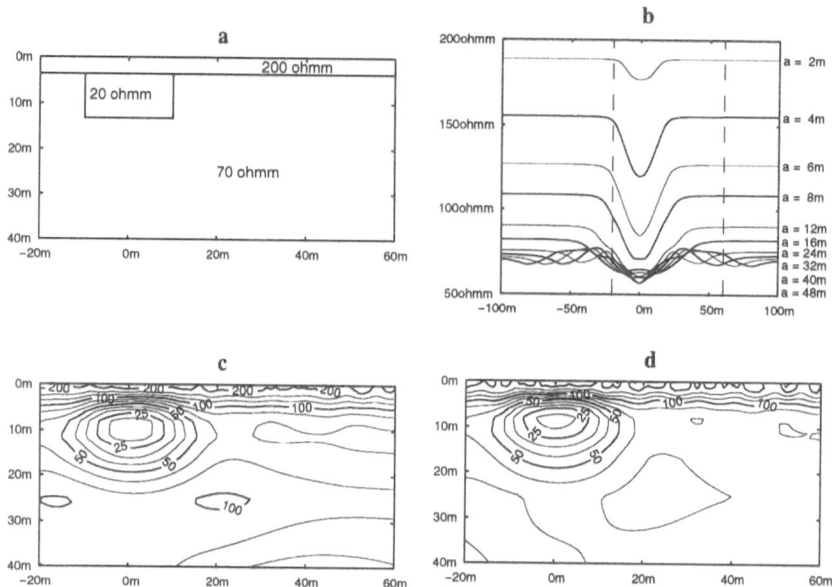

Fig. 2. a The true 2D model used for calculation of a forward response. **b** The forward response calculated using (2) for 10 Wenner configurations with spacings logarithmically distributed between 2m and 48m. The data are shown from -100m to 100 m, The dashed lines at -20m and 60m mark the interval displayed in the other figures. **c** Model estimate, when (5) is used as $\mathbf{C_x}$. **d** Model estimate, when (4) is used as $\mathbf{C_x}$. A logarithmic scale is used for the contouring in **c** and in **d**, as well as in fig. 4 and fig. 5.

long. Using (4) as $\mathbf{C_x}$ it turns out that estimates obtained for $L = 32$m and $L = 10,000$m look alike, so the choice of correlation length seems not to be critical.

Modelling Error as Data Error Covariance Matrix

When using the linearized equations (1) or (2) for forward calculations or inverse mapping an error is made. This modelling error can be taken into account in a pragmatic sense by adding a modelling error covariance matrix to the measuring error covariance matrix as $\mathbf{C_e}$ (Tarantola and Valette 1982).

Forward responses of the model in fig. 2a have been calculated using a finite element code (Labrecque 1992). The modelling error for each configuration is shown in fig. 3a. The curves for the middle configurations are most influenced by the conductive body, thereby containing more modelling errors.

Simulations are performed with a family of covariance functions

$$f(s) = \frac{1}{1 + (s/L)^r} \tag{6}$$

where L is the correlation length.

Fig. 3b shows simulations with 3 covariance functions with different L and r. The upper curves display functions where $L = 10$ and $r = 2$. The middle curves show the effect of a longer correlation length, now $L = 15$, and the bottom functions show the effect of a smaller exponent, r, which is now 1. All simulations are performed with a variance of 0.002. On inspection the curves in fig. 3b1 seem to simulate the actual modelling errors the best, and so the corresponding covariance function is used to generate the modelling error covariance matrix.

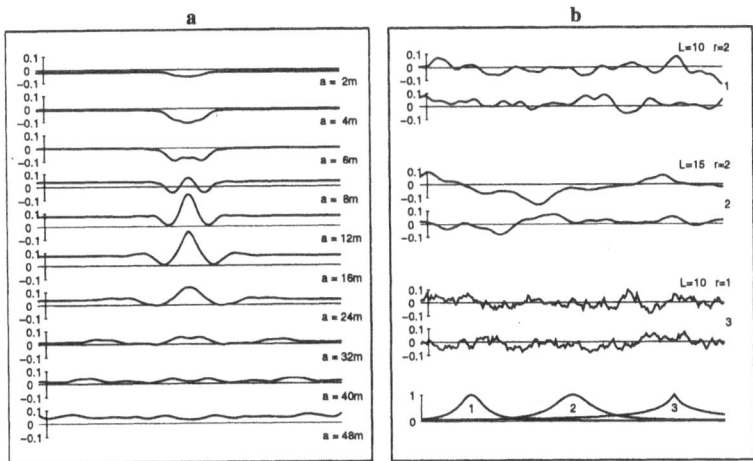

Fig. 3. a The modelling error due to the use of the Born approximation. The curves show for each configuration the relative difference between the forward response from the Born approximation and from a finite element modelling. **b** Simulations of the modelling error using the 3 different covariance functions of the bottom of the figure. The full length of the curves are in both **a** and **b** 200m.

The effect of adding a modelling error covariance matrix to the data covariance matrix when correct data are interpreted is shown in fig. 4. As before a 5% white gaussian noise is superimposed on the data, and all estimates are obtained using (4) as C_x with variance 0.25. The differences are in the C_e. The model estimate in fig. 4a is found by using the diagonal C_e with the variance 0.0025, which makes the regularization the same as in fig. 2d. This estimate is understabilized. Traditionally the modelling error is taken into account by using a higher measuring error variance. Simulations of the modelling error have shown that a variance of 0.002 is realistic. Fig. 4b shows an estimate, where the modelling error is traditionally treated, now the diagonal C_e has a variance of 0.0045. This gives a more stabilized estimate. Using the modelling covariance matrix with the variance 0.002 added to the diagonal measuring error covariance matrix with the variance 0.0025 as C_e, we get the model estimate in fig. 4c. The structures become more smooth, but artifacts seen in fig. 4b are removed.

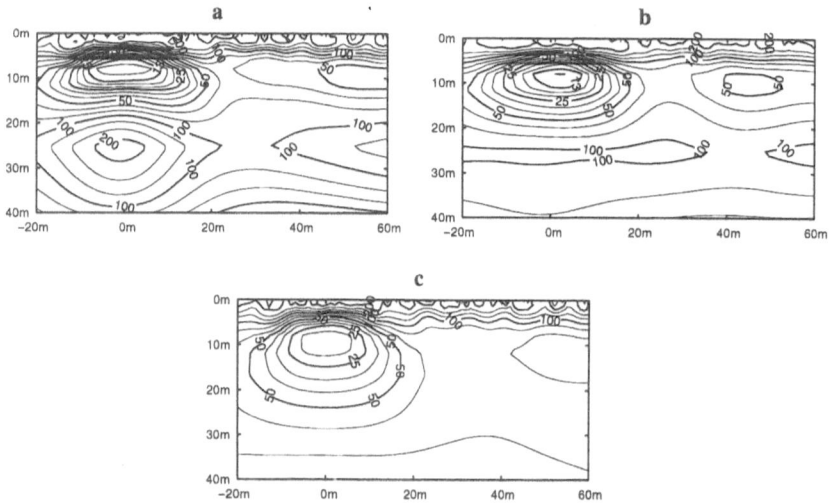

Fig. 4. a Model estimate of finite element calculated data of model in figure 2a using covariance matrices and variances as for model estimate in figure 2d. **b** Model estimate, when the modelling error is treated as measuring error. **c** Model estimate, when the modelling error covariance matrix is added to the measuring error in $\mathbf{C_e}$.

Field Example

In this field example data have been measured by the ABEM Lund Imaging System (Dahlin 1993) at Skejby test line near Aarhus. 10 Wenner configurations with spacings logarithmically distributed between 2 and 48 metres have been used. Fig. 5a shows the data as a data pseudo section. The data have been interpreted using the modelling error covariance matrix with variance 0.002 added to the diagonal measuring error covariance matrix with variance 0.0025 as $\mathbf{C_e}$ and (4) as $\mathbf{C_x}$ with variance 0.25, which was the regularization of the estimate in fig. 4c. The solution is displayed in fig. 5b.

Fig. 5c shows a result of the iterative quasi-Newton least-squares inversion program, RES2DINV. The inversion program (Loke and Barker, 1996) works in the space domain and performs a smoothness-constrained least-squares inversion using a flatness filter, which is a very sparse matrix with elements for which the amplitude is increased as the elements correspond to model parameters of increased depth.

Discussion and Conclusions

A 2D one-pass inversion algorithm for DC resistivity problems has been presented. The algorithm resolves the resistivity structures, though they become smoothed. The absolute resistivity values of the anomalies do not attain as high or

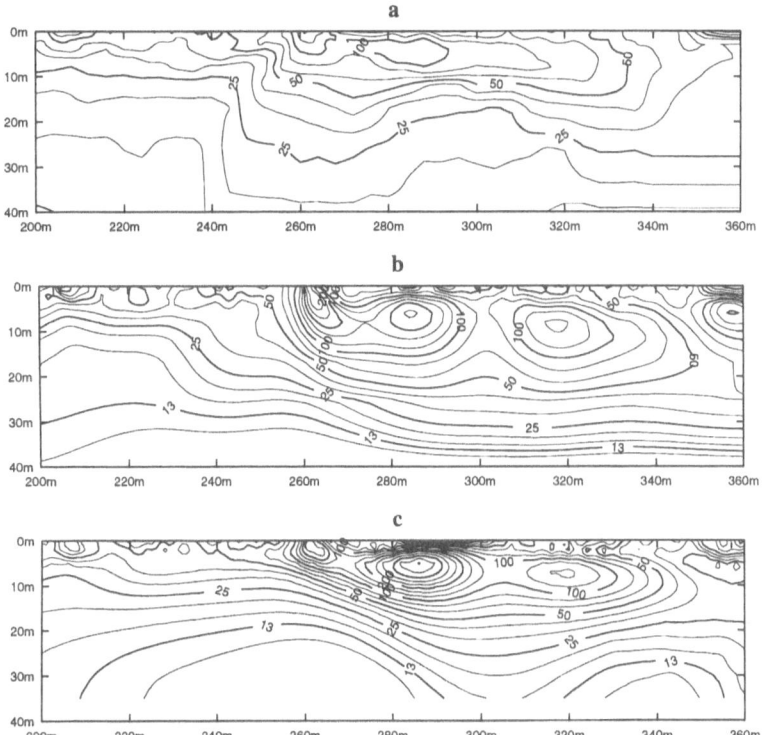

Fig. 5. a Data pseudo section showing the apparent resistivity. **b** Model estimate using covariance matrices and variances as in figure 4b. **c** Model estimate using the iterative quasi-Newton least-squares inversion program, RES2DINV.

low values as the true model or the resistivity values obtained from iterative least squares algorithms, when contrasts are high.

Regularization is performed using covariance matrices. A correlated model covariance matrix is used to describe variability on the a priori model. The data error covariance matrix is pragmatically specified as a combination of measuring error and nonlinear/numeric modelling errors.

This choice of regularization parameters leads to more extreme estimates of the amplitudes of the anomalies without amplifying the noise in the data.

This paper shows how pragmatic analyses through simulations of the covariance properties enable us to determine the type of covariance functions, the correlation lengths, and amplitudes.

The inversion algorithm can be very fast. If regularization is standardized, a family of generalized inverses can be calculated once and stored. Then the deconvolution algorithm consists of as many Fourier transforms as configurations, as many matrix-vector multiplications of small matrices, as there are data measuring positions, and as many inverse Fourier transforms as layers. This is performed by means of an ordinary PC in less than one second for 500–1000 measuring points and 10 configurations/layers. If the choice of variances is decided interactively, then the deconvolution is about one order of magnitude slower.

Solving the problem in the Fourier domain limits the demand for large memory capacity or sophisticated matrix household of big matrices. A 10,000-by-10,000 matrix in the space domain decouples into 1000 10-by-10 matrices in the Fourier domain.

This paper presents a limited analysis of the problem of specifying the covariance. Particularly the modelling error covariance matrix should be determined from analyses of several more complex models.

Acknowledgement

The authors thank Peter Duch for letting us use the finite element calculated forward responses of the model in fig. 2.

References

Dahlin, T., 1993. On the automation of 2D resistivity surveying for engineering and environmental applications. PhD thesis, Department of Engineering Geology, Lund Institute of Technology, Lund University, Sweden.

Jackson, D.D., 1979. The use of a priori data to resolve nonuniqueness in linear inversion. Geophysical Journal of the Royal Astronomical Society, 57, 137–157.

Jacobsen, B.H., 1993. Practical methods of a priori covariance specification for geophysical inversion. In: K.Mosegaard (ed.), Proc. Interdisciplinary Inversion Workshop 2, The Niels Bohr Institute for Astronomy, Physics and Geophysics, University of Copenhagen, 1–10.

Jacobsen, B.H., 1996. Rapid 2D inversion by multichannel deconvolution of geophysical profile data, in 'Proceedings of the Symposium on the application of Geophysics to Engineering and Environmental Problems', Keystone, Colorado 1996, 659–668.

Labrecque, D.J, Miletto, M., Daily, W. and Ramirez, A., Owen L, 1996. The effects of noise on Occam's inversion of resistivity tomography data. Geophysics, 61, 538–548.

Li, Y and Oldenburg, D.W., 1992. Approximate inverse mapping in DC resistivity problems, Geophysical Journal International, 109, 343–362.

Li, Y and Oldenburg, D.W., 1994. Inversion of 3-D DC resistivity data using an approximate inverse mapping, Geophysical Journal International, 116, 527–537.

Loke, M.H. and Barker, R.D., 1996. Rapid least-squares inversion of apparent resistivity pseudosections by a quasi-newton method, Geophysical Prospecting, 44, 131–152.

Oldenburg, D.W. and Ellis, R.G., 1991. Inversion of geophysical data using an approximate inverse mapping, Geophysical Journal International, 105, 325–353.

van Overmeeren, R.A and Ritsema, I.L., 1988. Continuous vertical electrical sounding, First Break,vol 6, no 10, p 313–324.

Snyder, D.D., 1976. A method for modeling the resistivity and IP response of two-dimentional bodies, Geophysics, 41, 997–1015.

Sørensen, K.I and Sørensen K.M., 1995. Pulled array continuous vertical sounding (PA-CVES), Proceedings of the SAGEEP'95 Orlando, Florida, p 893–897.

Tarantola, A. and Valette, B., 1982. Inverse Problems: Quest for information, Journal of Geophysics, 50, 159–170.

Imaging of Transient Electromagnetic Soundings Using a Scaling Approximate Fréchet Derivative

Niels B. Christensen

Department of Earth Sciences, Aarhus University, Aarhus, Denmark

1 Introduction

Transient electromagnetic (TEM) soundings have found widespread use in a variety of geophysical investigations in connection with environmental problems, hydrogeological investigations, and mineral prospecting (Spies and Frischknecht 1991). Modern strategies of measurement and instrument design emphasize dense measurements covering large areas resulting in vast amounts of data to be interpreted (Christensen and Sørensen 1994, Sørensen et al. 1995, Sørensen 1996). For airborne applications this is carried to the extreme.

Traditionally, TEM soundings have been interpreted with one-dimensional (1D) earth models using an iterative least squares inversion technique to determine layer conductivities and thicknesses (Boerner and West 1989, Raiche et al. 1985). This problem is highly nonlinear making the inversion task considerable even on modern PC's.

An algorithm for 1D approximate inversion, "imaging", of TEM soundings based on the Born approximation and the Fréchet kernel of the homogeneous halfspace is presented. The kernel is scaled according to the average conductivity of the earth for each time of measurement. To a large extent this scaling linearizes the TEM inversion problem, and the use of apparent conductivity as data makes it considerably less dependent on the measurement configuration. In practical calculations simple, piecewise linear approximations to the Fréchet kernel are used resulting in more or less damped inverse operators.

The computation time is approximately 0.5 sec/sounding/Mflop, which is 2-3 orders of magnitude faster than conventional least squares iterative inversion. The imaging produces models with many layers, typically 20-40, which fit the original data typically within 5-15% rms. No initial model is required, and the algorithm is therefore well suited for on-line and automatic inversion as well as for approximate inverse mapping schemes (AIM) (Oldenburg 1991). The imaging provides good initial models for a more rigorous least squares inversion.

The principle of a scaling Fréchet kernel can be extended to the 3D transient case as well as to other types of electromagnetic data.

2 Current distribution and sensitivity function

The current density distribution resulting from a step off at time zero from
a vertical magnetic dipole is shown in figure 1a. The current in the ground
has only an azimuthal component, and the current maximum diffuses
outwards and downwards with time, while the maximum broadens and
decreases in amplitude due to ohmic loss in the conductor. The current
density maximum follows a straight line, making an angle of
approximately 27° with the surface.

The sensitivity of the transient method, however, does not depend
directly on the current distribution at depth, but on the effect of the
current distribution on the measured field - in our case the vertical
magnetic field measured at the center. Using the usual formula for the H-
field on the axis of a circular current, the current element situated at the
point with the cylindrical coordinates (r, z) contributes to the magnetic
field at the surface as

$$dH(\sigma, t) = j(\sigma, t, r, z) \, dr \, dz \times \tfrac{1}{2} \frac{r^2}{\left(r^2 + z^2\right)^{3/2}} \tag{1}$$

where σ is the conductivity, t is time, and $j(\sigma, t, r, z)$ is he current density
at (r, z).

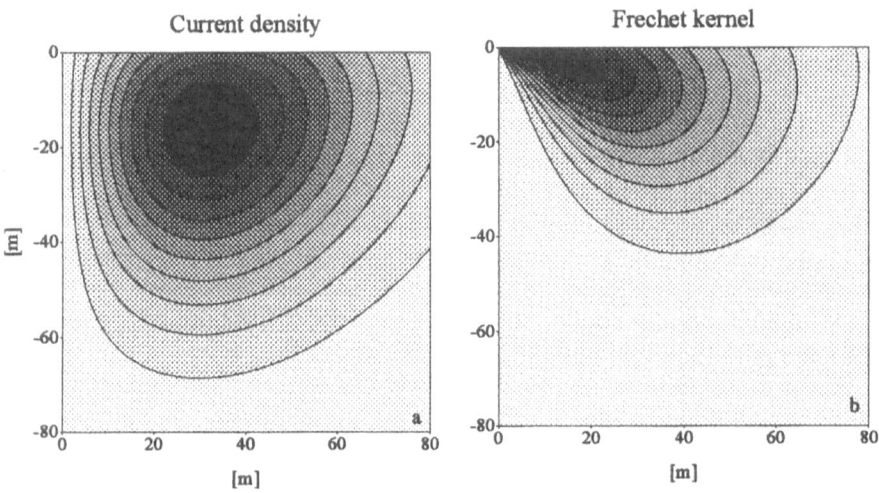

Fig. 1. a: Contoured plot of the current density in the ground 500 μs after
turn-off of a vertical magnetic dipole on a 1Ωm homogeneous halfspace. b:
Contoured plot of the current density weighted according to its contribution to
the vertical H-field at the center (equation 1). Contour intervals are 10% af
the maximum value.

Figure 1b shows the current distribution of figure 1a weighted according to its influence on the H-field. This plot reveals some interesting features of the transient method. The sensitivity of the method stays high at the surface close to the center at all times and can be described as a cone-shaped structure, which broadens outwards and downwards with time. The 'ridge' lies along a straight line, making an angle of approximately 15° with the surface. Note also that there is almost no sensitivity to the conductivity distribution within a cone with an opening angle of 30° directly under the source.

In the quasi-stationary case, which is the one considered here, the function described in figure 1b is the 3D rotationally symmetric sensitivity function of the vertical magnetic field with respect to conductivity for the homogeneous halfspace. The Fréchet kernel is identical to this function divided by the conductivity, σ, of the halfspace.

3 The 1D Fréchet kernel

The 1D Fréchet kernel for the homogeneous halfspace can be found by integrating the 3D Fréchet kernel. This results in an analytic expression (Christensen 1995):

$$F(\sigma, t, z) = \frac{M}{4\pi} \frac{1}{16\,\sigma\,\tau^4} \left\{ \frac{2u}{\sqrt{\pi}} (2u^2 + 1)\,e^{-u^2} - (4u^4 + 4u^2 - 1)\,\mathrm{erfc}(u) \right\} \quad (2)$$

$$= \frac{M}{30\,\pi^{3/2}\,\tau^3} \frac{1}{\sigma}\,\tilde{F}(\sigma, t, z) \ , \quad \tau = \sqrt{t/(\mu\sigma)} \quad \text{and} \quad u = z/(2\tau) \ .$$

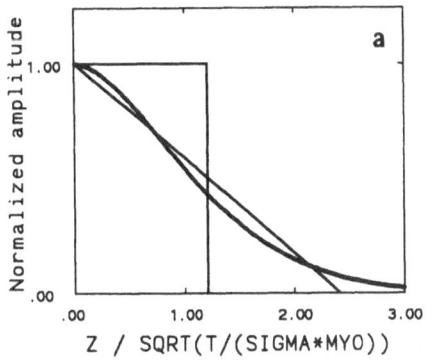
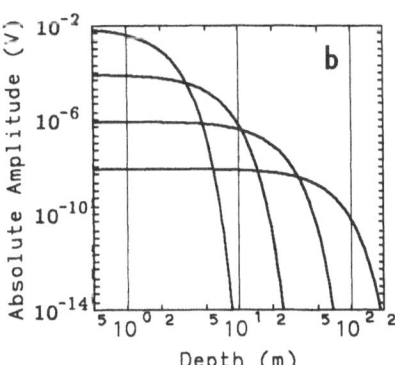

Fig. 2. a: The normalized 1D Fréchet kernel for the transient vertical magnetic dipole in the case of a homogeneous halfspace as a function of normalized depth (thick grey curve) together with the constant and the linear approximations to the Fréchet kernel. b: The unnormalized Fréchet kernel for a halfspace resistivity of $1\,\Omega m$ at the times $1\,\mu s$, $10\,\mu s$, $100\,\mu s$, and $1\,ms$.

Figure 2a shows a plot of $F(\sigma, t)$ on a linear scale and figure 2b shows $F(\sigma, t)$ on a logarithmic scale at four different times for a source dipole of unit moment. It is seen that the sensitivity is a bell-shaped function of z, which has its maximum at the surface at all times. Figure 2b shows how the amplitude decreases and the depth of diffusion increases as a function of time. The sensitivity drops very abruptly to zero with depth, decreasing as $u^{-3} \exp(-u^2)$.

4 The imaging algorithm

Imaging of transient electromagnetic soundings has been the subject of many investigations in the geophysical litterature (e.g. Macnae and Lamontagne 1987, Smith et al. 1994).

The imaging algorithm proposed here is based on the Born approximation and the Fréchet derivative of the homogeneous halfspace. However, instead of choosing a constant reference conductivity of the halfspace, a new reference conductivity is chosen for each time of measurement. This reference conductivity is chosen equal to the instantaneous apparent conductivity of the measurement, which by definition is equal to the halfspace conductivity, which would produce the actual field at the time of the measurement. In this way the slower diffusion of current (relative to the homogeneous halfspace) in good conductors and the faster diffusion through bad conductors is taken into account.

In the approximation that conductivity changes are small, the Born approximation describes the change in the response as a linear functional of the change in subsurface conductivity:

$$H_i \simeq H_i^{ref} + \int_0^\infty F\left(\sigma^{ref}(z), t_i, z\right) \left[\sigma(z) - \sigma^{ref}(z)\right] dz \qquad (3)$$

where

$\quad$ H_i is the measured H-field at the time t_i.

$\quad$ H_i^{ref} is the H-field of the reference model.

$\quad$ $F(\sigma^{ref}(z), t, z)$ is the Fréchet kernel of the reference model.

$\quad$ $\sigma(z)$ is the conductivity of the subsurface.

$\quad$ $\sigma^{ref}(z)$ is the conductivity of the reference model.

The response of the homogeneous halfspace with conductivity σ^0 is the integral of the Fréchet kernel multiplied with conductivity

$$H_i^0 = \int_0^\infty F(\sigma^0, t_i, z) \sigma^0 dz = \frac{M}{30} \left(\frac{\sigma^0 \mu}{\pi t}\right)^{3/2} = \frac{M}{30 \, \pi^{3/2}} \frac{1}{\tau^3} . \qquad (4)$$

For a measured $H_i(t_i)$ over any 1D structure the solution of equation (4) in terms of σ^0 will give us the apparent conductivity $\sigma^a(t_i)$.

In the case, where the reference model is a homogeneous halfspace, we find

$$H_i \simeq H_i^0 + \int\limits_0^\infty F\left(\sigma^0, t_i, z\right)\left(\sigma\left(z\right) - \sigma^0\right) dz = \int\limits_0^\infty F\left(\sigma^0, t_i, z\right) \sigma\left(z\right) dz \qquad (5)$$

For a layered 1D structure with L layers given by the layer boundaries z_i, $i=1, L+1$, $z_1 = 0$, $z_{L+1} = \infty$, the conductivity is constant within each layer and equation (5) becomes

$$H_i \simeq \sum_{j=1}^{L} \sigma_j \int\limits_{z_j}^{z_{j+1}} F\left(\sigma^0, t_i, z\right) dz \qquad (6)$$

The scaling of the Frechet kernel is obtained by substituting the apparent conductivity $\sigma^a(t_i)$ instead of the constant halfspace conductivity σ^0 in the expresion for the Fréchet kernel in equation (2), and substituting equation (4) in (2) we arrive at:

$$H_i \simeq \sum_{j=1}^{L} \sigma_j \int\limits_{z_j}^{z_{j+1}} F\left(\sigma^a(t_i), t_i, z\right) dz_i = \sum_{j=1}^{L} \sigma_j \frac{H_i}{\sigma^a(t_i)} \int\limits_{z_j}^{z_{j+1}} \tilde{F}\left(\sigma^a(t_i), t_i, z\right) dz$$

$$\Rightarrow \qquad \sigma^a(t_i) \simeq \sum_{j=1}^{L} \sigma_j \int\limits_{z_j}^{z_{j+1}} \tilde{F}\left(\sigma^a(t_i), t_i, z\right) dz \ . \qquad (7)$$

Equation (7) expresses the apparent conductivity as a weighted sum of the conductivities of each layer with easily calculated weights.

The problem of damping the inverse is adressed by using a constant and a linear approximation to the Fréchet kernel (fig. 2a) instead of using the exact kernel. The constant approximation results in a well damped inverse, whereas the linear approximation gives a more sensitive "spiked" inverse.

By choosing the surface amplitude of the approximations equal to that of the true kernel and by demanding the integral over the halfspace to be the same as that of the true kernel, the constant approximation is given by

$$F\left(\sigma, t, z\right) = F^0 = \frac{M}{4\pi} \frac{1}{16\,\sigma\,\tau^4} = H_i \cdot \frac{15\sqrt{\pi}}{32\,\tau} \frac{1}{\sigma} \qquad \text{for } z < z_D$$

$$F\left(\sigma, t, z\right) = 0 \quad \text{for } z > z_D \quad , \quad z_D = \frac{32}{15\sqrt{\pi}}\,\tau \qquad (8)$$

Choosing the layer boundaries z_j equal to the diffusion depths of the Fréchet kernel, z_D, gives a particularly simple version of equation (7)

$$\sigma_i^a = \sum_{j=1}^{i} \sigma_j \cdot \frac{h_j}{z_{Di}} \tag{9}$$

where h_j is the thickness of the j'th layer. Equation (9) may easily be inverted for the layer conductivities by forward substitution. It is a "stripping the earth" algorithm.

The above derivations have all assumed the source and receiver to be coincident vertical magnetic dipoles, which is not a practical measuring configuration. However, equation (7) may be used for other configurations as well, provided that a suitable definition of apparent conductivity can be found with which the Fréchet kernel can be scaled.

Among the many definitions of apparent conductivity, the all-time apparent conductivity based on the step response - the H-field itself - must be chosen. For the central loop and the coincident loop configuration it is unambiguously defined and exists always, as the H-field is a monotonically decreasing function of time, and it has a smooth transition between resistivity levels. The field quantity most often measured with transient equipments is dH/dt, but the apparent conductivity definitions based on dH/dt (Spies 1986) are not applicable. However, given dH/dt data we may calculate the H-field numerically, and from the H-field we can calculate the all-time apparent conductivity as a function of time (Christensen 1995).

The all-time apparent conductivity does not only serve the scaling purpose, but is also used as input data. This parameter is to a much larger extent independent on measuring configuration than the H-field, and so the procedure developed for the vertical magnetic dipole is also applicable for other configurations.

5 Results and discussion

Figure 3 shows the models resulting from the application of the imaging algorithm on synthetic noise-free data from four different models together with the true resistivity models: two 2-layer models (descending and ascending) and two 3-layer models (minimum and maximum). The results for the double descendeing and double descending 3-layer models a similar to the 2-layer models. It is seen that the imaging algorithm works very well with descending type models, but reacts slower to ascending resistivity models. The worst performance is seen with the maximum model, which is to be expected. From the asymptotic behaviour it is seen that the imaged models reach the true value of the resistivity very well.

The constant approximation to the Fréchet kernel results in very little undershoot and overshoot in the imaged models and is thus a more damped inverse, whereas such effects are seen to some extent, if the linear approximation is used.

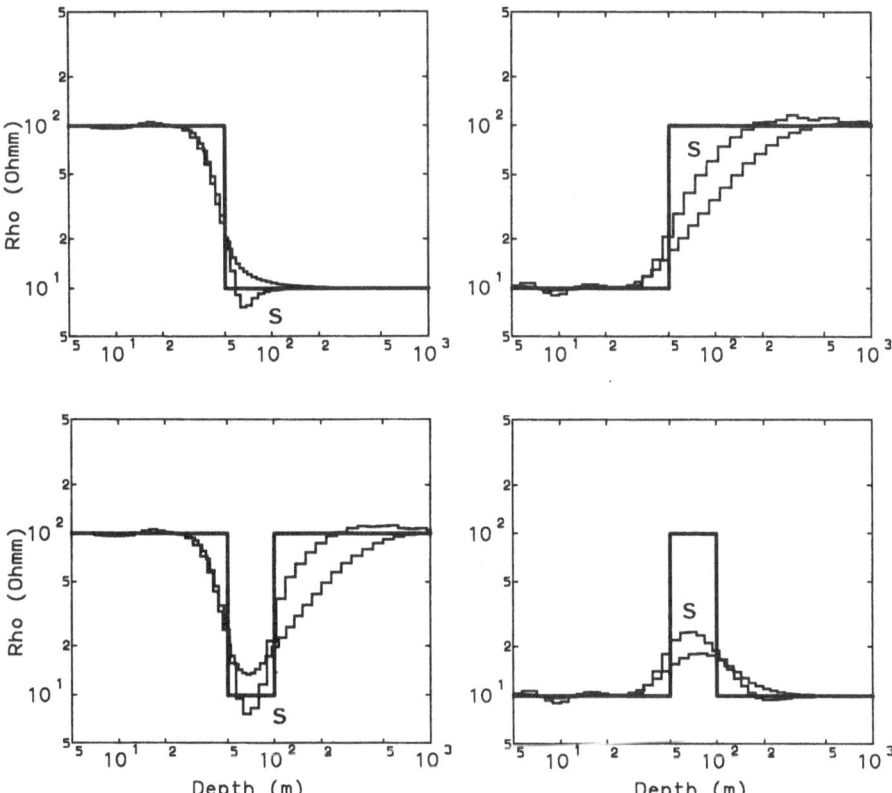

Fig. 3. The models obtained using the imaging procedures on 40×40 m^2 central loop soundings. The figure shows a comparison between the results obtained with the more damped constant approximation and the more sensitive linear approximation to the Fréchet kernel for four models. The true models are shown with thick grey curves. The sensitive inverse is marked with an "S".

The models produced by the imaging procedure fit the original data typically within 5-15% rms and are well-behaved without extreme discrepancies. This is very satisfactory for a one-pass algorithm working almost instantaneously. Two examples are shown in figure 4: the two layer descending and ascending models, representing the best and the worst case among the four models of figure 3, respectively. From dH/dt data the H-field and the all-time apparent conductivity have been calculated and the imaging applied using the constant approximation to the Fréchet kernel.

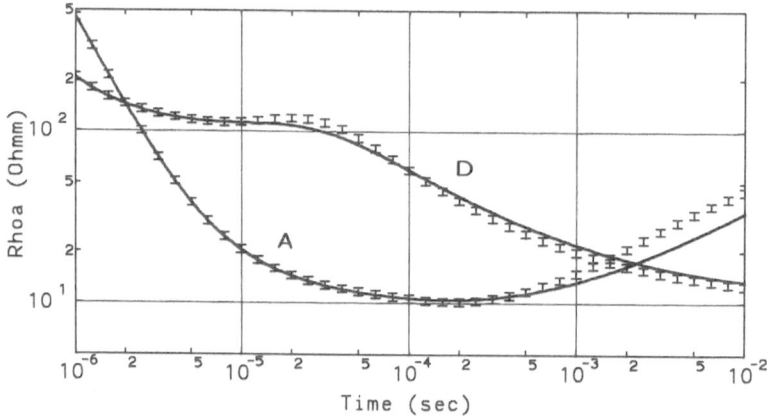

Fig. 4. Comparison between forward responses of imaged models and original data for the two-layer descending (D) and ascending (A) models of figure 3. The rms misfits are 6.9% and 13.3%, respectively.

Figure 5 shows the application of the imaging algorithm to a profile of transient soundings in the central loop configuration. The result of the imaging is displayed as a contoured model section. The profile is from the island of Rømø, Denmark, where it transects the NE corner of the island. The good conductor - the salt water horizon - is closer to the surface at the ends of the profile, where the distance to the coast is small and lies deeper in the middle of the profile, which is situated furher inland.

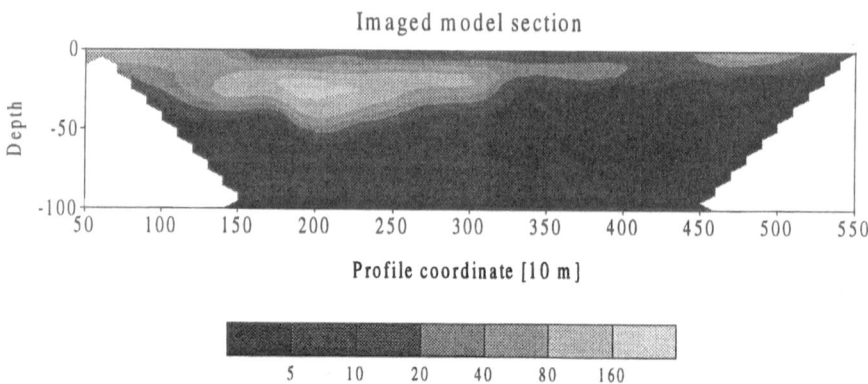

Fig. 5. An imaged model section from a TEM profile for the detection of the salt water horizon on the island of Rømø, Denmark.

6 Conclusion

The imaging algorithm is fast, robust, fully automated, and requires no initial model. The computation time, including the transformation from dH/dt to H-field, is approximately 0.5 sec/sounding/Mflop, which is 2-3 orders of magnitude faster than conventional least squares iterative inversion. The imaging produces models which fit the original data typically within 5-15%.

The imaged models may be used as very good input models to an iterative least squares inversion program, can be implemented as an on-line interpretation in TEM instruments, and the imaging procedure lends itself readily to AIM strategies (Oldenburg 1991). Contoured model sections based on imaged models from soundings along profile lines give a very fast insight in the subsurface conductivity distribution.

The principles of the imaging algorithm based on the Fréchet kernel can be extended to the 2D and 3D case of transient data. In fact, the idea of an instantaneous scaling of the Fréchet kernel according to some "average value" of the investigated parameter is applicable in other areas, as long as an "average parameter" can be defined and if the Fréchet kernel depends on the investigated parameter.

References

Boerner, D.E., and West, G.F., 1989. A generalized representation of the electromagnetic fields in a layered earth. Geophysical Journal, 97, 529-547.

Christensen N.B., 1995. Imaging of central loop transient electromagnetic soundings. Journal of Environmental and Engineering Geophysics, vol. 0, No. 1, 53-66.

Christensen, N.B., and Sørensen, K.I., 1994. Integrated use of electromagnetic methods for hydrogeological investigations. Proceedings of the Symposium on the Application of Geophysics to Engineering and Environmental Problems (SAGEEP), Boston, March 1994, p 163-176.

Macnae, J., and Lamontagne, Y., 1987. Imaging quasi-layered conductive structures by simple processing af transient electromagnetic data. Geophysics, 52, 545-554.

Oldenburg, D.W., and Ellis, R.G., 1991. Inversion of geophysical data using an approximate inverse mapping. Gephys. J. Int. 105, 325-353.

Raiche, A.P., Jupp, D.L.B., Rutter, H., and Vozoff, K., 1985. The joint use of coincident loop transient electromagnetic and Schlumberger sounding to resolve layered structures. Geophysics, 50, 1618-1627.

Smith, R.S., Edwards, R.N., and Buselli, G., 1994. An automatic technique for presentation of coincident-loop, impulse-response, transient, electromagnetic data. Geophysics, 59, 1542-1550.

Spies, B.R., and Eggers, D.E., 1986. The use and misuse of apparent resistivity in electromagnetic methods. Geophysics, 51, 1462-1471.

Spies, B.R., and Frischknecht, F.C., 1991. Electromagnetic Sounding. In: M.N. Nabighian (ed.), Electromagnetic methods in Applied Geophysics, Investigations in Geophysics, Vol. 2A, SEG.

Sørensen, K.I., Effersø, F., and Christensen, A.J.,1995. Pulled array transient electromagnetic method (PA-TEM). SAGEEP conference, Orlando, p 899-903.

Sørensen, K.I., 1996. Pulled array continuous electrical profiling, PA-CEP, First Break, in press.

A Case of Processing Design Using Inversion: Simple Gating Versus Iterative Reweighted Least Squares Inversion

M.S. Munkholm

Department of Earth Sciences, Laboratory of Geophysics, Aarhus University, Aarhus, Denmark

Introduction

The importance of TEM soundings in hydrogeophysics as a means to map the depth extent of an aquifer is increasing. New instrumentation, the Pulled Array Transient Electromagnetic Method, PA-TEM, (Sørensen, 1995) is planned with this purpose in mind. This study illustrates the role of inversion concepts in the phase of raw data processing for a new instrument of this type.

The theory of current induction due to a changing magnetic field is the physical basis of the TEM method. A switch-off of current in a transmitter loop induces eddy currents in the ground. The decay rate of the resulting secondary magnetic field is dependent on the conductivity structure of the subsurface and may be inferred from measurements of the induced voltage in a receiver coil (Consult Nabighian and Macnae (1991) for a review). The planned instrument samples the decay every 2 μs in the time interval from 2 μs to 8 ms, producing 4000 data points for each transient. As this corresponds to a data flow in the range of GBytes per hour, data reduction and on-line processing is essential.

TEM earth responses decay extremely rapidly ($t^{-5/2}$ over a halfspace) decreasing into the electromagnetic (EM) noise at some delay time. A comparison of the model response over the 3 layered earth model (see fig. 1.a) and the same model response with real EM noise added shows that the decay drowns in noise after 10-20 μs (see fig. 1.b). Although an increase in the signal to noise (S/N) ratio may be obtained by increasing the transmitter moment, hydrogeophysical surveys require small mobile systems, setting limits on the transmitter moment.

To extract the transient electromagnetic (TEM) earth response from TEM data it is essential to improve the S/N ratio of the raw data. Traditionally this has been achieved through simple averaging over time, so called gating in time windows, and summing of repeated measurements, termed stacking (e.g. Munkholm and Auken, 1995). In the following, the optimal window density for this procedure is investigated. Alternatively, for non-Gaussian noise contamination, known to be a problem in TEM data, robust methods can be superior. The source of the outliers is predominantly noise from local thunderstorms, an activity that has both a seasonal and geographic variation. Problems with outliers in TEM measurements have previously been treated successfully through use of robust stacking methods (e.g. Buselli and Cameron, 1992; Strack, 1992). One such method interpolates the signal by an iterative reweighted least squares, IRLS, inversion in which outlier data points are downweighted through the use of a Huber influence function. This paper presents a comparison of the signal extraction ability of gating and the more robust interpolation of the signal by an IRLS inversion scheme.

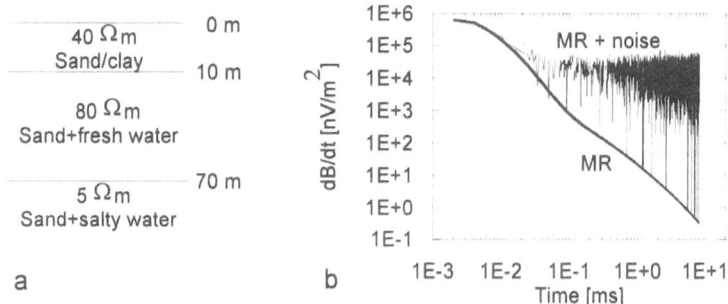

Fig. 1. a. 3 layered earth model. **b.** The model response (MR) over the earth model in a. (thick) and the MR contaminated by real EM noise (thin). The decay is seriously disturbed by noise at 10-20 µs and after 50-60 µs only noise is seen.

The signal is interpolated from the densely sampled PA-TEM data to the window center times used in the gating. A Huber influence function is used as weighting function (Hampel et al., 1986), rendering the method less sensitive to outliers. The IRLS method has been used in the extraction of various other geophysical signal types with outlier contamination, such as geomagnetic observatory data (Constable, 1988) and MT data (Sutarno & Vozoff, 1989).

The comparison of the signal extraction ability of the two schemes is carried out on model data contaminated by true EM noise measured with a prototype of the PA-TEM receiver. The results of signal extraction are analyzed to assess the degree of processing sophistication required in different noise environments.

Analysis of Gate Position

For Gaussian noise contamination it is well known that simple stacking increases the S/N ratio by a factor of the square root of the stack size. A further increase in the S/N ratio for TEM data may be obtained by exploiting the data redundancy and gating the data in windows. By choosing windows with widths increasing with delay time the effect of gating becomes more pronounced at later times when the S/N ratio is the poorest. At the same time, it reflects in fact the distribution of the information contained in a transient decay. There is however a trade-off between the improvements in S/N obtained by increasing the width of the windows and the resolution of the decay as it unfortunately also results in a decrease in the detail of the decay.

A TEM earth response, D(t), may be approximated by a set of time constants, τ, and amplitude coefficients, A, as $D(t) = \mathbf{R}(t,\tau)A$. Vector D contains the measurements at the times, t, vector A contains the amplitude coefficients to be determined, and $\mathbf{R}$ is the kernel matrix dependent on t, τ and the waveform of the EM system (Stoltz & Macnae, 1992). Through a SVD analysis of the kernels, $R(t,\tau)$ for different gating schemes, the trade-off for a given EM system may be investigated. The calculation of the kernels and the subsequent SVD analysis is carried out using the program WAVBASIS (Macnae, 1992).

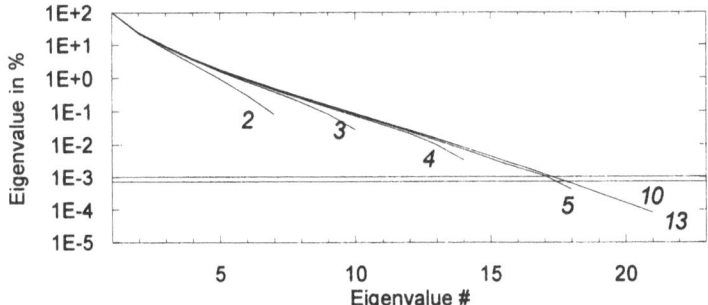

Fig. 2. Resolution of 40 τ values in the range from 2 μs to 20 ms (only 21 shown) for EM systems with different window density. The numbers refer to the number of windows per decade (WPD). Less than 5 WPD degrades the resolution. 5 or more WPD results in essentially the same τ–range resolution. The 10^{-3}% and $7 \cdot 10^{-4}$% levels are shown.

The decrease in resolution with increased gate width is investigated by studying the resolution of 40 τ values in a 4 decade time range from 2μs to 20 ms (see fig.2). A TEM system with a transmitter moment of 3000 Am^2, a repetition frequency of 31.25 Hz, a turn-on time of 2 ms, a turn-off time of 10 μs, and a resolution of eigenvalues over a range of 10^5 is defined to represent the instrumentation. 17 of the 40 eigenvalues are above the resolution limit when gating with 5 or more windows per decade, (WPD). However, when gating with less than 5 WPD fewer τ values may be singularly defined.

The increase in the S/N ratio obtained by gating in windows of double length (i.e. 5 in comparison to 10 WPD) is a factor of $\sqrt{2}$ for Gaussian noise. If 10 WPD can resolve singular values down to 10^{-3} % of the largest singular value, then 5 WPD should resolve down to $7 \cdot 10^{-4}$ %. These levels result in the resolution of essentially the same τ range. It is noteworthy though, that for non-Gaussian noise contamination the longer windows present a better opportunity to detect and reject outliers.

The Iterative Reweighted Least Squares Method

Interpolation by IRLS inversion presents an alternative to simple gating for extraction of the earth response from noise contaminated TEM data. In each iteration a local spline interpolation of a 3rd degree polynomial with a continuous

1st degree derivative is fitted to 4 local points. The weights are updated between iterations in accordance with the Huber influence function (Hampel et al., 1986)

$$\psi(e) = \{ \begin{array}{l} e \ , for \ e \leq t \\ 1 \ , for \ e > t \end{array} \tag{1}$$

For data with errors less than the scale parameter, t, the Huber influence function weights the data according to the L2 norm, whereas data with errors exceeding t the data are weighted according to the L1 norm (see fig. 3). This limits the influence of outliers. The scale parameter is updated between each iteration to be 1.4 times the standard deviation of the errors of a given iteration weighted according to the $\Psi(e)$. More detailed accounts of the method are given in i.e. Constable (1988), Sutarno & Vozoff (1989) and Munkholm et al. (1994).

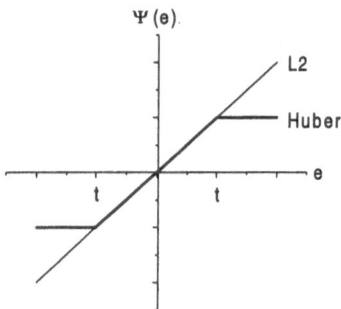

Fig. 3. The L2 and the Huber influence functions, $\Psi(e)$, with scale parameter t. For the L2 norm the weight on data increases with their error whereas when using the Huber influence function data with errors outside t are given equal weight.

One considerable drawback of the IRLS is its computational expense. Each iteration requires an update of three vectors and one scalar, namely the residual vector, e_k, the scaling parameter, t_k, the weights, w_k, and the model parameter vector, x_k. Equations 2 to 5 state the updates.

$$e_k = y - A x_k \tag{2}$$

$$t_k = 1.4 \ \sigma(e_k) \tag{3}$$

$$w_k = \Psi(e_k, t_k)/e_k \tag{4}$$

$$x_k = (A^T W_k A)^{-1} A^T W_k y \tag{5}$$

Furthermore, upon initialization of the algorithm the kernel must be calculated. As part of an IRLS standard processing tool **A** may be predefined for specified values of the measurement and interpolation times though. The most expensive operation is the calculation of the model parameters in equation 5 costing in the order of NM² operations in each iteration. For a system with 4096 samples per decay (N) and approximately 35 interpolation times (M) it amounts to 5 Mflop which is not feasible for online processing. However, exploitation of the spasity of the matrices should reduce this drastically bringing IRLS within the scope of online processing. IRLS algorithms for sparse matrices are presented in O'Leary (1990). In contrast simple gating is an N order operation.

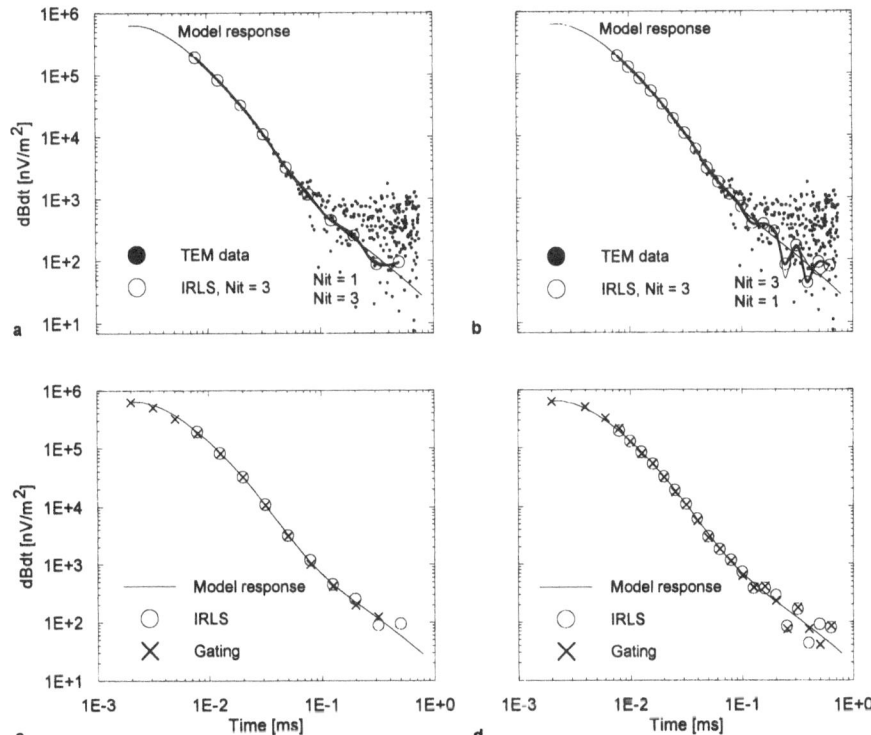

Fig. 4. a. The results after 1 and after 3 iterations of IRLS are shown and may be compared with the model response. The original 2 µs data are shown for comparison. 5 windows per decade, (WPD). **b.** As a. with 10 WPD. **c.** Comparison of the signal extraction ability of IRLS and gating, 5 WPD. **d.** As c. with 10 WPD.

Signal Extraction by Gating and IRLS

The performance of gating and IRLS is compared for a noise contaminated TEM response over the 3 layered model in fig. 1.a. The signal consists of the model response to which a stack of 800 noise measurements with a prototype of the PA-TEM receiver from an urban, culturally disturbed area have been added. At a speed of 1 m/s, a stack size of 800 corresponds to averaging of the response and thus the geology over a distance of 13 m. The stacked transient drowns in noise after only approximately 0.08 ms (see figs. 4.a and 4.b).

Although the analysis of the optimal window width showed that the same resolution should be obtained through use of 5 as 10 WPD, the comparison of gating and IRLS is carried out for both. In the IRLS the response is interpolated to the center times of the windows used in the gating. The interpolation is not carried out for the earliest time windows as these contain only one or no samples, rendering the interpolation unstable. At these early times the S/N is sufficiently high that the stacked samples themselves to be used. In both cases the signal is now recovered to delay times of 0.2 ms (see fig. 4.c-d), favoring neither of the methods over the other.

To illustrate the reweighting of the errors, consider the change in weights from one iteration to the next in the case of 10 WPD. Fig. 5.a shows the weights after the first iteration. Outlier observations are downweighted. Fig. 5.b shows the changes in the weights from the 2nd to the 3rd, the 3rd to the 4th, and the 4th to the 5th iteration - parallel shifted 0.05 to allow variations to be observed. After 3 iterations the difference in weights between iterations has become insignificant.

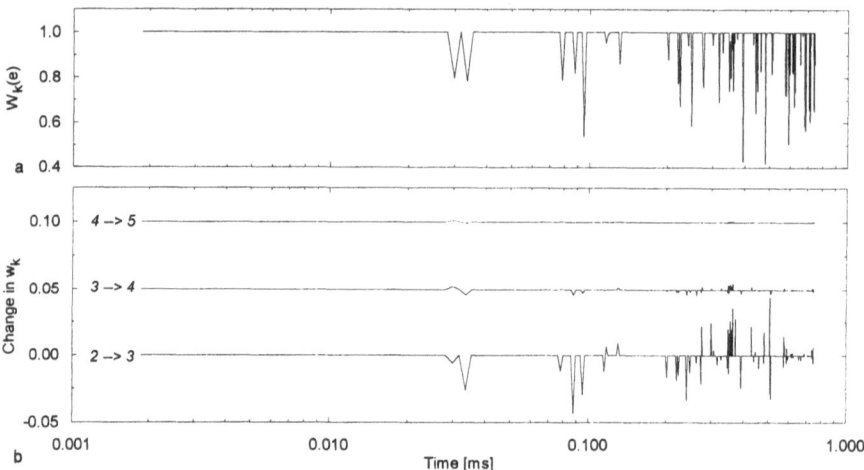

Fig. 5. a. The weights for the IRLS interpolation to 10 per decade times as a function of delay time for the first iteration. **b.** The change in weights between subsequent iterations.

Inversion of the Extracted Signals

To complete the comparison of the performance of the processing schemes, test inversion of the extracted signals is carried out using a layered earth inversion program (Christensen & Auken, 1992). Table 1 shows the 3 layered earth models resulting from inversion of the signals extracted by gating and IRLS for 5 and 10 WPD. The data could not be fitted with 2 layered models. The entire extraction segments have entered into the inversion with equal weight without any cutoff of data in an attempt to produce fully automatic processing.

Although the residual is superior for the 5 WPD, both the resistivity and the thickness of the first layer is unresolved. The sampling is not dense enough to resolve the first layer for these noise contaminated data. Furthermore, the resistivity of the conductive half space is underestimated. This is found to be the case for both the gating and the IRLS. For the 10 WPD the model parameters are well resolved and close to the initial model. The large residuals for the IRLS 5 per decade and the two 10 per decade data sets are due to noisy data at late delay times (see fig. 4.c-d). Inversion of the extracted signals shows that the 5 WPD sampling is not dense enough at early times to resolve the model parameters of the first layer. Thus, despite the results of the SVD analysis, 10 WPD is superior to 5 WPD for these data. It appears that an eigenvalue resolution of 10^6 is required in which case 10 WPD have 3 more resolvable eigenvalues than 5 WPD (see fig. 3). The eigenvectors corresponding to the extra eigenvalues contain significant contribution from large τ, not present in the 5 WPD eigenvectors.

	Model	5 per decade sampling				10 per decade sampling			
Meth./resid.		Gating, 0.9		IRLS, 5.5		Gating, 8.2		IRLS, 10.3	
		Para	Rel	Para	Rel	Para	Rel	Para	Rel
rho1 [Ωm]	40	22.5	0.9	23.2	1.1	34.8	0.3	41.7	0.1
rho2 [Ωm]	80	64.7	0.04	62.7	0.03	80.1	0.1	84.7	0.1
rho3 [Ωm]	5	1.4	0.1	2.1	0.1	4.6	0.1	6.0	0.1
t1 [m]	10	2.6	1.7	2.3	1.8	7.0	0.6	11.9	0.4
t2 [m]	60	67.4	0.1	65.5	0.1	65.4	0.1	56.2	0.1

Table 1: The model parameters (Para), their relative uncertainties (Rel) and the residuals (Resid) obtained from the inversion of the signals extracted by gating and IRLS.

Conclusions

A comparison of simple gating of the sampled transient in logaritmically spaced windows and interpolation by iterative reweighted least squares inversion has been carried out to assess the optimal signal extraction of the earth response to a transient electromagnetic input.

An SVD analysis to identify the optimal window positioning for a TEM system such as the one described here, showed that the optimal number of WPD in respect to both resolution and S/N considerations is 5. Both the IRLS extracted signal and the gated signal for 5 WPD seem to fit the model response better than the 10 WPD signal.

The severity of the outlier problem for these noise data is limited, and therefore robust approaches such as IRLS interpolation for signal extraction do not improve results. However, previous tests of the IRLS on model responses with synthetic noise containing outliers showed that the method is, indeed, robust towards such bursts when they occur (Munkholm et al., 1994). Thus, for data measured under noisier circumstances on days with outlier problems the IRLS is expected to be superior. It is indeed noteworthy that the results of IRLS inversion are not inferior to simple stacking under conditions of effectively Gaussian noise. Although this suggests that the cheaper gating should be pursued in the processing of these data rather than the expensive IRLS, one must remember that the noise is not always effectively Gaussian. As the statistics of the noise varies both for geographic location and for season less favorable conditions are to be expected. The desired automatic processing scheme ought to present at least some kind of diagnostic of the severity of the outlier problem and preferably be able to handle cases of extreme outliers for which simple stacking fails.

Acknowledgements

The author is grateful to Prof. J. Macnae, Macquarie University for the use of WAVBASIS and to the reviewers for their constructive suggestions to the manuscript. I would also like to thank Kurt Sørensen, Århus University for providing me with the PA-TEM noise measurements.

References

Buselli, J. and Cameron, M.A., 1992. Improved sferics noise reduction in TEM measurements. AMIRA Project 273R, CSIRO, Australia, 34 pp.

Christensen, N.B. and Auken, E., 1992. Simultaneous electromagnetic layered model analysis. In: Jacobsen, B.H. (ed) Proc. Interdisciplinary Inversion Workshop I, Geoskrifter 41, Aarhus University, Denmark, 49-56.

Constable, C.G., 1988. Parameter estimation in non-Gaussian noise. Geophysical Journal, 94, 131-142.

Hampel, F.R., Ronchetti, E.M., Rousseeuw, P.J. and Stahel, W.A., 1986. Robust statistics. The approach based on influence functions. Wiley.

Macnae, J., 1992 Arbitrary waveforms in EM -P378. Report for CRCAMET, Macquarie University, Australia.

Munkholm, M.S. and Auken, E., 1995. Electromagnetic noise contamination on transient electromagnetic soundings in culturally disturbed environments. Submitted to Journal of Environmental and Engineering Geophysics.

Munkholm, M.S., Jacobsen, B.H., and Sørensen, K.I., 1994. Extraction of the transient electromagnetic earth response for colored non-Gaussian noise contamination. In Sibani, P. (ed) Proc. Interdisciplinary Inversion Workshop 3, Odense, Denmark, 31-37.

Nabighian, M.N. and Macnae, J., 1991. Time domain electromagnetic prospecting methods. In Nabighian, M.N. (ed) Electromagnetic Methods in Applied Geophysics, SEG, Vol. 2, Part A, pp 427-514.

O'leary, D.P., 1990. Robust regression computation using iteratively reweighted least squares. SIAM J. Matrix Anal. Appl., 11, 466-480.

Sørensen, K.I., 1995. Pulled Array Transient Electromagnetic Method, PA-TEM. Proc. of the SAGEEP, 899-904.

Stolz, N. and Macnae, J., 1992. The mathematical basis for representation of arbitrary waveforms in EM. In: Arbitrary waveforms in EM - P378. CRCAMET, Macquarie University, Australia. 32 pp.

Strack, K.M-, 1992. Exploration with deep transient electromagnetics. Elsevier, Amsterdam.

Sutarno, D. and Vozoff, K., 1989. Robust M-estimation of magnetotelluric impedance tensors. Exploration Geophysics, 20, 383-398.

Inverting Magnetotelluric Data Using Genetic Algorithms and Simulated Annealing

Oliver Bäumer

Institut für Geophysik und Meteorologie, Braunschweig, Germany

1 Introduction

The goal of the inversion of magnetotelluric (MT) data is to obtain an electrical conductivity structure as accurately and unambiguously as possible. However, because only a finite number of frequencies are observed, the data can be explained by a whole class of models. An even greater variation of the models may be allowed by considering the errors of the data, because fitting the data better than within their error ranges means also fitting the noise in the data. Thus there exists not only one model but a population of valid models, and therefore two topics are of special interest: The common attributes of all or at least a large number of models fitting the data and the properties of extremal models.

The inversion task is mostly done by using linearized methods to minimize the sums of squared residuals. These methods suffer from the need for an initial model from which they start their search. Because the character of the search is only local, they will find a minimum only in the neighbourhood of the initial guess. Therefore this solution may be suboptimal and is strongly dependent on the initial model.

The commonly used methods can also be computationally slow: Calculating gradients may be time consuming in cases with a large number of parameters and data.

So one looks for a method which will do the minimization in a numerically easy way and which avoids becoming trapped in suboptimal minima by applying a global search concept.

2 Nature as a Prototype

Search and optimization concepts can be found in nature in very distinct forms. Some of them are adopted to derive "evolutionary algorithms" which overcome the problems of traditional methods.

2.1 Simulated Annealing

The cooling of melt minimizes the energy of the resulting crystal system. But this crystal system depends on the cooling scheme: Slow cooling gives a monocrystal while fast cooling yields a state resembling glass; the latter resides in a suboptimal energy minimum. By considering the system energy function, Metropolis *et al.* (1953) constructed an algorithm based on the Monte Carlo method (MC).

Unlike MC, which chooses a model M randomly in the parameter space, compares it with a given reference model $M^{(\text{ref})}$ and replaces $M^{(\text{ref})}$ by M only if M is better than $M^{(\text{ref})}$, the Simulated Annealing algorithm (SA) computes M as a (small) change of $M^{(\text{ref})}$ and accepts M as new reference model not only if M is better than $M^{(\text{ref})}$, but with a certain probability p also if M is worse than $M^{(\text{ref})}$. The acceptance probability

$$
p(\Delta E, t) = \begin{cases} 1 & ; \Delta E \leq 0 \\ \exp\left(\frac{-\Delta E}{kT(t)}\right) & ; \Delta E > 0 \end{cases} \tag{1}
$$

depends on the energy difference $\Delta E = E - E^{(\text{ref})}$ and a pseudo temperature $T(t)$ (which decreases with iteration number or time t):

1. In the beginning even large deteriorations will be accepted with a high probability. This leads to a widely spread sampling of the parameter space.
2. Later on the algorithm becomes more conservative, preserving a found minimum. But the allowance to do some steps for the worse permits the discovery of a lower minimum.
3. Finally even small deteriorations will scarcely be accepted, and only a local search is performed.

2.2 Genetic Algorithms

Genetic Algorithms (GA), introduced by Holland (1975), imitate the optimization strategy of biology: All features of an individual – i. e. appearance, action and behaviour – are coded in chromosomes which are chains of two kinds of base pairs (adenine–thymine, AT, and cytosine–guanine, CG). These features determine the individual survival fitness in a given environment: By "survival of the fittest" a population of superior models arises.

The first thing to do is to imitate the coding of the model's features: This can simply be done by a single bit string (by replacing AT↦0 and CG↦1), which therefore is a haploid representation, or – to take the orientation of the base pairs into consideration – by a diploid bit chain; e. g. the chromosome

A–C–A–T–G
T–G–T–A–C

may be represented either haploid,

0–1–0–0–1 ,

or diploid,

$$0\text{--}1\text{--}0\text{--}1\text{--}0$$
$$0\text{--}0\text{--}0\text{--}1\text{--}1$$

Coded in this way, a randomly initialized population of individuals (all their bits are set with a probability of $\frac{1}{2}$, therefore the models are widely spread in the parameter space) can be manipulated by operators applied in biological propagation to get the next generation:

1. In the "reproduction" phase, based upon the distribution of the fitness values of the current generation, the number of expected offsprings for each individual in the population is calculated and a new population is created with respect to these values. So the number of superior individuals increases while poor models vanish.
2. The individuals are randomly combined to pairs. Then, depending on the type of individuals, the following operation is performed:
   - The "crossover" (for a haploid GA) exchanges for each pair the bits between two randomly chosen sites in the strings.
   - The "recombination" (for a diploid GA) performs for each pair of individuals and each pair of diploid bit strings representing one physical parameter (each of them consisting of two haploid bit strings) the following procedure: For each partner one of the two haploid bit strings is randomly selected and exchanged with the partner's chosen haploid bit string.

   So new regions of the parameter space are sampled without losing too much information about already scanned good areas.
3. The "mutation" changes all bits with only a small probability $\simeq 10^{-4}$ to prevent a premature convergence. Higher mutation probabilities would result in a greater loss of information and would therefore decrease the performance.

So far, the translations were totally independent on the problem. The only problem specific tasks to do are

1. Decoding the haploid or diploid bit strings to physical model parameters:
   This can be done by e. g. simply considering the bit string for each parameter as an integer number with which the model parameter is discretized within a given interval. To evaluate the diploid bit string, it is considered as a parallel arrangement of two haploid individuals and a weighted mean is calculated.
2. Definition of the fitness function which has to be maximized:
   In addition to a term containing the residual to be minimized this fitness may depend e. g. on extremal criteria.

Notice that there is no need to calculate derivatives of the fitness function – only the values themselves are required.

3 Application on the Inversion of MT Data

In MT, the Fourier components $\mathbf{E}(\omega)$ and $\mathbf{H}(\omega)$ of the electric and magnetic field are determined at the earth's surface. In the case of a layered half space (horizontal isotropic homogenous layers) it is sufficient to determine the transfer functions $C(z, \omega)$ between horizontal orthogonal components at $z = 0$, e. g.

$$C(\omega) = \frac{E_x(\omega)}{i\omega\mu_0 H_y(\omega)} \; . \tag{2}$$

3.1 The Parameter Coding

The model parameters – depth and resistivity of the layers – are discretized logarithmically equidistant in an interval whose boundaries are derived from the extremal values of $\varrho^\star$ and $z^\star$ (these values are obtained by interpreting each observed complex $C(\omega)$ individually by a simple model with two parameters).

3.2 The Fitness Function

The model fit to the data is described by a normalized residual

$$R = \sqrt{\frac{1}{N_f} \sum_{d=1}^{N_f} \left(\frac{\left| C_d^{(\mathrm{obs})} - C^{(\mathrm{mod})}(\omega_d) \right|}{\delta \left| C_d^{(\mathrm{obs})} \right|} \right)^2} \tag{3}$$

with $C^{(\mathrm{mod})}(\omega_d)$ as the calculated transfer function, $C_d^{(\mathrm{obs})}$ as observed transfer function at ω_d, $d = 1, \ldots, N_f$, and $\delta \left| C_d^{(\mathrm{obs})} \right|$ as an error of its module. So $R = 1$ if the model fits the data just within their error ranges.

Best Fitting Model. The fitness (which is to be maximized) is dependent only on R (which is to be minimized), so a bounded fitness function may be written as

$$f = \left(1 + \left(\frac{1}{q} - 1 \right) R \right)^{-1} \qquad \in (0, 1] \tag{4}$$

with $f = q$ in case $R = 1$ ($q \in (0, 1)$ defined by the user).

Extremal Models. Considered is the extremalization of the conductance (integrated conductivity) τ in a given depth interval.

At first, a fitness $f \geq q$ (explanation of data within their error ranges, $R \leq 1$) must be achieved. For those models not fulfilling this criterion an extremalization of τ is senseless. So the fitness calculation for each individual is performed as following:

1. A fitness $\tilde{f}$ is calculated using (4).

2. Then the final fitness value is determined depending on the value of $\tilde{f}$:
   - $\tilde{f} < q$: The data are not explained within their error ranges, so the value of τ is discarded: $f = \tilde{f}$.
   - $\tilde{f} \geq q$: Now the data are explained, but only an explanation just within their error ranges is desired, which gives a portion q to f. From the remainder $1 - q$ a portion will be added with respect to the fulfilling of the extremal criterion: $\varrho^{(\min)} \leq \varrho \leq \varrho^{(\max)}$ leads to $\tau^{(\min)} \leq \tau \leq \tau^{(\max)}$ ($\tau^{(\min)} = d/\varrho^{(\max)}$ with d as thickness of the considered depth interval), and so you get in the case of minimization of τ (with $\bar{\varrho} = \frac{d}{\tau}$)

$$f = q + (1 - q) \frac{\ln\big(\bar{\varrho}/\varrho^{(\min)}\big)}{\ln\big(\varrho^{(\max)}/\varrho^{(\min)}\big)} = \begin{cases} q & ; \bar{\varrho} = \varrho^{(\min)} \\ 1 & ; \bar{\varrho} = \varrho^{(\max)} \end{cases} \qquad (5)$$

(similar in case of maximization of τ).

3.3 Results

Comparison. The tested methods (GA and SA) are nearly equally efficient applied on problems with a small number of parameters (~ 5). With an increasing number of parameters a superiority of GA becomes apparent: Due to the parallel sampling of the parameter space using a population instead of using only one model per iteration, GA needs significant less forward model calculations than SA to find a model fulfilling the desired fitness criteria.

Best Fitting Model. Used is a data set determined by Larsen and already interpreted by Parker and Whaler (1981). The D^+ models of Parker (1980) – consisting of thin conducting layers embedded in an insulating half space – are known to be the best fitting one dimensional models. Therefore they allow the assessment of models found by SA and GA. The D^+ model which achieves a residual $R_{D+} \simeq 0.74$ is shown in Table 1.

A model with only 2 layers (including the half space) is sufficient to achieve a residual $R \simeq 0.9$. So the data are explained, but you are still away from the best fitting D^+ model.

By increasing the number of layers a better fit is achieved. With 12 layers (now the number of model parameters is greater than the number of observed data, so the model parameters are not independent) the residual $R \simeq 0.75$ comes near to the optimum R_{D+}. Three of the four conducting layers of the D^+ model can be well reproduced, see Table 1 and Fig. 1. The population has well converged in $z < 700\,\mathrm{km}$, showing a good resolution of the parameters in this depth range, while the missing convergence at greater depths hints at a poor significance of the parameters there.

Extremal Models. Considered is the minimization of τ in the depth interval $100\,\mathrm{km} \ldots 300\,\mathrm{km}$. The model shown in Fig. 2 is an obviously (nearly) extremal model, containing a poorly conducting layer with its lower boundary close to that of the considered depth interval (the upper boundary of this layer is deeper than $100\,\mathrm{km}$ because of the necessity of fitting the data).

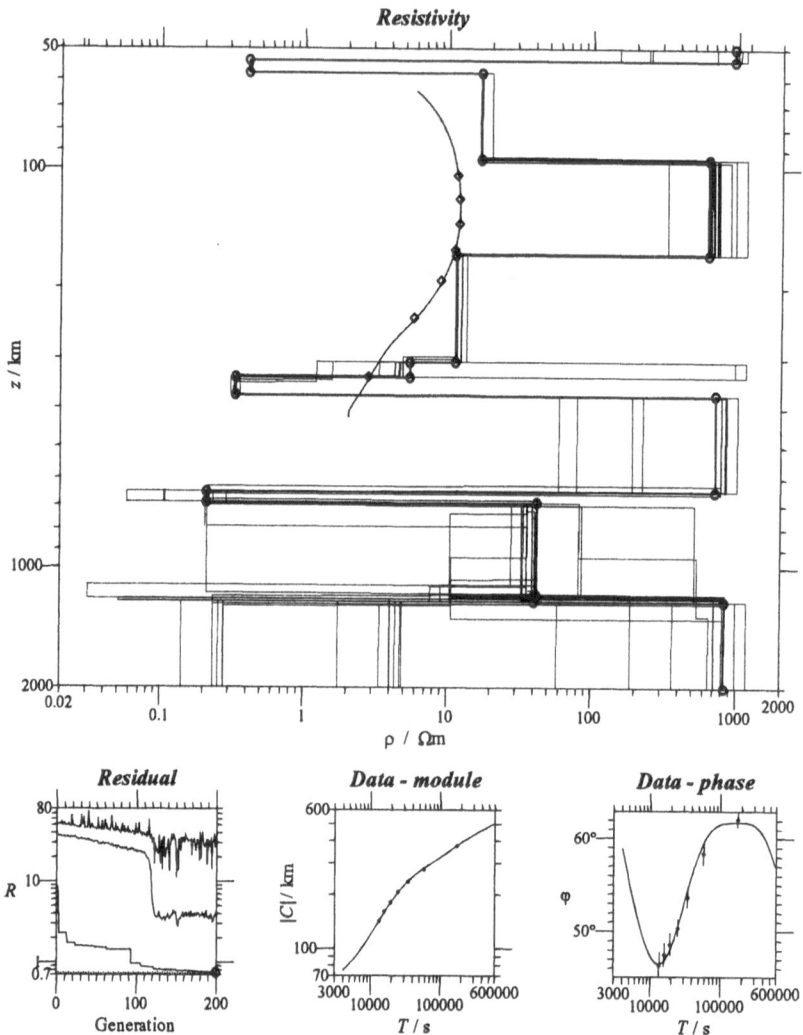

Fig. 1. Best fitting 12 layer model for Larsen's data set. At the top, the data are presented in terms of ϱ^*-z^* values (diamonds), and those 42 models of the final population of 200 models (with 9 bits per parameter) are shown, which achieve $f \geq q = 0.9$. For the best model (marked with circles), the $\varrho^*(z^*)$ curve is displayed. At the bottom left, the best (small R) and worst (large R) residual and the population's average residual for each generation are displayed. At the bottom center and right you can see the observed data with their 95 % error bars and the theoretical phases and moduli of the transfer function of the final best model.

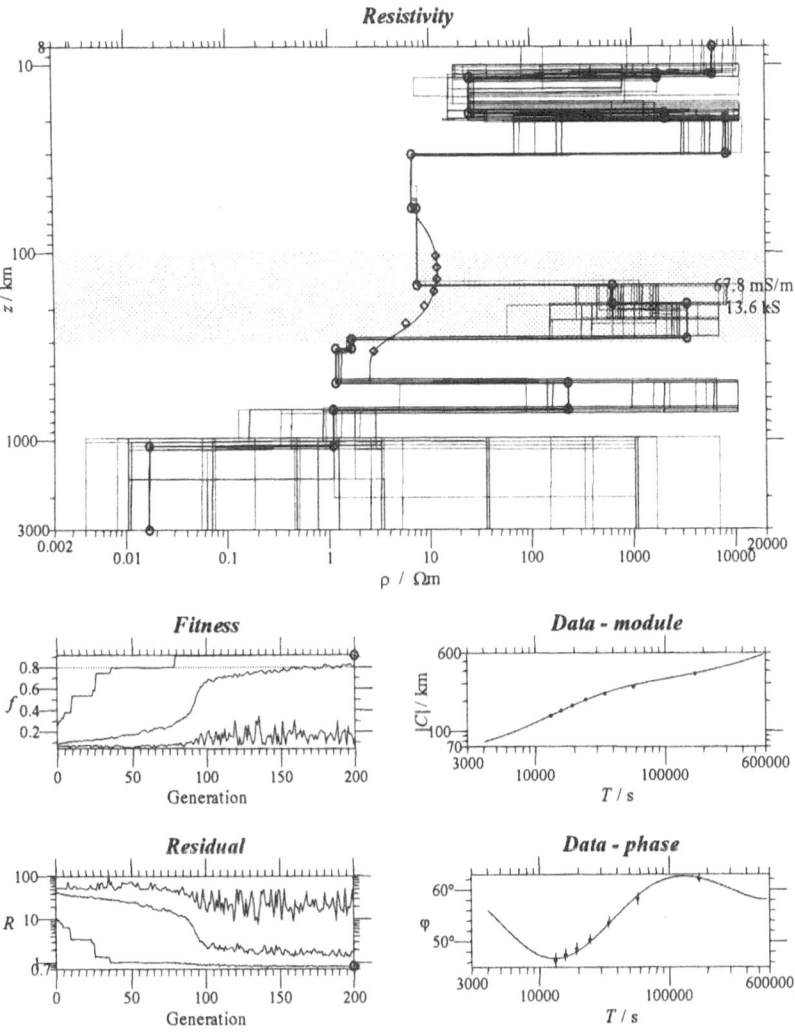

Fig. 2. Model minimizing τ (in depth interval $100\,km\ldots 300\,km$) for Larsen's data set. Same presentation as in Fig. 1. In addition, now also the fitness (best, average and worst value for each generation) is displayed, because now the fitness depends not only on R (the value of q is marked with a dashed line). Now 98 of the initial 200 models (with 8 bits per parameter) are finally valid.

Table 1. Larsen's data set: D^+ model (depth and integrated conductivity of well conducting sheets), the three best well conducting layers of a 12 layer model (upper and lower boundary and integrated conductivity of well conducting layers).

D^+ model		12 layer model		
z/km	τ/kS	z_{ub}/km	z_{lb}/km	τ/kS
0.0	0.5152			
64.42	13.33	53.7	57.6	9.7
318.6	85.65	333.8	369.8	107.2
559.7	159.65	646.9	689.0	198.5

4 Conclusions

Nonlinear methods like GA and SA, which sample the model space not only locally but widely spread, have an increased chance of finding the global optimum. Therefore the initial model has a smaller influence on the final solution. These algorithms also permit the consideration of constraints in an easy way. Manipulating a population of models with GA, a very efficient search is achieved due to the implicit parallelism in sampling the model space. The distribution of the final models in the parameter space also gives hints on the confidence intervals of the parameters.

Parker's D^+ model as the best fitting 1D model can be well reproduced. Extremal models can be calculated in an easy way.

The global character of the performed search leads to a decreased efficiency in the final iterations. There only a local search is demanded which can be better done by e. g. gradient methods. So the most efficient way is to start with a GA for calculating a population of nearly optimal models, which then are used as initial models for local search methods.

References

Goldberg, D. E. (1989). *Genetic Algorithms in Search, Optimization, and Machine Learning*. Addison Wesley.

Holland, J. H. (1975). *Adaption in Natural and Artificial Systems*. The University of Michigan Press.

Metropolis, N., Rosenbluth, A. W., Rosenbluth, M. N., Teller, A. H., and Teller, E. (1953). Equation of state calculations by fast computing machines. *J. Comp. Phys.*, **21**(6), 1087–1092.

Parker, R. L. (1980). The inverse problem of electromagnetic induction: Existence and construction of solutions based on incomplete data. *J. Geophys. Res.*, **85**, 4421–4428.

Parker, R. L. and Whaler, K. A. (1981). Numerical methods for establishing solutions to the inverse problem of electromagnetic induction. *J. Geophys. Res.*, **86**, 9574–9584.

Ionospheric Tomography

Manabu Kunitake

Communications Research Laboratory, Tokyo, Japan

1 Introduction

Computerized tomography techniques are being applied to reconstruct two-dimensional electron density distributions in a vertical plane of the ionosphere from the measurment of the line integral of the electron density. This application is called ionospheric tomography. The tomographic reconsruction system for the ionosphere was first proposed by Austen et al. (1986). The system uses a moving satellite as a transmitter and a chain of several ground stations as receivers. Total electron content (TEC), which is the line integral of the electron density, can be measured by an inexpensive receiving equipments.

Before they proposed ionospheric tomography, TEC observations made by using radio beacon waves from moving satellites were used to study the latitudinal structure of electron distributions. However, this method is not sufficient since the earth's ionosphere has not only latitudinal but also vertical structures. Two-dimensional density distribution information is not only important for ionosphere physics but also for satellite-ground communication and navigation by satellite because the electron distribution affects the propagation conditions for radio signals in the ionosphere. The technique to obtain two-dimensional density distribution by utilizing the receiving equipments had been desired.

After their proposal, various algorithms for ionospheric tomography were proposed and their feasibilities were examined by numerical simulations (for a review of algorithms see, Raymund 1994). The incompleteness of observations owing to the geometrical constraints of the data acquisition system, such as limited viewing angles, makes tomographic reconstruction an ill-posed problem. The limitations and the resolution degradations originating from the geometrical constraints were investigated in detail (Yeh and Raymund 1991; Na and Lee 1994; Raymund et al. 1994b).

The observations for ionospheric tomography were actually carried out(e.g. Andreeva et al. 1990, 1992; Kunitsyn and Tereschenko 1992; Pryse and Kersley 1992; Kersley et al. 1993; Pryse et al. 1993; Raymund et al. 1993; Bust et al. 1994; Foster et al. 1994; Kersley and Pryse 1994; Kunitsyn et al. 1994; Kronschnabel et al. 1995; Kunitake et al. 1995; Pakula et al. 1995). Many chains of receivers

were and have been operated for ionospheric tomography at different longitudes.
The electron distributions reconstructed from the observations showed many
interesting structures of the ionosphere.

In this paper, we explain the experimental method and basic formulation
of ionospheric tomography, and briefly review inversion algorithms focusing on
non-iterative algorithms, which have been developed recently. Finally, recent
results by Kunitake et al. (1995) are presented as a concrete example of actual
applications.

2 Experimental Method

Ionospheric tomography is an inversion problem in which an unknown electron
density distribution is reconstructed from a known ray path distribution and a
set of the line integrated electron contents along the ray path (Fig. 1). The system
consists of radio transmissions from a polar orbiting satellite and several receiving stations aligned latitudinally on the ground. The American NNSS satellite
system or the Russian CICADA satellite system are used for ionospheric tomography. The satellite emits two coherent radio waves (150 MHz and 400 MHz) at
an altitude of about 1100 km. When the two waves pass through the ionosphere
and reach the ground, the waves have phase shifts originating from the ionospheric electron density. This is caused by the refractive index in the ionosphere
being slightly different than one. The line integral of the electron density can be
obtained from the difference between phase shifts in the two waves (Leitinger
et al. 1975). The line integral is called the total electron content (TEC), and is
given by

$$TEC = \int_p N(x, y) ds \qquad (1)$$

where $N(x, y)$ expresses the electron density distribution and p specifies a ray
path between the satellite and receiver. At 150 MHz or 400 MHz, each ray path
is approximated by a straight line. Because the satellite systems broadcast accurate information about satellite location, we can know the ray path distribution
precisely. The electron density distribution is assumed to be stable during the
satellite passing (about 20 minutes).

3 Basic Formulation and Algorithm for Reconstruction

Generally, linear inversion algorithms are used for ionospheric tomography. The
basic equation is:

$$\mathbf{A}x = b . \qquad (2)$$

The whole set of TEC measurements is expressed by a column vector b with
m elements. The $\mathbf{A}$, which is an $m \times n$ matrix, is a mapping operator between
unknown electron density distribution x and known b.

In many cases, the two-dimensional density distribution is discretized into a
column vector x with n elements (e.g. Fig. 1). The imaged region is divided into

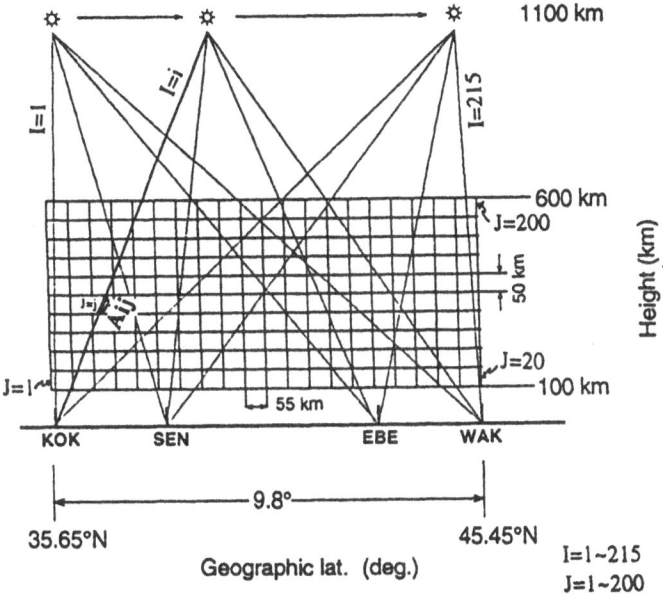

Fig. 1. The observation system and reconstruction geometry for ionospheric tomography. This is the case in Kunitake et al. (1995)

small areas called pixels. Within each pixel, the electron density is assumed to be constant. Matrix **A** is determined by the reconstruction geometry that results from the ray path distribution and pixel division. In these cases, the element A_{ij} corresponds to the length of the path-pixel intersection for the ith ray path in the jth pixel.

In other cases, the two-dimensional density distribution is assumed to be approximated by a column vector x which is a set of basis functions (e.g. Fourier functions) or model realizations.

In ionospheric tomography, geometorical limitations (no horizontal ray path, incomplete viewing angles, a limited number of receiving stations etc.) of the observation system make the available data insufficient for ideal reconstruction (Na and Lee 1994; Raymund et al. 1994b). The reconstruction of ionospheric tomography is in the category of an ill-posed inversion problem. Thus a priori information is incorporated into inversion algorithms to compensate for the incomplete observations.

Formally, iterative algorithms such as the algebraic reconstruction technique (ART) (Austen et al. 1988), the simultaneous iterative reconstruction technique (SIRT) (e.g. Pryse and Kersley 1992; Afraimovitch et al. 1992), and the multiplicative algebraic reconstruction technique (MART) (Raymund et al. 1990) were applied to solve (2). In any iterative algorithm, an initial guess is used, and generally, the estimated image depends on the initial guess.

Recently, non-iterative algorithms have been developed. In ill-posed inversion problems, small perturbations (error or noise) in b in equation (2) tend to lead to extremely large perturbations in the solution x. To reduce the perturbations in the solution, regularization techniques have been used in the algorithms. Fremouw et al. (1992, 1994) applied the "weighted, damped, least squares" technique of stochastic inversion on the assumption that the distribution is composed of empirical orthogonal functions for vertical variations and Fourier components (with red power low) for horizontal variations. Lehtinen et al. (1994) applied stochastic inversion with generarized Tikohnov regurarization. Fehmer (1994) applied the technique of non-iterative optimization with quadratic constraint. Fougere (1995) applied the maximum entropy method. Raymund et al. (1994a) proposed another algorithm. A model vector x_0 with n elements was used in this Raymunds' algorithm:

$$x_0 = Sm + w \tag{3}$$

where S is an $n \times l$ matrix, whose columns are stacked realizations of the parameterized ionospheric model (Daniell 1991), the vector m is an n length vector of coefficients, and w is an additive deviation. From all x minimizing $\|Ax - b\|_2$, Raymund et al. searched x minimizing $\min_m \|Sm - x\|_2$. Modified truncated singular value decomposition (MTSVD) (Hansen et al. 1992; Hansen 1993) was applied to ionospheric tomography by Kunitake et al. (1995). The MTSVD technique looks for the solution of the following problem:

$$\min_{x \in S} \|Lx\|_2, \quad S = \{x \mid \|A_k x - b\|_2 = \min\} \tag{4}$$

where A_k is the truncated singular value decomposition (TSVD) of A and $k < n$. A_k is obtained by truncating the singular value decomposition expansion of A "at k", in other words, by discarding $n - k$ smallest singular values of A. The k is the parmeter for the truncation. TSVD is effective in filtering out the contributions to the solution corresponding to the $n - k$ smallest singular values. In (4), L is a $p \times n$ matrix and $p \leq n$. Introducing the minimization of $\|Lx\|_2$ realizes the incorporation of a priori conditions such as smoothness and nonnegativeness. When smoothness is incorporated, L is a matrix expression of the derivative operators.

4 Application

Kunitake et al. (1995) applied MTSVD to ionospheric tomography. First of all, numerical simulations of tomographic reconstruction were carried out using several model distributions of the electron density in the vertical plane. The reconstruction geometry is shown in Fig. 1. The reconstruction region was from $35.65°N$ to $45.45°N$ in geographic latitude and from 100 km to 600 km in altitude. The total number of pixels was 200 (20×10), and the total number of ray paths was 215. In the simulation, TECs generated from model distributions were used as b. In applying the MTSVD to tomography, key points for good reconstructions are "what kind of a priori conditions are incorporated?" and "how many singular values are discarded?" (or "what is the optimum k-value?"). Two a priori conditions were incorporated in **L**. The first condition was that the whole distribution is smooth. The second condition was that the density values in the top row and the bottom row of pixels are zero. A comparison between the model distribution and reconstructed distributions with various k values gives us information about the optimal k. Figure 2. a shows a model distribution. Figure 2. b shows a reconstructed distribution for a particular k. An appreciable discretization error should arise due to the small total number of pixels in their simulation. The solution for a distribution with a larger k value had larger perturbations owing to the discretization error. On the other hand, the solutions for small k values did not reconstruct sharp structures. The optimal range of k is determined from a compromise between these two situations.

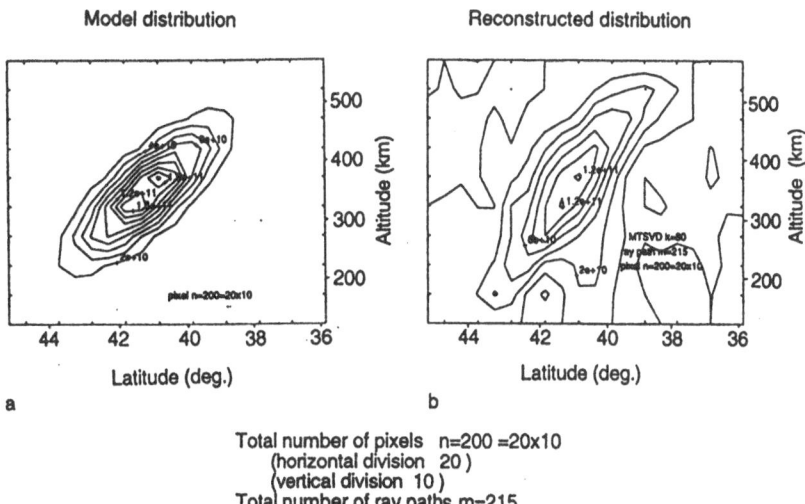

Total number of pixels   n=200 =20x10
(horizontal division 20)
(vertical division 10)
Total number of ray paths m=215

Fig. 2. a Model input of electron density distribution, **b** Reconstructed distribution from TEC generated from the model distribution. $k = 80$ was applied

The MTSVD with the optimal k determined above was then applied to the reconstruction from the TEC observed at four receiving stations, Wakkanai (WAK), Ebetsu (EBE), Sendai (SEN), and Kokubunji (KOK), aligned near the 140 °E meridian. The reconstruction geometry **A** (Fig. 1) and a priori conditions **L** were the same as those used in the simulations. An optimal value of k was determined to be 100 based on careful examination of the results of the simulation. Figure 3 shows the electron density distribution reconstructed from TEC observations around 1955 UT (LT=UT+9h) on 2 May 1994. There are two peaks and one valley between them. One subpeak can be seen at 39.55 °N at a height of 200 km. The maximum electron densities observed by ionosonde (see pages 89–98 of Davies (1990) for the technique) at Wakkanai and Kokubunji agree with the reconstruction.

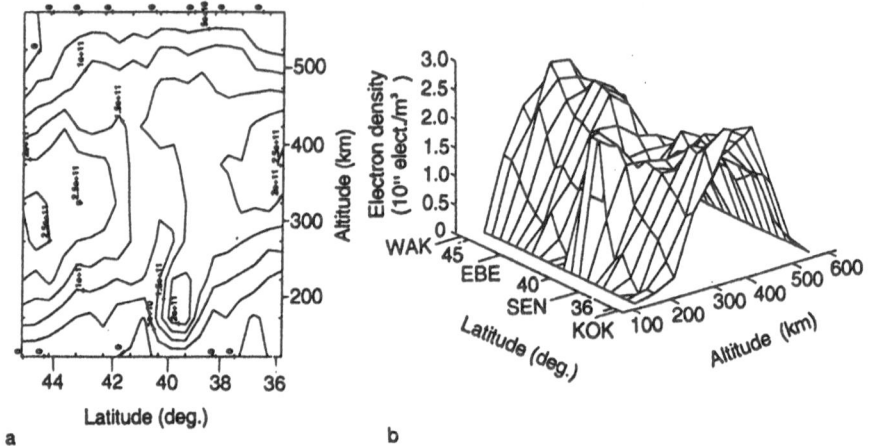

Fig. 3. Reconstructed image of ionospheric electron density from TEC measured at four stations in Japan around 1955 UT on 2 May 1994: a contour plot, b three-dimensional plot

5 Summary

In this paper, it is shown how computerized tomography technique has been applied to reconstruct two-dimensional electron density distribution of the ionosphere. The geometrical constraints of ionospheric tomography make the inversion an ill-posed problem. To compensate for incomplete measurements, incorporation of a priori information is necessary. Further, regularization techniques are effective. Reconstructions from actual TEC observations proved the usefulness of algorithms. In many cases, the reconstructed distributions were compared with other observations by using ionosondes or an incoherent scatter radar (see pages 106–110 of Davies (1990) for the technique). The reconstructions agreed reasonably with other observations.

The observation system to acquire data for ionospheric tomography is much less expensive than other observation systems. Moreover, the receivers are transportable. Thus many observation chains for tomography were established and are planning to be established at different longitudes. Cooperation in analyzing data obtained from observation chains which were (or will be) simultaneously operated at different longitudes would provide new information on the large-scale structures and dynamics of the ionosphere.

References

Afraimovitch,E., Pirog,O. and Terekhov,A., 1992. Diagnostics of large-scale structures of the high-latitude ionosphere based on tomographic treatment of navigation-satellite signals and of data from ionospheric stations. J. Atmos. Terr. Phys, 54, 1265–1273.

Andreeva,E., Galinov,A., Kunitsyn,V., Mel'nichenko,Y., Tereshchenko,E., Filimonov,M. and Chernyakov,S., 1990. Radiotomographic reconstruction of ionization dip in the plasma near the Earth. J. Exp. Theor. Phys. Lett., 52, 145–148.

Andreeva,E., Kunitsyn,V. and Tereshchenko,E., 1992. Phase-difference radiotomography of the ionosphere. Ann. Geophysicae, 10, 849–855.

Austen,J., Franke,S. and Liu,C., 1988. Ionospheric imaging using computerized tomography. Radio Sci., 23, 299–307.

Austen,J., Franke,S., Liu,C. and Yeh,K., 1986. Applications of computarized tomography techniques to ionospheric research. In: A.Tauriainen (ed.), Radio beacon contribution to the study of ionization and dynamics of the ionosphere and corrections to geodesy. Oulensis Universitas, Part 1, 25–35.

Bust,G., Cook,J., Kronschnabel,G., Vasicek,J. and Ward,S., 1994. Application of ionospheric tomographyw to single-site location range estimation. Int. J. Imag. Sys. Tech., 5, 160–168.

Daniell, R., 1991. Parametarized real-time ionospheric specification model: PRISM version 1.0. Tech. Rep. PL-TR-91-2299, Phillips Lab., Hanscom AFB, Mass. U.S.A.

Davies, K., 1990. *Ionospheric Radio*, Peter Peregrinus Ltd., London, United Kingdom.

Fehmers, G., 1994. A new algorithm for ionospheric tomography. In: L.Kersley (ed.), Proc. International Beacon Satellite Symposium, University of Wales, 52–55.

Foster,J., Buonsanto,M., Holt,J., Klobuchar,J., Fougere,P., Pakula,W., Raymund,T., Kunitsyn,V., Andreeva,E., Tereshchenko,E. and Khudukon,B., 1994. Russian-American tomography experiment. Int. J. Imag. Sys. Tech., 5, 148–159.

Fougere, P., 1995. Ionospheric radio tomography using maximum entropy 1. Theory and simulation studies. Radio Sci., 30, 429–444.

Fremouw,E., Secan,J., Bussey,R. and Howe,B., 1994. A status report on applying discrete inverse theory to ionospheric tomography. Int. J. Imag. Sys. Tech., 5, 97–105.

Fremouw,E., Secan,J. and Howe,B., 1992. Application of stochastic inverse theory to ionospheric tomography. Radio Sci., 27, 721–732.

Hansen, P., 1993. Regularization Tools. Danish Computing Center for Research and Education, Lyngby, Denmark.

Hansen,P., Sekii,T. and Shibahashi,H., 1992. The modified truncated SVD method for regularization in general form. SIAM J. Sci. Stat. Comput., 13, 1142–1150.

Kersley,L., Heaton,J., Pryse,S. and Raymund,T., 1993. Experimental ionospheric tomography with ionosonde input and EISCAT verification. Ann. Geophysicae, 11, 1064–1074.

Kersley, L. and Pryse, S., 1994. Development of experimental ionospheric tomography. Int. J. Imag. Sys. Tech., 5, 141–147.

Kronschnabel,G., Bust,G., Cook,J. and Vasicek,J., 1995. Mid-America Computerized ionospheric tomography experimennt (MACE '93). Radio Sci., 30, 105–108.

Kunitake,M., Ohtaka,K., Maruyama,T., Tokumaru,M., Morioka,A. and Watanabe,S., 1995. Tomographic imaging of the ionosphere over Japan by the modified truncated SVD method. Ann. Geophysicae, (accepted).

Kunitsyn,V., Andreeva,E., Tereshchenko,E., Khudukon,B. and Nygrén,T., 1994. Investigations of the ionosphere by satelite radiotomography. Int. J. Imag. Sys. Tech., 5, 112–127.

Kunitsyn, V. and Tereshchenko, E., 1992. Radio tomography of the ionosphere. IEEE Antennas Propag. Mag., 34, 22–32.

Lehtinen,M., Markkanen,M., Henelius,P., Vilenius,E., Nygrén,T., Tereshchenko,E. and Khudukon,B., 1994. Bayesian approch to satellite radio tomography. In: L.Kersley (ed.), Proc. International Beacon Satellite Symposium, University of Wales, 80–83.

Leitinger,R., Schmit,G. and Tauriainen,A., 1975. An evaluation method combining the differential Doppler measurements from two stations that enables the calculation of the electron content of the ionosphere. J. Geophysics, 41, 201–213.

Na, H. and Lee, H., 1994. Resolution degradation parameters of ionospheric tomography. Radio Sci., 29, 115–125.

Pakula,W., Fougere,P., Klobuchar,J., Kuenzler,H., Buonsanto,M., Roth,J., Foster,J. and Sheehan,R., 1995. Tomographic reconstruction of the ionosphere over North America with comparisons to ground-based radar. Radio Sci., 30, 89–103.

Pryse, S. and Kersley, L., 1992. A preliminary experimental test of ionospheric tomography. J. Atmos. Terr. Phys., 54, 1007–1012.

Pryse,S., Kersley,L., Rice,D., Russell,C. and Walker,I., 1993. Tomographic imaging of the ionospheric mid-latitude trough. Ann. Geophysicae, 11, 144–149.

Raymund, T., 1994. Ionospheric tomography algorithms. Int. J. Imag. Sys. Tech., 5, 75–85.

Raymund,T., Austen,J., Franke,S., Liu,C., Klobuchar,J. and Stalker,J., 1990. Application of computerized tomography to the investigation of ionospheric structures. Radio Sci., 25, 771–789.

Raymund,T., Bresler,Y., Anderson,D. and Daniell,R., 1994a. Model-assisted ionospheric tomography: A new algorithm. Radio Sci., 29, 1493–1512.

Raymund,T., Franke,S. and Yeh,K., 1994b. Ionospheric tomography: its limitations and reconstruction methods. J. Atmos. Terr. Phys., 56, 637–657.

Raymund,T., Pryse,S., Kersley,L. and Heaton,J., 1993. Tomographic reconstruction of ionospheric electron density with European incoherent scatter radar verification. Radio Sci., 28, 811–817.

Yeh, K. and Raymund, T., 1991. Limitations of ionospheric imaging by tomography. Radio Sci., 26, 1361–1380.

Application of Inversion to Global Ocean Tide Mapping
- preliminary results constrained by observations in the Baltic Sea

Ole B. Andersen

Kort & Matrikelstyrelsen, Copenhagen, Denmark

1. Introduction

Ocean tide modeling has improved vastly during the last two years as a direct result of the launch of the French-American TOPEX/POSEIDON (T/P) satellite, (e.g. Andersen 1994, 1995a,b; Eanes 1994; Egbert et al. 1994; Le Provost et al. 1994; Ray et al. 1994). All of the new ocean tide models are very similar and accurate in the deep ocean. However, Andersen et al. (1995a) pointed out, that major differences still exist along the shelf regions of the oceans. On these shelves the wave velocity reduces the horizontal scale of tidal features, and increases the vertical amplitude of the tidal signal dramatically. Therefore measurements from the T/P are inadequate in order to describe the tidal field and other sources of information, like the laws of physics, must be used in order to make an accurate description of the ocean tides.

The tides are constrained by two different types of information. The first is that obtained from the laws of physics, and the second is data collected from tide gauges and satellite altimetry. Prior to the launch of the T/P satellite, data was limited to sea surface height measurements from coastal tide gauges as well as a small number of pelagic tide gauge readings. However, with the launch of the T/P satellite the first global set of homogenous high quality sea surface height observations has become available.

The code has been derived at Oregon State University by Gary Egbert, Andrew Bennett, Rodney James and Mike Foreman. Evidently large part of the theory follows closely that by Bennett (1992) and Egbert et al. (1994). However, there are substantial differences in the model setup and in the use of data. Therefore, the theory will be briefly reviewed in the context of inversion.

2. Hydrodynamic equations, model setup and observations.

The ocean tides arise from the influence on the ocean from the sun and the moon. The ocean tide field at a position (ϕ,λ) is in the following denoted $u(\phi,\lambda)$. The ocean tide field is a complex vector field consisting of a number of constituents $u_i(\phi,\lambda)$ like

$$u(\phi,\lambda) = \left| u_1(\phi,\lambda),\ldots,u_n(\phi,\lambda) \right| \ with \ u_i(\phi,\lambda) = (u_x(\phi,\lambda),u_y(\phi,\lambda),h(\phi,\lambda)) \qquad (1)$$

where $u_x(\phi,\lambda)$ and $u_y(\phi,\lambda)$ is the transport component in the north and east directions respectively and $h(\phi,\lambda)$ is the elevation. All individual constituent fields making up the ocean tide field $u(\phi,\lambda)$ are complex vector fields - this way 3n complex values are required to completely describe the ocean tide field at each location, with the total ocean tide field arising from n constituents. Hence, solving for the major four constituents (M_2, S_2, K_1, O_1) would give a total of 12 complex parameters to be described at each location (ϕ,λ).

To a good approximation the tidal field, at a frequency ω, satisfies the linearized shallow water equations, with parameterised corrections for the effects of ocean self attraction and tidal loading (Hendershott, 1988). Furthermore, the tidal field will be subject to boundary condition on the boundary ∂O of the ocean domain like

$$Su = f_o \ \wedge \ \left| \begin{matrix} u \cdot n = 0 \ on \ \partial O_{coast} \\ h = h_o \ on \ \partial O_{open} \end{matrix} \right. \qquad (2)$$

where O is the full domain represented by the global oceans, f_o is the astronomical forcing. The rigid boundary condition is no flow normal to the coast.
Along the open boundary the condition is specified through elevation h_o at each grid cell using an existing ocean tide model.

The Laplace tidal operator S for linearized shallow water equations is given by

$$S = \left| \begin{matrix} i\omega+K & -f & gHG*\nabla_x \\ f & i\omega+K & gHG*\nabla_y \\ \nabla_x\cdot & \nabla_y\cdot & i\omega \end{matrix} \right| \qquad (3)$$

where H is ocean depth, f is the Coriolis acceleration, ω is the constituent frequency, $\nabla\cdot$ and ∇ represents the two dimensional divergence and gradient operators on a spherical earth, K represents dissipation and $G*$ represents a convolution with the Green's function for tidal loading and ocean shelf attraction. Details can be found in Andersen et al. (1995b)

The information about the ocean tide field from the hydrodynamic equations and from the observations can be represented like,

$$\left| \begin{matrix} S \\ L \end{matrix} \right| u = \left| \begin{matrix} f_o \\ d \end{matrix} \right| \qquad (4)$$

Here the knowledge about the hydrodynamics is included through the equation $Su = f_o$. The information from observations is represented by the equation $d = Lu$. Here d is the data vector and L is a measurement functional, equivalent to a height measurement like $d = L(u)+\varepsilon$, where ε is an additive measurement noise. For the present model, measurements were harmonic constants observed from tide gauge sites and harmonic constants derived from measurements at T/P crossover locations. Details on the processing of data can be found in Andersen et al. (1995b)

3. Global ocean tide inversion.

The inverse approach attempts a minimization of a quadratic penalty functional $P[d,u]$ or equivalently a cost function like

$$P[d,u] = (d-L u)^t C_e^{-1} (d-L u) + (\int\int [Su - f_0]^t C_f^{-1} [Su - f_0] \, d^2x \tag{5}$$

here C_e is data error covariance matrices and C_f and C_e are the covariances of the forcing error and the observations respectively. The use of a penalty functional enables a normalization of the two sources of errors. However, it also enables a tradeoff between fitting the dynamics and fitting the data, as the observations and the dynamics can be weighed through the choices of C_f and C_e.

Following Bennett (1992) it can be shown, that the weighted least-squares fit to data and dynamics found by minimizing P, is essentially the minimum variance best linear unbiased estimator (BLUE estimator) of the tidal field (Jackson 1973; Jacobsen 1982), which is also known as the Gauss-Markov smoother. If the dynamical and data errors are furthermore assumed to have a normal distribution, then the estimate is also the maximum-likelihood estimate.

In order to solve this generalized inverse problem we minimize the penalty functional or cost function P. In theory minimization of P is straightforward, and can be treated as a large linear least square problem. However if the ocean tides are to be represented with reasonable resolution, the number of unknown parameters becomes prohibitively large. For instance, our global test model has a resolution of 0.3 degrees which is equivalently to a global array with [1024 x 512] grid points. Hence, solving for one constituent, would require around one million parameters.

We will therefore take the Hilbert space approach derived by Bennett (1992) for oceanography. This approach uses a Hilbert space approach similar to what is used extensively in solid earth geophysics (Parker et al. 1987). The derivation is based on generalized inversion using the Euler Lagrange approach and details can be found in Bennett (1992) section 5.3 - 5.4.

Initially we define the full space of possible tidal states by $\mathfrak{S}$. Hereby any tidal field u (ϕ,λ) is in this space $\mathfrak{S}$. The measurement functional or "point elevation" L of the tidal field is therefore a functional acting on the elements in this state space. The penalty functional is a positive definite quadratic form. It can therefore be used to define an inner product on the state space like

$$< u_1, u_2 > = \int_0 [Su_1](x)^* [C_f^{-1} Su_2](x) \, d^2x \tag{6}$$

which defines a norm on the state space $\mathfrak{S}$. The subset for which $<u,u> < \infty$

corresponds to a weak assumption on C_f (Moritz, 1980). Hence, we have a Hilbert space. For each (reasonable) functional in this Hilbert space there exists an element r_i in the state space $\mathfrak{S}$ so that $L_i |u| = <r_i, u>$.

The element r_i is called the Riez representer of the measurement functional L. Following Bennett (1992), P attains its minimum for a unique element in this state space. This element has the form

$$u = u_0 + \sum_{k=1}^{K} b_k r_k \tag{7}$$

where u is the "complete" description of the tidal state best fitting dynamic and data, b_k is the coefficient of the representer r_k, K is the total number of representers, and u_0 is the model satisfying the Laplace tidal equation subjected to rigid boundary conditions.

The coefficients b for the representers satisfies the $K \times K$ system of linear equations

$$(R+C_\varepsilon) b = d - L[u_0] \wedge R_{ij} = <r_i, r_j> = L_i[r_j] \tag{8}$$

where R is the representer matrix which has elements derived from interpolation in the representer fields r_k. This approach massively reduces the number of unknowns to its real dimension equivalently to one parameter set (3n complex parameters) for each (independent) observation. The inversion of the representer matrix is now possible within the limitations of many computers. However, the calculations involved in this approach are still quite formidable, because we have to compute, the representer fields, the coefficients of these representer fields (equation 8), and finally the updated solution (equation 7).

Representers can be computed by solving a set of equations for the forward and adjoint model as described in Egbert et al. (1994) or Andersen et al. (1995b). One representer field is required for each (independent) observation. The set of equations is solved by running a full hydrodynamic model of array size [1024x512] backwards and forwards using 10.000 time steps in an Arakawa-C grid. These calculations are by all means the most computational intensive of the project. However, the code for deriving representers is highly parallel and therefore extremely well suited for massive parallel computers. It has been run on the Edinburgh CM-200 connection machine at Edinburgh Parallel Computing Center, However, the calculation of ONE representer field for the major four constituents still takes around 15

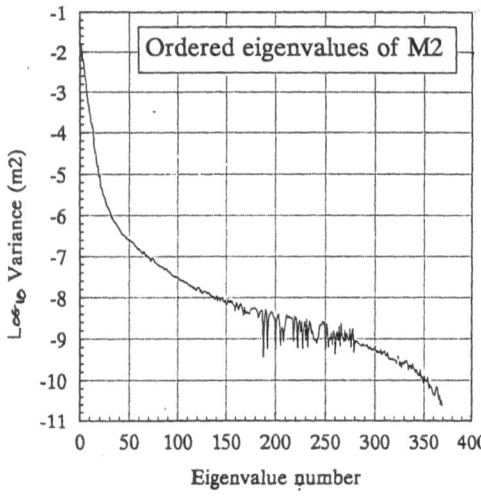

Figure 1 Ordered eigenvalues of the representer matrix for the M_2 test solution (400 elements).

minutes to compute using all 16.000 available processors. Hence, our test model consisting of 400 representers took around 100 hundred CPU hours to compute. Recently the code has been implemented on the CRAY T3D. The DRAY T3D is presently equipped with around 20Gb of memory. Hence, it can fit models of even higher resolution, which will be needed in the future.

Generally the representers are very broadly peaked with a maximum in the vicinity of the location of the applied forcing depending on the local dynamical conditions. It is also evident that the representers have relatively large amplitude throughout the basin where it is located. The representer fields are actually triplets with two transport components in addition to the elevation component.

The representer can be understood in both a physical and a mathematical sense. The physics of the representer fields is relatively hard to decipher, but the representer field indicates, how the entire hydrodynamic ocean tide model reacts to applied forcing at the site of observation.

In a mathematical sense, the representers can be viewed as covariance functions for the dynamical model. This is so, because the evaluation potential is bounded. Hence Moritz (1989) showed that the Hilbert space actually has a reproducing kernel (Tscherning 1986). If the Hilbert space has a reproducing kernel $K(x,y)$ then the representer for the functional L has a very simple form

$$r(x) = L^y K(x,y) \qquad (9)$$

where L^y denote that L acts on the reproducing kernel as a function of y. This reproducing kernel might be interpreted as covariance function for the dynamical residuals. The representer will have its maximum at or near to the location of the representer depending on the local dynamic conditions. Similarly the representer matrix R may then be viewed as the covariance of $L[u-u_0] = d-L[u_0]$ $(=Rb)$ in the absence of measurement errors. Hereby R contains information on how the dynamic errors affect the measurement deviations from our prior model u_0.

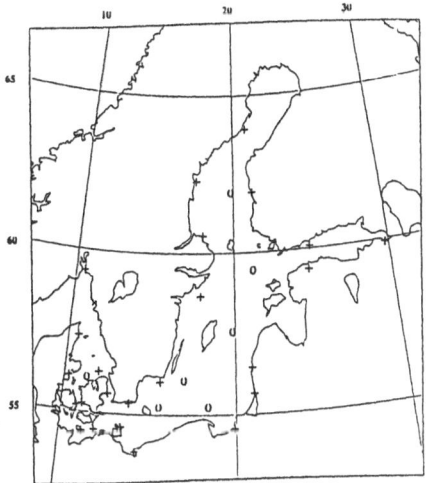

Figure 2 Position of representers. (+) tide gauge sites, (o) T/P crossover locations.

The second task is the estimation of the coefficients (b_k) of the representers. This task might also involve heavy computations (Egbert et al. 1994) if the number of unknowns is very large (> 10.000).

However, by constraining the solution to the Baltic Sea for this initial test, we could easily limit the number of unknown to around 100. The inversion could therefore be treated using standard inversion software as implemented in the GRAVSOFT software package (Tscherning et al. 1994).

In order to reduce the number of unknowns, it is often very informative to have a look at the singular value decomposition (SVD) of the representer matrix. Computing the SVD can be viewed as a rotation of the data vectors and the measurement functionals. Subsequently the inverse solution may be expressed as

$$u(x) = u_0(x) + \sum_{k=1}^{K} \frac{d_k^*}{\lambda_k + \sigma^2} r_k^*(x) \tag{10}$$

where r* and d* are the rotated functionals, and λ and σ^2 are the eigenvalues of the representer matrix and the variance of the measurement errors respectively. It is obvious that if $\lambda \ll \sigma^2$ then the eigenvalues are much smaller than the measurement noise and this datum can safely be ignored.

Practically this means, that the summation can be truncated at a level K without major loss of accuracy. The solution estimated from the truncated set of eigenvalues is often referred to as the regularised least square solution of the problem. For a test model involving representers in all of the Northeast Atlantic, the ordered eigenvalue spectra can be seen in Figure 1. From the estimated coefficients the improved solution was finally obtained through a direct summation over the representer fields using the coefficients b_k, by directly forming the summation as in equation 7 or equation 10.

4. Ocean tide model setup and initial results

The global model was constrained to observations from tide gauges and T/P altimetry in the Baltic Sea using a global regular grid with a size of 1024 x 512 elements. This corresponds to a resolution of 0.35° x 0.3° which in turn corresponds to 17 km x 30 km. Antarctica was used as southern boundary whereas the northern boundary was chosen to be the 82°N parallel. Along the northern boundary the ocean tide model by Kowalik & Proshutinsky (1993) was used as boundary conditions. The Baltic Sea M_2 model was, due to future considerations, derived as a global model constrained to thirty available representers in the Baltic Sea as shown in Figure 2. Twentythree of these representers were estimated at tide gauge locations, whereas the remaining seven were calculated at T/P crossover locations.

The M_2 constituent of the test model in the Baltic Sea can be found in Andersen, 1995b. The general assumption about the Baltic Sea is, that the ocean tide signal can be ignored, as the amplitude is less than 10 cm. However, with the accuracy of satellites like T/P, such a signal can be observed and modeled, and hence, taken into account. Comparisons to an existing model in press by Kantha (1995), showed nice agreement. However it became evident, that this model does not have high enough resolution in coastal areas like this, and that we must consider increasing the resolution in future versions.

5. Summary - future work

The preliminary ocean tide test model, constrained to thirty representers the Baltic Sea, presented in this report demonstrates the high potential of the global inversion method for the study of ocean tides in particular, and ocean science in general. Furthermore, the analyses of the eigenvalue spectrum indicated that we can easily truncate the solutions. Hence, it is possible to run much larger solutions based on several thousands of representers in the future (e.g. north Atlantic ocean, global ocean). On the other hand it is obvious that the resolution is critical for further improvement, which will put even higher demands on computer resources.

However, there is no doubt that this approach, which combines hydrodynamic and data from tide gauges and satellite observations, seems to provide the basis for further improving the accuracy of the ocean tide models both in the global ocean but also in the Northwest European shelf region.

Acknowledgment

Most of this work was done during a seven month stay at the Proudman Oceanographic Laboratory in Bidston UK. The author wishes to thank P. L. Woodworth, M. Ashworth and S. Wilkes as well as staff and colleagues there. Furthermore the author is thankful to G. Egbert at Oregon State University and his colleagues for providing the software for global ocean tide inversion.

References

Andersen, O.B., 1994. Ocean tides in the northern North Atlantic Ocean and adjacent ᴣas from ERS 1 altimetry, J. Geophys. Res., 99(C11), 22,557-22,573.

Andersen, O.B., 1995a Global ocean tides from ERS 1 and TOPEX/POSEIDON altimetry, J. Geophys Res.100 (C12), 25,249-25,259.

Andersen, O.B., 1995b. New ocean tide models for loading computations, Bull. Int. de Maree terreste, Belgium, 102, 9256-9264.

Andersen, O.B., P.L. Woodworth and R.A. Flather, 1995a. Intercomparison of recent global ocean tide models, J. Geophys. Res., 100 (C12), 25,261-25,282.

Andersen, O.B., M. Ashworth and S. Wilkes, 1995b. Global Ocean tide inversion, Preliminary results & parallel implementation, Proudman Oceanographic Laboratory, POL Internal rapport 41, 52pp.

Bennett, A.F., 1992. Inverse methods in Physical Oceanography Monographs on mechanics and applied mathematics, Cambridge University press, Cambridge, 346pp.

Eanes, R.J., 1994. Diurnal and semidiurnal tides from TOPEX/POSEIDON altimetry, (abstract), EOS, 75(16), 108.

Egbert, G.D., A.F. Bennett, and M.G.G. Foreman, 1994. TOPEX/POSEIDON tides estimated using a global inverse model, J. Geophys. Res. 99(C12), 24821-

24852.

Hendeshott, M., 1988. Long Waves and Ocean Tides. Evolution of Physical Oceanography. MIT Press, Cambridge.

Jackson, D.D., 1973. The use of a priori data to resolve non-uniqueness in linear inversion. Geophys.. J. R. astr. Soc., 57, 137-157

Jacobsen, B.H., 1982. Inversionsteori, Grundlag, teknik og anvendelse, Geoskrifter 21, Geofysisk Afd. University of Aarhus, 256 pp.

Kantha, L., 1995. Barotropic tides in the Global ocean from a nonlinear tidal model assimilating altimetric tides., J. Geophys. Res., in press.

Kowalik, Z. and A.Y. Proshutinsky, 1993. Diurnal tides in the Arctic Ocean. J. Geophys. Res., 98 (C9), 16449-16468.

Le Provost, C., M.L. Genco, F. Lyard, P. Vincent, and P. Canceil, 1994. Spectroscopy of the world ocean tides from a finite-element hydrodynamic model. J. Geophys Res. 99(C12), 24777-24797.

Moritz, H. 1989. Advanced Physical Geodesy, Wichmann, Karlsruhe, Germany.

Parker, R., L. Shure, and J. A. Hilderbrand, 1987. The Application of Inverse Theory to Seamount Magnetism., Rev. Geophys 25, 17-40.

Ray, R.D., B. Sanchez, and D.E. Cartwright, 1994. Some extensions to the response method of tidal analysis applied to TOPEX/POSEIDON. (abstract), EOS, 75(16) 108.

Tscherning, C.C., 1986. Functional methods for gravity field approximation, In Mathematical and numerical techniques in physical geodesy. Lecture Notes Earth Sci. Ser., vol. 7, edited by Hans Sunkel pp. 3-49, Springer Verlag, New York.

Tscherning, C.C., R. Forsberg, P. Knudsen, 1992. The GRAVSOFT package for geoid determination, Proc. First workshop on the European Geoid, Prague.

Inversion of STEP-Observation Equation Using Banach's Fixed-Point Principle

Wolfgang Keller

Geodetic Institute, Stuttgart University, Stuttgart, Germany

1 Introduction

Satellite gradiometry is a subject of geodetic research for several years. A satellite gradiometry mission called STEP, is expected to be launched at the beginning of the next century. Although the final mission parameters are not yet fixed, an one-axial gradiometer with an accuracy of 10^{-4} E.U. Paik et al (1991) is prefered. This gradiometer is supposed to measure the out-of-plane component V_{zz} of the *Marussi* -tensor in an orbit of 350 km height and an orbital plane inclination of $96°.85$.

Consequently, two problems have to be solved:

(i) to determine the gravitational potential V from the tensor component V_{zz} measured in the orbit

(ii) to investigate, how the measurement-noise contained in the data V_{zz} propagates into the solution V

Both problems will be treated first for the simplifying assumption of a *polar orbit*. The extension to a nonpolar orbit will be discussed subsequently.

2 Mathematical Model

Let us first discuss the determination of the gravitational potential V from the measured tensor component V_{zz}.

As an idealization of the real situation we assume a sphere of radius $R + h$ to be continuously covered by the gradiometer data V_{zz}.

What we are looking for is a function V, which fulfills the following conditions:

$$V_{rr} + \frac{1}{r^2}V_{\vartheta\vartheta} + \frac{2}{r}V_r + \frac{1}{r^2\sin^2(\vartheta)}V_{\lambda\lambda} + \frac{\cot(\vartheta)}{r^2}V_\vartheta = 0 \quad , \quad r > R \qquad (1)$$

$$\lim_{r\to\infty} V = 0 \qquad (2)$$

$$V_{zz}(\lambda, \vartheta) = \left(\frac{1}{r^2 \sin^2(\vartheta)} V_{\lambda\lambda} + \frac{1}{r} V_r + \frac{\cot(\vartheta)}{r^2} V_\vartheta \right) \Bigg|_{r=R+h} \tag{3}$$

These three equations form a *three-dimensional* elliptic boundary value problem with one particularity: the boundary condition is not prescribed at the boundary of the domain of harmonicity but in its interior. Consequently, standard methods of elliptic boundary value problems cannot be applied directly. If we combine equations (1) and (3), we obtain

$$V_{zz}(\lambda, \vartheta) = \left(-V_{rr} - \frac{1}{r^2} V_{\vartheta\vartheta} - \frac{1}{r} V_r \right) \Bigg|_{r=R+h} \tag{4}$$

which is valid only at the surface of the sphere of radius $r = R + h$. This means, instead of a boundary value problem in a part of the threedimensional *Euclidian* space we now have a partial differential on a two-dimensional curved and closed manifold. This equation has to be solved for $\bar{V}$ the restriction of V to this manifold. Once having determined $\bar{V}$ the gravitational potential V can be obtained by harmonic upward and downward continuation.

The solution of equation (4) is to be obtained by *Banach's* fixed-point principle. Therefore this equation has to be reformulated as an operator equation for operators mapping between different function spaces.

3 Reformulation

In order to study existence and uniqueness of a solution of (4), it has to be written as an operator equation with an operator **A** acting between suitable chosen function spaces.

As we will see later, these spaces turn out to be certain *Sobolev* spaces and the operator **A** will be a pseudodifferential operator (PDO) on these spaces.

Definition 1. A set $H_s(S_r) \subset L_2(S_r)$ is called **Sobolev** space of order s on the sphere S_r of radius r, if for every $u \in H_s(S_r)$ with the spherical harmonics expansion

$$u = \sum_{n \geq 0} \sum_{|m| \leq n} u_{nm} \bar{Y}_{nm} \left(\frac{R}{r} \right)^{n+1} \tag{5}$$

the following condition holds:

$$\sum_{n \geq 0} \sum_{|m| \leq n} (n + \frac{1}{2})^{2s} \left(\frac{R}{r} \right)^{2n+2} u_{nm}^2 < \infty \tag{6}$$

In the paper by Svensson (1983) the following property of the space $H_s(S_r)$ is proved:

Lemma 2. $H_s(R_r)$ *is a Hilbert space with respect to the scalar product*

$$< u, v >= \sum_{n \geq 0} \sum_{|m| \leq n} \left(n + \frac{1}{2}\right)^{2s} \left(\frac{R}{r}\right)^{2n+2} u_{nm} v_{nm} \qquad (7)$$

We now want to investigate mappings between different *Sobolev* spaces

Definition 3. A linear continuous mapping $p : H_s(S_R) \rightarrow H_{s-\mu}(S_r)$ with the property

$$p\bar{Y}_{nm} = p_n \bar{Y}_{nm} \left(\frac{R}{r}\right)^{n+1} \qquad (8)$$

is called invariant PDO of order μ with the spherical symbol

$$\tilde{p} = \{p_n\} \qquad (9)$$

The importance of invariant PDO lies in the fact that operator equations with this type of operators can be solved easily. In order to take advantage of this property, we now want to reformulate (4) as a PDO equation.

Lemma 4. *Let be S_R and S_r spheres with the radii R and $r = R+h$, respectively. For a function u being harmonic outside S_R the mapping*

$$\mathbf{A_0} : \begin{cases} H_2(S_R) \rightarrow H_0(S_r) \\ u \qquad \rightarrow \left(-\frac{1}{r}\frac{\partial u}{\partial r} - \frac{\partial^2 u}{\partial r^2}\right) \end{cases} \Bigg|_{r=R+h} \qquad (10)$$

is an invariant PDO with the spherical symbol

$$\tilde{A_0} = \{\frac{(n+1)}{R^2} \left(\frac{R}{r}\right)^{n+3} - \frac{(n+1)(n+2)}{R^2} \left(\frac{R}{r}\right)^{n+3}\} \qquad (11)$$

If we investigate the symbol of $\mathbf{A_0}$, we see that:

$$\tilde{A_0} = 0 \quad \Leftrightarrow \quad 1 = n + 2, \qquad (12)$$

which never occurs. Herefrom the existence of $\mathbf{A_0}^{-1} : H_0(S_r) \rightarrow H_2(S_r)$ can be concluded.

But it can be easily seen by comparing (10) and (4) that the operator $\mathbf{A_0}$ cannot represent the partial differential equation (4) completely. A second operator $\mathbf{A_1}$ is still necessary.

Lemma 5. *The operator*

$$\mathbf{A_1} : \begin{cases} H_2(S_R) \rightarrow H_0(S_r) \\ u \qquad \rightarrow \left(-\frac{1}{r^2}\frac{\partial^2 u}{\partial \vartheta^2}\right) \end{cases} \Bigg|_{r=R+h} \qquad (13)$$

is a noninvariant PDO.

Using these two operators the partial differential equation (4) gets the following form

$$V_{zz} = (\mathbf{A_0} + \mathbf{A_1})V \tag{14}$$

Because we do not know anything about the inverse of $\mathbf{A_1}$, an inverse of $(\mathbf{A_0} + \mathbf{A_1})$ cannot be given. We have to look for an iterative approach to solve (14).

4 Banach's Fixed-Point Principle

Using the fact that $\mathbf{A_0}^{-1}$ exists we obtain an equivalent form of (14) by

$$V = \mathbf{A_0}^{-1}(V_{zz} - \mathbf{A_1}V), \tag{15}$$

which leads to the following fixed-point iteration:

$$V^{(0)} = 0 \tag{16}$$
$$V^{(n+1)} = \mathbf{A_0}^{-1}(V_{zz} - \mathbf{A_1}V^{(n)}) \quad , \quad n = 0, 1, 2, ... \tag{17}$$

In order to give an interpretation of the formulae describing this iteration, we will indicate here which tasks have to be done in each step.

1. generation of the field $\mathbf{A_1}V^{(n)}$ on the sphere S_r by a spherical harmonics synthesis complete up to a certain degree and order.
2. correction of the data V_{zz} by the synthetic data generated in the previous step.
3. spherical harmonics analysis of the residual field $V_{zz} - \mathbf{A_1}V^{(n)}$, complete up to a certain degree and order.
4. application of $\mathbf{A_0}^{-1}$ which is known in the frequency domain, giving an improved field $V^{(n+1)}$ on the sphere S_R.

The question arises whether this iteration is convergent and if so whether the limit solves (14).
With the help of Banach's fixed point principle the following sufficient condition can be given:

Lemma 6. *Let be* $\|\mathbf{A_0}^{-1}\mathbf{A_1}\| \leq \gamma < 1.$ *then the following holds:*

(i) $V_{zz} = (\mathbf{A_0} + \mathbf{A_1})V$
    has a unique solution V^*.
(ii) $\lim_{n\to\infty} V^{(n)} = V^*$
(iii) $\|V^{(n+1)} - V^*\| \leq \gamma\|V^{(n)} - V^*\|$

An exact analytic determination of the bound γ was not possible. Only an upper estimate $\bar{\gamma}$ of this bound could be given, which unfortunately was larger than one.

Consequently, it cannot be expected that the field V can be *completely* determined by the iteration process . At least there are some spectral components of V which cannot be determined.

On the other hand this result does not contradict the fact that the remaining spectral components can be determined very well by this iteration approach.

5 Numerical Investigations

Because it is not possible to determine the value $\|A_0^{-1}A_1\|$ analytically, it has to be estimated during the iteration process. This estimation is based on the following result :

Lemma 7. *Let λ be the largest eigenvalue of $A_0^{-1}A_1$, then*

$$\lim_{n \to \infty} \frac{\|V^{(n)} - V^{(n+1)}\|}{\|V^{(n)} - V^{(n-1)}\|} = |\lambda| \tag{18}$$

holds.

This leads to the following modified iteration scheme :

$$V^{(0)} = 0 \tag{19}$$

$$\left. \begin{array}{l} V^{(n+1)} = A_0^{-1}(V_{zz} - A_1 V^{(n)}) \\[2mm] |\lambda^{(n+1)}| = \frac{\|V^{(n)} - V^{(n+1)}\|}{\|V^{(n)} - V^{(n-1)}\|} \end{array} \right\} \quad , \quad n = 0, 1, 2, ... \tag{20}$$

In order to verify the theoretical results numerically, the following numerical experiments were carried out

(i) generation of a synthetic gradiometer data set V_{zz} using the Earth model OSU91A
(ii) execution of the modified iteration process for $n = 1, ..., N$, where $N = 10, 20, 30$.

For the characterization of the iteration process the following quantities were observed in each iteration step :

- $|\lambda^{(n)}|$ as a function of n
- for each $N = 10, 20, 30$
  - the absolute errors $(V^{(N)} - V^*)$
  - the relative errors $(V^{(N)} - V^*)/V^*$
  - the changes between consecutive iteration steps $V^{(N+10)} - V^{(N)}$
  all in the frequency domain.

The following results were obtained:

- after a short initialization phase $|\lambda^{(n)}|$ quickly tends to the value of 0.88 (figure 1).

This not only verifies the existence and uniqueness of a solution of (4) but also indicates a comparatively high speed of convergence.

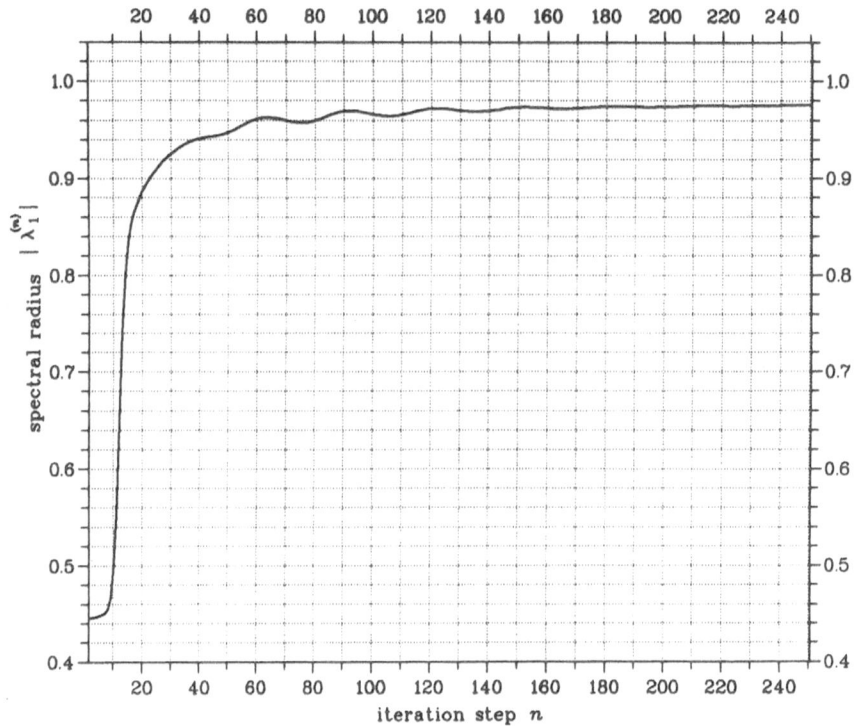

Fig.1. Estimated largest Eigenvalue

- The relative error $(V^{(10)} - V^*)/V^*$ after 10 iteration steps shows a typical *cone shaped* pattern.
  The coefficients inside the cone, i.e. the zonal coeficients and coefficients with a low order are badly determined. The coefficients outside the cones are well determined .
  Because the data V_{zz} contained no noise , this picture shows the truncation error.
- an increase of the number of iterations narrows the cone. This means the truncation error is more and more concentrated around the zonal coefficients

– after 30 iterations the following *rule of thumb* can be establishe :

Coefficients C_{nm} with $|m| > \frac{n}{3}$ can be determined by the iteration.

The determination of the coefficients by iteration is not only influenced by the truncation error but also by the propagation of the data errors. In order to investigate this effect a random noise of 10^{-4} E.U. was added to the data V_{zz}, giving the noisy data $\bar{V}_{zz}$.

The same experiments were performed with $\bar{V}_{zz}$. Here the *cone shaped pattern* changes into a *butterfly pattern*. Additionally to the coefficients which are close to the zonal axis, now the coefficients of high degree **and** high order cannot be determined . This can be explained by the fact that these coefficients have the smallest magnitude and therefore are influenced most strongly by the data noise. An increasing number of iterations reduces the number of not detectable coefficients but the *butterfly pattern* remains unchanged.

As a last interesting comparison the difference after 30 iterations between the results obtained from the noisy data and from the error- free data can be discussed .

This difference reflects only the influence of the data errors. The truncation error is approximately the same in both coefficient sets and cancels out. Again we have the *butterfly pattern* which is now very much concentrated around the zonal axis.

6 Summary

– The application of Banach's fixed point principle generates an iteration process which is convergent exept for a small band around the zonal axis.
– The convergence is fast, i.e. after 30 iterations the truncation error can be neglected.
– Exept of this narrow band around the zonal axis the gravitational field can be resolved from STEP gradiometer data at least up to degree and order 168.

References

Brovelli M. and Sanso F. *Gradiometry: the study of the V_{yy} component in the B.V.P. approach* Manuscripta Geodaetica **15**(1990)

Paik et al*Laboratory demonstrations of superconducting gravity and inertial sensors for space and airborne gravity measurements* In: **Colombo** ed. :*From Mars to Greenland: Charting Gravity With Space and Airborne Instruments*, pp 191-201, Wien, August 20 1991 , IAG Symposium No. 110, Springer Verlag

Svensson L. *Pseudodifferential Operators-a New Approach to the Boundary Problems of Physical Geodesy* manuscripta geodetica Vol. 8 (1983) , pp 1-40

Rummel R. and Colombo O. *Gravity field determination from satellite gradiometry,* Bull. Geod. **59**(1985)

Improving the Planar Isotropic Covariance Models

Gabriel Strykowski

Kort & Matrikelstyrelsen, Copenhagen, Denmark

Abstract *In this paper a general approximative method for dealing with the low frequency problem in local planar covariance models is proposed. The low frequency problem manifests itself in the equivalent frequency domain expression as non-zero values for frequencies corresponding to wavelengths larger than the size of the local area. After the correction for the effects of the low frequency problem, the value of the associated power spectral density model for these frequencies is approximately zero. The method is illustrated by a relevant example.*

1. Introduction

Planar Covariance Model: In stochastic methods of local and regional gravity field modelling (Moritz 1980), in planar approximation (Nash and Jordan 1978, Moritz 1980, Forsberg 1984, Forsberg 1987, Schwarz *et al.* 1990), the isotropic (i.e. azimuth-independent) and homogenous covariance models $C(s)$, where s is the Euclidean distance in a plane, are chosen among a limited class of functions or as a linear combination of such functions (Meier 1981).

In applications, the spatial planar covariance model $C(s,h_1,h_2)$ is being used, i.e a cross-process covariance model for two isotropic and homogenous stochastic processes in the respective heights h_1 and h_2 above the reference plane. The two stochastic processes must not necessarily be associated with the same physical quantity. Denoting by x and y the two types of physical quantities in heights h_1 and h_2 respectively, the notation for the spatial covariance model is $C_{xy}(s,h_1,h_2)$.

Cross-Power Spectral Density Model: The cross-power spectral density model (CPSD) $\Phi_{xy}(q,h_1,h_2)$ is the Hankel transform H of $C_{xy}(s,h_1,h_2)$:

$$\Phi_{xy}(q,h_1,h_2)=H[C_{xy}(s,h_1,h_2)]\equiv 2\pi\int_0^\infty C_{xy}(s,h_1,h_2)J_0(2\pi qs)sds \qquad (1.1)$$

The inverse relation is given by the inverse Hankel transform H^{-1}:

$$C_{xy}(s,h_1,h_2)=H^{-1}[\Phi_{xy}(q,h_1,h_2)]\equiv 2\pi\int_0^\infty \Phi_{xy}(q,h_1,h_2)J_0(2\pi qs)qdq \qquad (1.2)$$

where J_0 is the zero order Bessel function and q is the radial frequency

$$q=\sqrt{u^2+v^2} \qquad (1.3)$$

where u and v are the frequencies of the 2-dimensional Fourier transform (see e.g. Bracewell 1965).

Covariance Propagation: For linear (or linearized) operators L_X and L_Y, $X=L_X(x)$ and $Y=L_Y(y)$, the covariance model C_{XY} can be derived from C_{xy} using the law of *covariance propagation* (Moritz 1980):

$$C_{XY}=L_XL_Y[C_{xy}] \qquad (1.4)$$

The propagation law is also valid in the frequency domain (see e.g. Forsberg 1984). In physical geodesy a number of most important operators depend on the "vertical parameters" h_1 and/or h_2. This includes: (i) multiplication with a scalar depending on h_1 and/or h_2, (ii) multiple differentiation/integration with respect to h_1 and/or h_2, and (iii) any linear combination of the operations (i)-(ii). For such operators the Hankel transformation and propagation can be interchanged, i.e.

$$\Phi_{XY}=H[C_{XY}]=H[L_XL_YC_{xy}]=L_XL_YH[C_{xy}]=L_XL_Y\Phi_{xy} \qquad (1.5)$$

Fundamental CPSD models: In physical geodesy the CPSD models are constructed as linear combinations of *the fundamental CPSD models* (Forsberg 1987):

$$\Phi_{xy}(q,h_1,h_2)=2\pi\sqrt{C_x(0)C_y(0)}(2\pi q)^n e^{-2\pi q[\alpha_x+h_1+\alpha_y+h_2]} \qquad (1.6)$$

where $n\in\{...,-2,-1,0,1,2,...\}$, α_x,α_y are positive constants, and $C_x(0)$, $C_y(0)$ denote the variances of the stochastic processes x and y for $h_1=h_2=0$. The differentiation/ integration with respect to h_1 and/or h_2 applied to any fundamental CPSD model yields a fundamental CPSD model. However, the power term n changes by $+1$ for differentiation and by -1 for integration. Furthermore, the expressions for the inverse Hankel transforms of these functions are known.

Tscherning-Rapp covariance model: In spherical Earth approximation the spatial covariance model for the anomalous gravity potential T has the following form:

$$C_{TT}(\psi,r,r')=\sum_{k=2}^{\infty}\sigma_k(\frac{R^2}{rr'})^{k+1}P_k(\cos\psi) \qquad (1.7)$$

where r and r' are the radial distances from the centre of the Earth, ψ is the spherical distance, R is the mean Earth's radius, σ_k is the degree-variance for the harmonic degree k, and P_k is the Legendre polynomial of degree k (Moritz 1980, eq.(13-7)). Tscherning and Rapp (1974) found a closed expression for the above infinite series of Legendre polynomials.

2. The Low Frequency Problem of the Fundamental Planar CPSD Models

Data Reductions and the Consistent CPSD Model: In local gravity field modelling it is customary to centre the measurements by subtracting the estimated mean value and, if necessary, the tilt. This is interpreted as the subtraction of the low frequency part of the signal, i.e. the frequencies corresponding to wavelengths larger than the size of the local area. Consequently, the CPSD value for these frequencies should be zero.

The fundamental CPSD models have, however, a non-vanishing value for these frequencies. This value may even become ∞ for $q=0$. Fig. 2.1 shows the fundamental CPSD models for $n=-2,-1,0,1,2$. Throughout this paper this inconsistency between the fundamental CPSD model and the assumptions in data reduction is referred to as *the low frequency problem*.

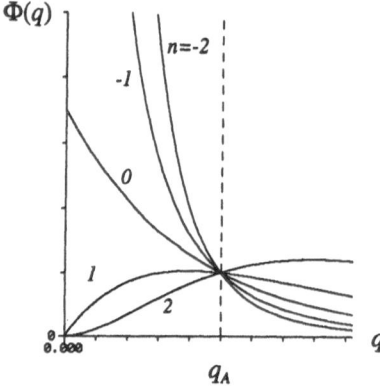

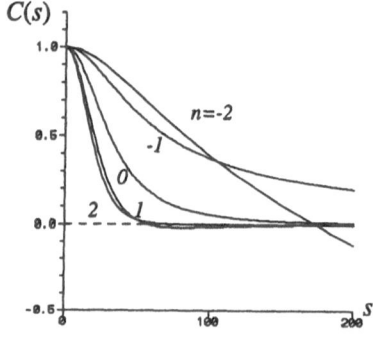

Fig. 2.1. Fundamental normalized [i.e. $\Phi(q_A)=1$] CPSD models for $n=-2,-1,0,1,2$. q_A is the lowest resolvable frequency in the local area.

Fig. 2.2 Normalized [i.e. $C(0)=1$] planar covariance models corresponding to the fundamental CPSD models for $n=-2,-1,0,1,2$.

Local Covariance Models in Spherical Approximation: The covariance models in spherical Earth approximation do not suffer from *the low frequency problem*. In eq.(1.7) the low frequency part of the signal can directly be associated with the low values of k. The local spherical covariance model is:

$$C_{TT}(\psi,r,r') = \sum_{k=K}^{\infty} \sigma_k \left(\frac{R^2}{rr'}\right)^{k+1} P_k(\cos\psi) \qquad (2.1)$$

where K is the harmonic degree corresponding to the lowest frequency which can be resolved by data given in a local area. Closed expressions for the above infinite series exist, e.g. a local version of the Tscherning-Rapp model.

"The Damped Oscillating Tail" and the Low Frequency Problem: Tscherning-Rapp covariance model exhibit a characteristic feature, "the damped oscillating tail", i.e. the covariance model as a function of distance has alternating positive and negative values and the amplitudes decrease in magnitude for increasing distance.

This feature is in agreement with the above interpretation of the data reduction. A subtraction of the low frequency signal from local data results in negative and positive residuals throughout the whole area. Thus, the alternating negative and positive covariances as a function of distance are expected. The decrease of the amplitude for increasing distance is consistent with the anticipated low value of the correlation coefficient for data far away from each other.

Fig. 2.2 shows the normalized, i.e. $C(0)=1$, planar covariance models for $n=-2,-1,0,1,2$. None of these models exhibits the "damped oscillating tail". For $n=-2$ (logarithmic model, see Forsberg 1984) the covariance value becomes negatively infinite for increasing distance, i.e. $C(s)\rightarrow-\infty$ for $s\rightarrow\infty$.

3. Correcting for the Effect of the Low Frequency Problem in Planar Covariance Models

The proposed general approximative method for dealing with *the low frequency problem* in planar covariance models consists of two stages: (i) stabilizing the model for $q=0$, and (ii) approximating and subtracting the power spectral density for low frequencies, i.e. for $q\in[0,q_A[$.

Stabilizing the Model for $q=0$: Eq.(1.6) shows that for $n<0$: $\Phi(q)\rightarrow\infty$ for $q\rightarrow0$, i.e. the model value becomes infinite for the DC-component ($q=0$). This is also indicated on Fig. 2.1. The method of stabilizing the model for $q=0$ is a method of obtaining a finite value for this frequency without affecting significantly the CPSD values for $q\geq q_A$, where q_A is the lowest resolvable radial frequency in a local area. Thus, the stabilizing of the model for $q=0$ concerns the fundamental CPSD models for $n<0$.

A series expansion of the fundamental CPSD model (based on Taylor expansion of the exponential function) yields:

$$\Phi_{xy}(q,h_1,h_2)=[2\pi\sqrt{C_x(0)C_y(0)}]\times(2\pi q)^n\times$$
$$\times\{1 + [\alpha_x+h_1+\alpha_y+h_2](-2\pi q) + \frac{[\alpha_x+h_1+\alpha_y+h_2]^2}{2!}(-2\pi q)^2 + ...\} \tag{3.1}$$

Thus, we obtain an expression in which different powers of $-2\pi q$ occur.

The next is to choose a new fundamental CPSD model $\Phi'(q,h_1,h_2)$, $n=-2$, with a new set of parameters α_x' and α_y'. As seen from eq.(1.6), the rate at which a fundamental CPSD model decreases depends on the parameters α_x and α_y. If the parameters of the new model are chosen in such a way that $\alpha_x'>>\alpha_x$ and $\alpha_y'>>\alpha_y$, the value of $\Phi-\Phi'$ for $q\in[q_A,\infty[$ is approximately equal the value of Φ, i.e. a subtraction of Φ' from Φ will not affect the CPSD values for these frequencies. The series expansion of the difference between the two models is obtained by subtracting the corresponding terms of the series expansions of the individual CPSD models:

$$[\Phi-\Phi']_{xy}(q,h_1,h_2)=[2\pi\sqrt{C_x(0)C_y(0)}]\times(2\pi q)^n\times\{\ [\alpha_x-\alpha'_x+\alpha_y-\alpha'_y](-2\pi q)\ +$$
$$+\ \frac{[\alpha_x+h_1+\alpha_y+h_2]^2-[\alpha'_x+h_1+\alpha'_y+h_2]^2}{2!}(-2\pi q)^2\ +\ ....\} \tag{3.2}$$

For $n=-2$, a comparison of eq.(3.1) and eq.(3.2) shows that the rate at which right hand side converges to infinity for $q\to0$ is different. Assuming that the value of q is very small, $[\Phi-\Phi'](q)\sim1/q$ in eq.(3.2) and $\Phi(q)\sim1/q^2$ in eq.(3.1).

In order to stabilize the model for $q=0$ ($n=-2$) two new CPSD models Φ'' and Φ''' with the appropriate set of parameters can be chosen. The sufficient conditions are $\alpha_x''-\alpha_x'''=\alpha_x-\alpha_x'$ and $\alpha_y''-\alpha_y'''=\alpha_y-\alpha_y'$. The series expansion for $[(\Phi-\Phi')-(\Phi''-\Phi''')](q,h_1,h_2)$ is:

$$[(\Phi-\Phi')-(\Phi''-\Phi''')]_{xy}(q,h_1,h_2)=[2\pi\sqrt{C_x(0)C_y(0)}]\times(2\pi q)^n\times$$
$$\times\{\ \frac{([f(\alpha_x,\alpha_y)]^2-[f(\alpha'_x,\alpha'_y)]^2)-([f(\alpha''_x,\alpha''_y)]^2-[f(\alpha'''_x,\alpha'''_y)]^2)}{2!}(-2\pi q)^2\ +\ ....\} \tag{3.3}$$

where the function f, $f(\alpha,\beta)\equiv\alpha+h_1+\beta+h_2$, is introduced to make the expressions shorter. For $n=-2$ the expression in eq.(3.3) converges to a constant value for $q\to0$. Furthermore, if $\alpha_x'',\alpha_x'''<<\alpha_x$ and $\alpha_y'',\alpha_y'''<<\alpha_y$ the value of the resulting CPSD model $(\Phi-\Phi')-(\Phi''-\Phi''')$, and for $q\in[q_A,\infty[$, is almost unaffected as compared to Φ.

In order to illustrate the method a number of parameters for the fundamental CSPD model for $n=-2$ were chosen (see Fig. 3.1): (i) the maximal radial wavelength that can be resolved in a local area is *300 km*; (ii) $h_1=h_2=0$ *km*; (iii) $\alpha_x=\alpha_y=20$ *km*; (iv) $\alpha_x'=\alpha_x+180$ *km* and $\alpha_y'=\alpha_y+180$ *km*; and (v) $\alpha_x''=\alpha_x'+1$ *km* and $\alpha_y''=\alpha_y'+1$ *km*. Fig. 3.2 shows the original fundamental CPSD model Φ (dashed line) and $(\Phi-\Phi')-(\Phi''-\Phi''')$ given by eq.(3.3) (solid line). As indicated above, $(\Phi-\Phi')-(\Phi''-\Phi''')$ converges to a finite value for $q\to0$ and is almost identical with Φ for $q\geq q_A$.

For $n=-3$ the corresponding procedure involves 8 fundamental CPSD models, for $n=-4$ it involves 16 fundamental CPSD models,... e.t.c. The conditions for the relation between the parameters α_x and α_y for these cases ($n<-2$) are more complicated than for $n=-2$. In general, the dependence on h_1 and h_2 do not enter these conditions. The explicit formulation of these conditions lies, however, outside the scope of the present paper.

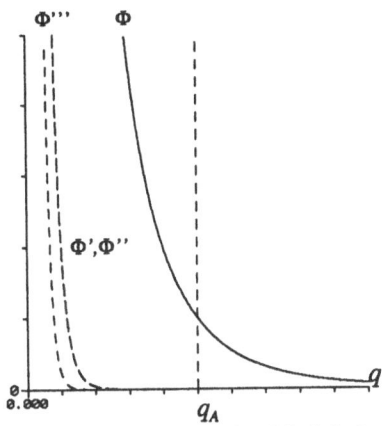

Fig. 3.1 *Stabilizing the Model for q=0. Example (n=-2)*: Fundamental CPSD models: Φ (solid line); Φ',Φ" and Φ''' (dashed line). [Φ' and Φ" are almost (but not quite) identical].

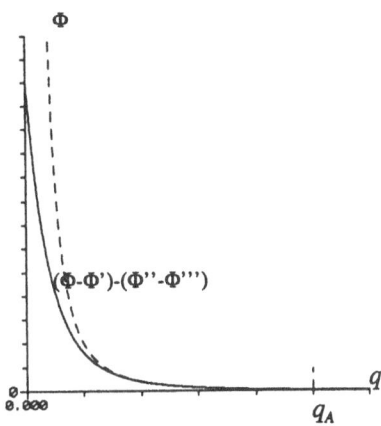

Fig. 3.2 *Stabilizing the Model for q=0. Example (n=-2)*: CPSD model Φ (dashed line); (Φ-Φ')-(Φ"- Φ''') (solid line). (Φ-Φ')-(Φ"-Φ''') converges to a constant for q→0 and (Φ-Φ')-(Φ"-Φ''')≈Φ for $[q_A, \infty[$.

Using the property of the Hankel transform operator (see Sect. 1) on a linear combination of the fundamental CPSD models (i.e. $\sum a_i \Phi_{xy,i}$) the following is valid:

$$C_{xy}(s,h_1,h_2)=H^{-1}[\sum_i a_i \Phi_{xy,i}(q,h_1,h_2)]=\sum_i a_i H^{-1}[\Phi_{xy,i}(q,h_1,h_2)]=$$
$$=\sum_i a_i C_{xy,i}(s,h_1,h_2) \tag{3.4}$$

where $C_{xy,i}\equiv H^{-1}[\Phi_{xy,i}]$. Thus, the resulting covariance model is simply a linear combination of known covariance models.

Approximating and Subtracting the Power for Low Frequencies: Bracewell (1965, Table 12.2) lists a number of functions and their Hankel transforms:

$$\frac{\sin(2\pi q_0 s)}{s} \leftrightarrow \frac{\Pi(q/2q_0)}{\sqrt{q_0^2-q^2}} \tag{3.5}$$

$$\frac{q_0^2 J_2(2\pi q_0 s)}{\pi s^2} \leftrightarrow (q_0^2-q^2)\Pi(q/2q_0) \tag{3.6}$$

$$\frac{q_0 J_1(2\pi q_0 s)}{s} \leftrightarrow \Pi(q/2q_0) \tag{3.7}$$

where

$$\Pi(q/2q_0)=\begin{cases} 1 & \text{for } |q|<q_0 \\ \frac{1}{2} & \text{for } |q|=q_0 \\ 0 & \text{for } |q|>q_0 \end{cases} \tag{3.8}$$

and where q_0 is a constant, J_1 and J_2 are the Bessel functions of order 1 and 2.

The frequency domain expressions in eqs.(3.5)-(3-7) can be used to approximate the (stabilized) CPSD model for $q \in [0,q_A[$, i.e. for the low frequencies. The method resembles the approximation of a definite integral by a finite sum of rectangular elements. Eq.(3.7) yields such a "rectangular element" in the radial frequency domain.

The resulting CPSD model is a linear combination of CPSD models with known inverse Hankel transform expressions so that eq.(3.4) applies also in this case. Consequently, the expression for the resulting covariance model is also known. However, not all of these models are fundamental CPSD models. Fig. 3.3. shows the resulting CPSD model after the technique has been used on the stabilized CPSD model on Fig. 3.2. Fig. 3.4 shows the covariance models corresponding to different stages of model construction. The planar covariance model C on Fig. 3.4 exhibits, as expected, "the damped oscillating tail".

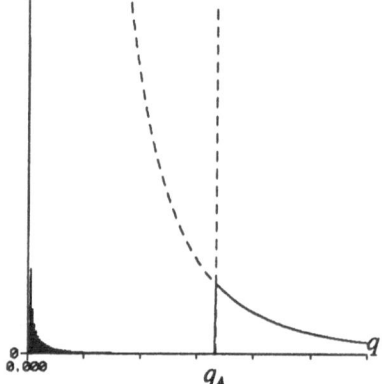

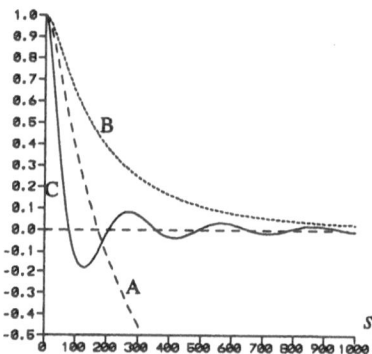

Fig. 3.3 *Residual CPSD Model.* *Example (n=-2)*: Residual CPSD model after the correction for the effect of *the low frequency problem* (solid line); original CPSD model (dashed line).

Fig. 3.4 *Planar Covariance Models.* *Example (n=-2)*: (A) original planar logarithmic model; (B) after stabilization for $q=0$; (C) corrected for the effect of *the low frequency problem*. (The size of the local area is *300 km*).

4. Upward Continuation of the Planar Covariance Model

The upward continuation of a fundamental CPSD model to the new heights h_1+H_1 and h_2+H_2 above the reference plane yields (see e.g. Forsberg 1984):

$$\Phi_{xy}(q,h_1+H_1,h_2+H_2)=\Phi_{xy}(q,h_1,h_2)e^{-2\pi qH_1}e^{-2\pi qH_2}=$$
$$=2\pi\sqrt{C_x(0)C_y(0)}(2\pi q)^n e^{-2\pi q[\alpha_x+(h_1+H_1)+\alpha_y+(h_2+H_2)]} \qquad (4.1)$$

Thus, the propagated CPSD model is also a fundamental CPSD model of the same type, but with a new set of parameters $\alpha_x+h_1 \rightarrow \alpha_x+h_1+H_1$ and $\alpha_y+h_2 \rightarrow \alpha_y+h_2+H_2$.

Using eq.(4.1) and eq.(3.4) the expression for the upward continued stabilized CPSD model can be obtained. The propagated CPSD model is also a stabilized linear combination of the fundamental CPSD models and it converges to the same constant value for $q \rightarrow 0$.

This nice property is not valid for the CPSD models in eqs.(3.5)-(3.7) which were used for correcting for the low frequency problem of the stabilized CPSD model. However, if for some frequency band the value of a CPSD model is zero, the upward continued value for these frequencies will also be zero.

Consequently, the problem of upward continuation of the CPSD model corrected for the low frequency problem should be understood as an approximation to the ideal zero CPSD value for low frequencies. Thus, in order to obtain approximately consistent set of planar CPSD models corrected for the low frequency problem one should: (i) upward continue the stabilized linear combination of the fundamental CPSD models; (ii) correct independently the stabilized CPSD models for the low frequency problem.

5. Conclusions

In this paper a general approximative method of correcting for the effect of the low frequency problem in planar covariance models is proposed. The principles of upward continuation of such corrected planar covariance models are also addressed.

References

Bracewell, R., The Fourier Transform and its Application, McGraw-Hill, New York, 1965.

Forsberg, R., Local Covariance Functions and Density Distributions, Rep.356, Dep. of Geod. Sci., Ohio State Univ., Columbus, 1984.

Forsberg, R., A New Covariance Model for Inertial Gravimetry and Gradiometry, J. Geophys. Res., 92, 1305-1310, 1987.

Meier, S, Planar Geodetic Covariance Functions, Reviews of Geophysics and Space Physics, vol. 14, 673-686, 1981.

Moritz, H., Advanced Physical Geodesy, Herbert Wiechnmann Verlag; Karlsruhe, Germany, 1980.

Nash, R.A. and S.K. Jordan, Statistical Geodesy-An Engineering Perspective, Proc. IEEE, 66, 532-550, 1978.

Schwarz, K.P., M.G. Sideris and R. Forsberg, The Use of FFT Techniques in Physical Geodesy, Geophys. J. Int., 100, 485-514, 1990.

Tscherning, C.C. and R.H. Rapp, Closed Covariance Expressions for Gravity Anomalies, Geoid Undulations, and Deflections of the Vertical Implied by Degree-Variance Models, Rep. 208, Dept. of Geod. Sci., Ohio State Univ., Columbus, 1974.

Optimal Linear Estimation Theory for Continuous Fields of Observations

F. Sacerdote and F. Sansò

Dipartimento di Ingegneria Idraulica, Ambientale e del Rilevamento, Politecnico di Milano, Italy

1 Introduction

Linear estimation methods - namely least-squares adjustment and least-squares collocation - are widely used in geodetic sciences. In both procedures a finite number of data is used to obtain estimates of a finite number of quantities (that can possibly be chosen among the infinite values assumed by functions defined on continuous sets); therefore they can be classified as discrete methods, even though the involved quantities may be continuously defined.

   This is for example the case of gravity field measurements and geoid height evaluations, that are carried out at discrete points, whereas they are taken from physical quantities defined on continuous point sets, whose functional properties (for example the fact that the gravity field is the gradient of a potential) are fundamental in the theoretical developments of physical geodesy.

   In least-squares adjustment a redundant number of measurements, affected by observational noise, is used to filter the noise and to estimate the parameters that describe a linear manifold to which the vector of the observed quantities is constrained.

   In least-squares collocation the measured quantities are functionally related to a stochastic field and are used to determine the estimates of a number of arbitrary functionals of the field. In this case, in addition to the observational noise with its statistical properties, the observed and the estimated quantities too are equipped with a stochastic structure (described in terms of covariance functions), which provides some a-priori information on the solution (Moritz 1980). This feature enables in some cases to define a solution for intrinsically singular problems or to stabilize the solution of ill-conditioned problems.

   It is remarkable that in the collocation method the relation between the field and the observed quantities involves operators acting on functions defined on continuous point sets, as differentiations or integrations; for example in geodesy gravity anomaly and geoid height are both functionally related to the disturbing potential. Nevertheless, the procedure is essentially finite-dimensional, and all continuous operations involve only the covariance functions, from which covariance matrices are obtained. This feature characterizes the classical theory of stochastic properties, where continuity, differentiability and so forth are typically defined on the covariance function (cf. for example (Lamperti 1977)).

It may appear that the discrete approach covers all possible physical situations; indeed, one has always to deal with a finite number of measurements. Yet, this number may be very high, and the measurement density may be larger than the required resolution. To be more precise, assume that the variation of the deterministic part of the measured quantity $y(t)$ inside a small domain T_i containing a single measurement at the point t_i is very small with respect to the observational noise, i.e.

$$|y(t_i)\mu_i - \int_{T_i} y(t)dt| << \sigma_\nu \mu_i \qquad (1.1)$$

where σ_ν is the standard deviation of the observational noise, and μ_i is the Euclidean measure of T_i . In this case, the amount of information carried by all individual point measurements is certainly redundant for a good global knowledge of the signal, and, on the other hand, an enormous computational effort may be required to process the whole data set. For instance in collocation theory a linear system has to be solved with the same dimension as the number of observations, the relevant matrix being in general full and with no special structure.

Then, algorithms derived from the continuum properties of the quantities involved, suitably discretized for numerical computation according to the required resolution, may be a powerful tool to preserve all the richness of information contained in the data.

The deterministic model is defined by introducing linear operators relating the functions representing the observables to a field defined on a continuous domain. The estimable quantities are linear functionals of this field; in many cases they are not point values, but involve field values on continuous sets (as averages, or Fourier coefficients).

A natural framework for this approach is the theory of random processes with values in Hilbert spaces, that has been developed in a number of papers, starting from Franklin's (1970) pioneering work (see also Rozanov 1971; Skorohod 1974; Mandelbaum 1984; Lehtinen et al. 1989). The problems on which the attention is focused in the present paper are characterized by the fact that their stochastic nature comes out from the presence of observational noise, whereas the field itself is a sort of (infinite-dimensional) deterministic parameter introduced to describe the observed quantities as its functionals. Therefore they can be considered as an infinite-dimensional generalization of least-squares adjustment; their solution relies on the redundance of the available data sets, that is used to filter out the observational noise. The

Wiener measures are the mathematical tool introduced to properly describe an uncorrelated noise in a continuous context.

Examples of these problems can be found in different topics of geodesy, geophysics and surveying sciences:

- geodetic overdetermined boundary-value problems, characterized by redundant boundary conditions;

The field $x(t)$ is the disturbing gravity potential, that satisfies the Laplace equation in the exterior of the earth; the operators B_i represent boundary conditions, whose number is larger than necessary to yield a unique solution, for example $B_1 = \partial/\partial r$, i.e. gravity disturbance, $B_2 = \partial^2/\partial r^2$, i.e. radial component of the gravity gradient tensor. The simplest way to express the harmonic function $x(t)$ is a spherical harmonic expansion; analytical computations are particularly simple if the operators B_i are diagonal with respect to spherical harmonics.

- tracking of satellites by a combination of different techniques, as for example, range and range-rate measurement in PRARE;

The observables are range and range-rate of the satellite; the field is the range itself. Therefore $B_1 = I$; $B_2 = d/dt$. If the observable space is equipped with the L^2-norm, it is natural to use for the field space the Sobolev H_1-norm, defined as $\|u\|_{H_1} = \|u\|_{L^2} + \|\dot{u}\|_{L^2}$; indeed in this case both u and its derivative are required to belong to L^2 . It turns out that, since H_1 functions of a 1-dimensional variable are continuous, point evaluation functionals are estimable. A detailed description is given in section 5.

- recovering of a digital terrain model from a redundant number of photogrammetric images.

The model has two components: the radiation reflected by the terrain and its topographic height. The observables are the grey intensities of different plates.

- geophysical prospecting by different kinds of signals (for example, gravity and tomography);

The field is the mass density in a layer, the observables are surface gravity and time-delay of seismic waves. in order to apply the method, it is necessary to assume a linear dependence of the time-delay on density.

The first two examples are illustrated in detail in (Sansò and Sona 1995). For the last example, see (Barzaghi and Sansò 1994).

2   Finite-dimensional case: a review

The problem of determining a "best" estimate of a k-dimensional vector x from an n-dimensional vector y of observations satisfying

$$y = Ax + \nu \tag{2.1}$$

where A is an $n \times k$ matrix and ν is an observational noise, leads to the well-known least-squares procedure, which can be as well formulated in a deterministic context, by requiring the minimization of the norm of the difference $y - Ax$. A norm in the Euclidean space $\mathbf{R}^n$ can be defined as

$$\|v\|^2 = (v, Kv) \tag{2.2}$$

where K is an arbitrary $n \times n$ positive-definite matrix and $(\ ,\ )$ is the Euclidean scalar product: $(u, v) = \sum u_i v_i$. The minimization of $(y - Ax, K(y - Ax))$ leads to the equation

$$A^T K Ax = A^T K y \tag{2.3}$$

and consequently to the well-known least-squares solution

$$\hat{x} = (A^T K A)^{-1} A^T K y \tag{2.4}$$

provided that $A^T K A$ is invertible, that occurs if $n > k$ and A is full-rank. Otherwise, the solution of the minimization problem is not unique, and the choice of the estimate $\hat{x}$ requires additional conditions.

For example, a minimum requirement may be imposed to $\|x\|$ (x satisfying (3)), with a norm in $\mathbf{R}^k$ defined by a $k \times k$ positive-definite matrix K' :

$$\|x\|^2 = (x, K'x) \tag{2.5}$$

Alternatively, a hybrid norm principle can be introduced, by requiring the minimization of $Q(x) = \|y - Ax\|^2 + \|x\|^2$, that leads to the equation

$$(A^T K A + K')x = A^T K y \tag{2.6}$$

Now, the introduction of $\|x\|^2$ regularizes the problem; indeed, the matrix on the left-hand side is certainly invertible, so that this principle leads anyway to a unique solution:

$$\hat{x} = (A^T K A + K')^{-1} A^T K y \tag{2.7}$$

This technique can be adopted even if (2.3) is not singular, but its solution is unstable because of the large condition number of $A^T K A$; in this case the introduction of $\|x\|^2$ in the minimum principle has the role of stabilizing rather than regularizing the problem. Obviously, the choice of K and K' determines the relative weights of the two terms in $Q(x)$.

In order to introduce a stochastic interpretation of (2.4), where ν is a random vector with zero mean and covariance matrix C, one may choose $K = C^{-1}$; in this case, assuming normal distributions, the adopted minimum principle can be viewed as arising from the maximum likelihood principle. On the other hand, x is not stochastic, and is simply interpreted as a parameter describing a k-dimensional subspace of an n-dimensional Euclidean space. Consequently, y is a random variable with mean Ax and covariance matrix C. It is well-known that in this case the estimate (4) with $K = C^{-1}$ has the minimum variance property with respect to estimates of the same form (2.4) with different K (that are all unbiased); this is usually called the BLUE (best linear unbiased estimation) property in statistical literature.

For the stochastic interpretation of (2.7), it is assumed that x is a stochastic vector too, with zero mean and covariance matrix C_{xx} (whereas in this case the covariance matrix of ν is denoted by $C_{\nu\nu}$). Then, the estimate (2.7), with $K = C_{\nu\nu}^{-1}$ and $K' = C_{xx}^{-1}$, coincides with the estimate that minimizes the mean square estimation error (MSEE), $E\{(x - \hat{x})^2\}$ among all linear estimates $\hat{x} = Ly$ (that are automatically unbiased, since $E\{y\} = AE\{x\} + E\{\nu\} = 0$, implying $E\{\hat{x}\} = LE\{y\} = 0 = E\{x\}$).

A more familiar expression for (2.7) is

$$\hat{x} = C_{xx}A^T(AC_{xx}A^T + C_{\nu\nu})^{-1}y \tag{2.8}$$

The equivalence of (7) and (8) arises from the well-known identity

$$(A^T C_{\nu\nu}^{-1} A + C_{xx}^{-1})C_{xx}A^T = A^T C_{\nu\nu}^{-1}(AC_{xx}A^T + C_{\nu\nu}) \tag{2.9}$$

(2.7) or (2.8) can be interpreted as representing a situation in which some a-priori knowledge on x is assumed, so that its treatment as a stochastic vector is justified; the expression (2.7) shows that (2.7) tends to (2.4) when C_{xx} becomes large, i.e. when the a-priori knowledge on x becomes negligible, which is referred to as the non-informative case in Bayesian theory.

On the other hand, (2.7) can be used in a more general context, in which the dimension of x is not necessarily smaller than the dimension of ν. The case (2.4) will be denoted as *partially stochastic approach*, the case (2.7) as *fully stochastic approach*.

A particular aspect, whose importance will be clear in the sequel, is the estimation of continuous functionals, i.e. the determination of a functional f in $\mathbf{R}^n$ (represented by the scalar product by an $\mathbf{R}^n$ vector, that will be denoted with the same symbol, f) that, applied to the data vector y, gives an estimate, optimal in some sense, of a given functional of x, represented by the $\mathbf{R}^k$ vector g:

$$\widehat{(g,x)} = (f,y) \tag{2.10}$$

From the equation

$$(f,y) = (f,Ax) + (f,\nu) = (A^T f, x) + (f,\nu) \tag{2.11}$$

it is clear that, if $A^T f = g$, (f,y) is an unbiased estimate of (g,x), as $E\{(f,\nu)\} = 0$.

It can be easily proved, as illustrated below, that, looking for the linear unbiased estimate of (g,x) that minimizes the variance in the partially

stochastic case and the MSEE in the fully stochastic one, results equivalent
to the ones obtained for the direct estimation of x are retrieved.

The partially stochastic approach, as seen earlier, can be adopted if
$n > k$ and A is full-rank; consequently, owing to the very structure of the
matrix A^T, the equation $A^T f = g$ is not uniquely solvable for f. The
optimal f may be chosen as the one which minimizes the variance

$$E\{(f, y)^2\} = (f, C_{\nu\nu} f) \quad ; \quad \text{(2.12)}$$

the equation $A^T f = g$ plays the role of a constraint. The application of
the Lagrange multiplier technique leads to the minimization of the function

$$\frac{1}{2}(f, C_{\nu\nu} f) + (A^T f - g, \lambda) \quad \text{(2.13)}$$

where λ is a vector of Lagrange multipliers. The solution has the
expression

$$f = C_{\nu\nu}^{-1} A (A^T C_{\nu\nu}^{-1} A)^{-1} g \quad \text{(2.14)}$$

i.e.

$$\widehat{(g, x)} = (C_{\nu\nu}^{-1} A (A^T C_{\nu\nu}^{-1} A)^{-1} g, y) = (g, (A^T C_{\nu\nu}^{-1} A)^{-1} A^T C_{\nu\nu}^{-1} y) \quad \text{(2.15)}$$

that corresponds exactly to (2.4).
In the fully stochastic approach, the MSEE

$$E\{|(f, y) - (g, x)|^2\} = E\{|(f, Ax + \nu) - (g, x)|^2\} =$$
$$= E\{|(A^T f - g, x) + (f, \nu)|^2\} = (A^T f - g, C_{xx}(A^T f - g)) + (f, C_{\nu\nu} f) \quad \text{(2.16)}$$

is minimized with respect to f. The result is

$$f = (A C_{xx} A^T + C_{\nu\nu})^{-1} A C_{xx} g \quad \text{(2.17)}$$

that leads to

$$\widehat{(g, x)} = (A C_{xx} A^T + C_{\nu\nu})^{-1} A C_{xx} g, y) = (g, C_{xx} A^T (A C_{xx} A^T + C_{\nu\nu})^{-1} y) \quad \text{(2.18)}$$

corresponding to (2.8).

When these estimation problems are generalized to the case of functions
defined on continuous sets, that belong to infinite-dimensional vector spaces,
procedures similar to the ones leading to (2.15) and (2.18) can be adopted, but
a straightforward generalization of (2.4) and (2.8) is generally not possible.
A detailed discussion of these procedures is presented in the sequel.

3   Generalization to infinite-dimensional spaces

As remarked in the introduction, the theory of infinite-dimensional linear estimation is naturally formulated in the frame of Hilbert spaces.

Equation (2.1) is again considered, but in the present case the Hilbert spaces H_X and H_Y are introduced, and A is a linear operator from H_X to H_Y. The problem of defining a suitable space for the observational noise is left open at the moment. For example, in many situations it is natural to choose H_X and H_Y as subspaces of L^2, but a noise uncorrelated between different points, namely a white noise, cannot be properly described by such regular functions.

Anyway, the realizations of the "stochastic process" ν may be considered as stochastic functionals on H_Y; indeed, for example, as illustrated in section 4, the classical Wiener theory gives a precise meaning to functionals of the form $I(f) = \int f(t)\nu(dt)$, characterized by the stochastic behaviour $E\{I(f)\} = 0$, $E\{I^2(f)\} = \int f^2(t)dt$. A continuous, self-adjoint, positive-definite linear operator C on H_Y is assumed to exist, such that, for any pair f_1, $f_2 \in H_Y$,

$$E\{< f_1, \nu >_Y < f_2, \nu >_Y\} = < f_1, Cf_2 >_Y \qquad (3.1)$$

(cf. for example (Franklin 1970)). This operator is called *covariance operator* and is the generalization of the covariance matrix widely used in the previous section.

What is remárkable is that, even in the case that ν belongs to H_Y with probability 1, as a direct consequence of the definition, it turns out that the quadratic form $(\nu, C^{-1}\nu)_Y$ is infinite with probability 1, as recalled in (Sansò and Sona 1994); therefore, a straightforward generalization of the minimum principle introduced in (2.2) is not possible.

Consequently, minimum variance and minimum MSEE have to be adopted in order to obtain optimal estimates.

Usually the problem that has to be solved is the estimation of a certain class of functionals on the Hilbert space H_X; by virtue of Riesz's theorem they can be represented by scalar products in H_X : $g(x) \to < g, x >_X$.

In the sequel it will be shown that, under suitable conditions, it is possible to define an estimation procedure whose result formally looks as a generalization of the estimate (2.14) obtained in the finite-dimensional case. Only the partially stochastic approach, that is systematically discussed in (Sansò and Sona 1994) and has been adopted for the geodetic applications mentioned in the introduction, will be described in full detail. Most of the mentioned papers (for example, Franklin 1970, Mandelbaum 1984, Lehtinen et al. 1989), on the contrary, deal with the fully stochastic approach.

It turns out that, if the process ν is such that $E\{\|\nu\|_Y^2\} < \infty$, C is nuclear, i.e. its trace, $\sum < Ce_k, e_k >_Y = E\{\|\nu\|_Y^2\}$ is finite (and independent of the particular choice of the orthonormal basis $\{e_k\}$).

Consequently, the range of C is dense in H_Y, but does not coincide with H_Y, and its inverse C^{-1} is unbounded. Therefore the new norm $\|y\|_{Y'} = <y, Cy>_Y^{\frac{1}{2}} = \|C^{\frac{1}{2}}y\|_Y^{\frac{1}{2}}$ is not equivalent to $\|y\|_Y = <y, y>_Y^{\frac{1}{2}}$, and H_Y, equipped with the norm Y', is not complete. Let its completion be denoted by $H_{Y'}$; any element $y' \in H_{Y'}$ can be expressed as $y' = C^{-\frac{1}{2}}y, y \in H_Y$, and the operator $C^{-\frac{1}{2}}$ is an isometry from H_Y to $H_{Y'}$. As a particularly interesting consequence, if the process $\epsilon = C^{-\frac{1}{2}}\nu$ is introduced, it turns out that $E\{\|\epsilon\|_{Y'}^2\} = E\{\|\nu\|_Y^2\} < \infty$.

In order to determine the covariance structure of ϵ, first it has to be remarked that any $y' = C^{-\frac{1}{2}}y \in H_{Y'}$ may be viewed as a functional acting on the space $H_{Y''} = \{y'' = C^{\frac{1}{2}}y, y \in H_Y\}$ according to the coupling rule $<y_2', y_1''> = <C^{-\frac{1}{2}}y_2, C^{\frac{1}{2}}y_1> = <y_2, y_1>_Y$. In other words, the Hilbert space $H_{Y'}$ is the dual of the space $H_{Y''}$, equipped with the norm $\|y''\|_{Y''} = \|C^{-\frac{1}{2}}y''\|_Y$, with the coupling above defined. Furthermore, the embeddings $H_{Y'} \subset H_Y \subset H_{Y''}$ are dense. This procedure is classical in functional analysis, where it is called *pivoting* (Aubin 1972):

Hence, for any $h \in H_{Y''}, h = C^{\frac{1}{2}}f, f \in H_Y$, $<h, \epsilon>_Y$ is defined almost surely, and $<h, \epsilon>_Y = <C^{\frac{1}{2}}f, C^{-\frac{1}{2}}\nu>_Y = <f, \nu>_Y$. Consequently

$$E\{<h, \epsilon>_Y^2\} = E\{<f, \nu>_Y^2\} = <f, Cf>_Y = \|h\|_Y^2 \qquad (3.2)$$

$<h, \epsilon>_Y$ is a random variable in the Hilbert space H_ϵ generated by $\{<h, e_i>\}$, where $\{e_i\}$ is a basis in $H_{Y''}$, with norm $\|\cdot\|^2 = E\{|\cdot|^2\}$. (3.2) shows that the operator $<\cdot, \epsilon>_Y$ is an isometry from the set of the elements of $H_{Y''}$, viewed as a subspace of H_Y with the H_Y-norm, to H_ϵ. Therefore the coupling $<\cdot, \epsilon>_Y$ can be extended to the whole H_Y in the sense of the convergence in H_ϵ:

$$<f, \epsilon>_Y = \lim_{H_\epsilon} <h_n, \epsilon>_Y \quad \text{where} \quad f = \lim_{H_Y} h_n \qquad (3.3)$$

From (3.2) and the related equality

$$E\{<h_1, \epsilon>_Y <h_2, \epsilon>_Y\} = <h_1, h_2>_Y \qquad (3.4)$$

compared with (2.19), it appears that the covariance operator for ϵ is the identity operator I. Consequently, $\|\epsilon\|_Y^2$ cannot have finite mean; furthermore, it can be proved (cf. Sansò and Sona 1994) that the realizations of ϵ do not belong to H_Y with probability 1.

Equation (3.4), applied to vectors e_i, e_j of an orthonormal basis, shows that the Fourier components of the noise ϵ are uncorrelated and have all the same variance. This is a typical character of the white noise, that is particularly relevant for applications; therefore the estimation procedure will be described in detail in this context.

Starting from an equation of the form (2.1) in a Hilbert space context, as illustrated at the beginning of the present section, first to both sides of the equation the operator $C^{-\frac{1}{2}}$ is applied, obtaining an expression of the form

$$w = Bx + \epsilon \qquad (3.5)$$

where $w = C^{-\frac{1}{2}}y$, $B = C^{-\frac{1}{2}}A$; all terms in (3.5) are in $H_{Y'}$. Now, it has been shown that the scalar product $< f, \epsilon >_Y$ is meaningful for any $f \in H_Y$ in the sense illustrated by equation (3.3). Assume in addition that B maps continuously H_X into H_Y (or, equivalently, that A is continuous from H_X into $H_{Y''}$), that can be done by suitably restricting the space H_X and introducing a sufficiently strong norm, as for example

$$\|x\|_X = \|Bx\|_Y = \|Ax\|_{Y''} \qquad (3.6)$$

or anyone equivalent. Then the scalar product $< f, Bx >_Y$ is meaningful too, and it is possible to write the equation

$$< f, w >_Y = < f, Bx >_Y + < f, \epsilon >_Y = < B^{\dagger}f, x >_X + < f, \epsilon >_Y \qquad (3.7)$$

On the analogy of the finite-dimensional case, the optimal unbiased estimate of a functional $< g, x >_X$ of the field x by means of a functional $< f, w >_Y$ of the "observable" w is determined by minimizing the variance of $< f, \epsilon >_Y$, that,as previously remarked, is equal to $\|f\|_Y^2$, under the constraint $B^{\dagger}f = g$.

The usual technique leads to the solution

$$f = B(B^{\dagger}B)^{-1}g \qquad (3.8)$$

The invertibility of $B^{\dagger}B$ is ensured if a norm equivalent to (3.6) is adopted for H_X . From the equalities

$$< f, w >_Y = < B(B^{\dagger}B)^{-1}g, C^{-\frac{1}{2}}y >_Y = < C^{-\frac{1}{2}}B(B^{\dagger}B)^{-1}g, y >_Y \qquad (3.9)$$

replacing B with its expression $C^{-\frac{1}{2}}A$, an equality analogous to the first one in (2.15) is obtained, namely

$$< \widehat{g, x} >_X = < C^{-1}A(A^{\dagger}C^{-1}A)^{-1}g, y >_Y \qquad (3.10)$$

4   Linear estimation in function spaces

Assume now that the Hilbert spaces involved in the linear estimation problem are spaces of functions defined on a continuous set of points in a finite-dimensional space or manifold; in particular, let H_Y be equipped with the well-known L^2 scalar product:

$$< f_1, f_2 >_Y \rightarrow \int f_1(t)f_2(t)dt \tag{4.1}$$

As previously illustrated, in the case of white noise, for which the co-variance operator is proportional to the identity, the realizations of ϵ are with probability 1 not functions in L^2 ; they are generally represented by stochastic measures, $< f, \epsilon >_Y \rightarrow \int f(t)\mu_\epsilon(dt)$, and

$$E\{\int f_1(t)\mu_\epsilon(dt) \int f_2(t)\mu_\epsilon(dt)\} = \sigma_0^2 \int f_1(t)f_2(t)dt \tag{4.2}$$

The constant σ_0^2 is in some way related to the amount of uncertainty arising from the presence of an observational noise; the fact that it is a constant means that the uncertainty is uniform over the whole domain. More generally, (4.2) may be modified by introducing a variable weight:

$$E\{\int f_1(t)\mu_\epsilon(dt) \int f_2(t)\mu_\epsilon(dt)\} = \int \frac{f_1(t)f_2(t)}{\rho(t)} dt \tag{4.3}$$

in order to take into account a non-uniform uncertainty corresponding, in the discrete case, to different accuracies or different densities of the individual measurements.

Formulas (4.2) and (4.3) describe the stochastic properties of Wiener measures (Sansò 1988); they can be obtained starting from discrete measurements with independent observational noises and taking the limit for indefinitely increasing number of measurements, preserving their relative density and the amount of uncertainty per unit area. The latter requirement is fulfilled by increasing the variance of individual measurements proportionally to their density.

This trick has obviously no physical meaning. Observations of physical reality are discrete, with a finite number of measurements, whose density and accuracy are assumed to be known and contribute to the global information in a given area. They are therefore used to determine the constant σ_0 in (4.2) or the function $\rho(t)$ in (4.3); the continuum language is just a device to construct, with simple procedures, estimation algorithms, and its role is similar to the one of differential and integral calculus in deterministic classical physics.

To describe in a more specific case the estimation procedure, assume that a functional of a field $x(t)$ defined on a finite-dimensional domain T (for example, a region in space or a portion of a surface) has to be estimated; let this functional be expressed as $\int_T g(t)x(t)dt$. The observables $y_i(\tau)$ are defined on domains D_i (for example, in boundary-value problems T is the solution domain and D is its boundary) and are related to the field by suitable operators: $y_i = B_i x$. As the noise is not represented by a function, but rather by a measure, the observation equations cannot be expressed pointwise, but can be written for example in infinitesimal domains:

$$\mu_{Y_i}(d\tau) = B_i x(\tau) d\tau + \mu_{\epsilon_i}(d\tau) \tag{4.4}$$

where $\mu_{\epsilon_i}(dt)$ are mutually uncorrelated, and each of them fulfils a relation similar to (4.2) or (4.3). Equivalently, the variance of the noise can be differentially expressed as

$$E\{\mu_{\epsilon_i}(d\tau)\mu_{\epsilon_i}(d\tau')\} = \frac{d\tau\, d\tau'}{\rho_i(\tau)}\delta(\tau - \tau') \tag{4.5}$$

What has to be determined is an estimate linear with respect to the observables:

$$\int_T \widehat{g(t)x(t)}dt = \sum_i \int_{D_i} f_i(\tau)\mu_{Y_i}(d\tau) \tag{4.6}$$

The optimality criterion is given by the minimization of the quantity

$$E\{[\sum_i \int_{D_i} f_i(\tau)\mu_{\epsilon_i}(d\tau)]^2\} = \sum_i \int_{D_i} f_i(\tau)^2 d\tau \tag{4.7}$$

The constraint is a uniform unbiasedness condition:

$$\sum_i \int_{D_i} f_i(\tau)B_i x(\tau)d\tau = \int_T g(t)x(t)dt \tag{4.8}$$

that expresses the fact that, if the data are noiseless, the estimate gives the exact value of the functional.

The procedure to obtain the estimation formulas follow essentially the lines illustrated in the previous sections. In some cases, for particular forms of the operators B_i and of the domains D_i, it is possible to obtain analytical formulas for the solution, as in the previously mentioned examples.

5   An example

In order to clarify the procedure previously described, a simple example, with interesting practical applications, is here illustrated. Assume that a quantity $x(t)$ depending on a continuous parameter - for example, the distance from a satellite to a receiver as a function of time - is measured in some time interval, say $[0,1]$, together with its derivative $\dot{x}(t)$. This is schematically the principle of a PRARE receiver, already mentioned in the introduction. The system (1.1) of observation equations has in this case the form

$$\begin{pmatrix} w_1 \\ w_2 \end{pmatrix} = \begin{pmatrix} x(t) \\ \dot{x}(t) \end{pmatrix} + \begin{pmatrix} \epsilon_1(t) \\ \epsilon_2(t) \end{pmatrix} = \begin{pmatrix} I \\ \frac{d}{dt} \end{pmatrix} x(t) + \begin{pmatrix} \epsilon_1(t) \\ \epsilon_2(t) \end{pmatrix} \tag{5.1}$$

or, following (4.4),

$$\begin{pmatrix} \mu_{W_1}(dt) \\ \mu_{W_2}(dt) \end{pmatrix} = \begin{pmatrix} I \\ \frac{d}{dt} \end{pmatrix} x(t)dt + \begin{pmatrix} \mu_{\epsilon_1}(dt) \\ \mu_{\epsilon_2}(dt) \end{pmatrix} \qquad (5.2)$$

The natural function space for $x(t)$ is the space $H^1([0,1])$ of functions belonging to $L^2([0,1])$ together with their first derivatives; then the vector function $w(t)$ belongs to $L^2([0,1]) \times L^2([0,1])$. The norms in H_X and H_Y are defined by

$$\begin{aligned} \|x\|_X^2 &= \int_0^1 (x^2(t) + \dot{x}^2(t))dt \\ \|w\|_Y^2 &= \int_0^1 (w_1^2(t) + w_2^2(t))dt \end{aligned} \qquad (5.3)$$

The operator $B = (I \quad d/dt)^T$ (where $(\cdot)^T$ denotes the algebraic transpose) is then clearly an isometry.

Assume further that the quantity to be estimated is simply the evaluation functional $x(\bar{t})$ at time $\bar{t}$, that is a continuous functional in H^1, as t is a one-dimensional variable. The covariance operator in $L^2([0,1]) \times L^2([0,1])$ is a 2×2-matrix of operators in $L^2([0,1])$, that in this case is assumed to be diagonal, with diagonal elements proportional to I; let their coefficients be σ_{01}^2 and σ_{02}^2. It can be proved that the evaluation functional is represented, as required by Riesz's theorem, by the H_X element

$$g(t) = -\frac{1}{2}e^{|t-\bar{t}|} + \frac{\cosh(1-\bar{t})}{2\sinh 1}e^t + \frac{e\cosh \bar{t}}{2\sinh 1}e^{-t} \qquad (5.4)$$

Before computing explicitly the solution, it is useful to illustrate in detail how the operator $B^\dagger$ acts. The equality $u = B^\dagger w$ in the present case means

$$\int_0^1 (w_1 x + w_2 \dot{x})dt \equiv < w, Bx >_Y = < u, x >_X \equiv \int_0^1 (ux + \dot{u}\dot{x})dt \qquad (5.5)$$

for any $x \in \mathcal{D}_B$, or equivalently

$$\int_0^1 (w_1 - u)x dt = -\int_0^1 (w_2 - \dot{u})\dot{x} dt \qquad (5.6)$$

(5.6) is fulfilled if $u(t)$ is so chosen that it satisfies the boundary-value problem

$$\begin{cases} w_1 - u = \frac{d}{dt}(w_2 - \dot{u}) & \text{in } [0,1] \\ \dot{u}(0) = w_2(0) \\ \dot{u}(1) = w_2(1) \end{cases} \qquad (5.7)$$

Now, the problem to be solved, in order to obtain an optimal estimate of $< g, x >_X$ in terms of a functional of the observations, $< f, w >_Y$, is the minimization of the quadratic expression $< f, Cf >_Y$, with f satisfying $B^\dagger f = g$. The application the method of Lagrange multipliers leads to the minimization of the expression

$$\frac{1}{2} < f, Cf >_Y - < B^\dagger f, \lambda >_X \tag{5.8}$$

that yields $Cf - B\lambda = 0$; then

$$(B^\dagger C^{-1} B)\lambda = g \tag{5.9}$$

the result f is obtained using the formula $f = C^{-1}B\lambda$. In view of (5.7), solving (5.9) for λ is equivalent to determining the solution of the boundary-value problem

$$\begin{cases} \frac{d}{dt}(\dot{g} - \frac{1}{\sigma_2^2}\dot{\lambda}) = g - \frac{1}{\sigma_1^2}\lambda & \text{in } [0,1] \\ \dot{g}(0) - \frac{1}{\sigma_2^2}\dot{\lambda}(0) = 0 \\ \dot{g}(1) - \frac{1}{\sigma_2^2}\dot{\lambda}(1) = 0 \end{cases} \tag{5.10}$$

that can be expressed in the simpler form

$$\begin{cases} \ddot{\gamma} - \alpha^2\gamma = (1 - \alpha^2)g & \text{in } [0,1] \\ \dot{\gamma}(0) = 0 \\ \dot{\gamma}(1) = 0 \end{cases} \tag{5.11}$$

where $\gamma = g - (1/\sigma_2^2)\lambda$; $\alpha^2 = \sigma_2^2/\sigma_1^2$. Hence the solution of (5.10) can be expressed as

$$\lambda(t) = \sigma_2^2(g(t) - \gamma(t)) \tag{5.12}$$

and $\gamma(t)$ is explicitly expressed by

$$\gamma(t) = (1 - \alpha^2) \int_0^1 G(t, \tau; \alpha)g(\tau)d\tau \tag{5.13}$$

where $G(t, \tau; \alpha)$ is Green's function of (5.11):

$$G(t, \tau; \alpha) = -\frac{1}{2\alpha}e^{\alpha|t-\tau|} + \frac{\cosh(1 - \tau)}{2\alpha \sinh \alpha}e^{\alpha t} + \frac{e^\alpha \cosh \tau}{2\alpha \sinh \alpha}e^{-\alpha t} \tag{5.14}$$

Then, using for f the formula following (5.9), the explicit analytical solution of the original problem can be written as

$$< \widehat{g, x} >_X = \alpha^2 \int_0^1 \left(g(t) - (1 - \alpha^2) \int_0^1 G(t, \tau; \alpha)g(\tau)d\tau \right) \mu_{W_1}(dt) + $$
$$+ \int_0^1 \left(\dot{g}(t) - (1 - \alpha^2) \int_0^1 \frac{\partial G}{\partial t}(t, \tau; \alpha)g(\tau)d\tau \right) \mu_{W_2}(dt) \tag{5.15}$$

As one can see, all that might appear as a pure analytical exercise, yet the very practical problem stated at the beginning could now find a practical solution by simply discretizing formula (5.15); this procedure has actually been adopted and seems to work in a number of examples.

References

Aubin, J.P., 1972. Approximations of Elliptic Boundary-Value Problems, Wiley Interscience.

Barzaghi, R. and F.Sansò, 1994. A continuous approach to the integration of tomographic and gravimetric signals, preprint.

Forlani, G., F. Sansò and S. Tarantola, 1994. Digital Photogrammetry: Experiments with the Continuous Approach, in Proc. of the Symp. "Primary Data Acquisition and Evaluation", Como, Int. Arch. of Photogrammetry and Remote Sensing, **30**, Part 1, 229-236.

Franklin, J.N., 1970. Well-Posed Stochastic Extensions of Ill-Posed Linear Problems, Jour. of Math. Analysis and Appl., **31**, 682-716.

Lamperti, J., 1977. Stochastic Processes - A Survey of the Mathematical Theory, Appl. Math. Sci. **23**, Springer-Verlag.

Lehtinen, M.S., L. Päivärinta and E. Somersalo, 1989. Linear inverse problems for generalised random variables, Inverse Problems, **5**, 599-612.

Mandelbaum, A., 1984. Linear Estimators and Measurable Linear Transformations on a Hilbert Space, Z. Wahrscheinlichkeitstheorie verw. Gebiete, **65**, 385-397.

Moritz, H., 1980. Advanced Physical Geodesy, H. Wichmann Verlag.

Rozanov, Yu. A., 1971. Infinite-Dimensional Gaussian Distributions, Proc. of the Steklov Inst. of Math., **108**, English transl. American Math. Soc.

Sacerdote, F. and F. Sansò, 1985. Overdetermined boundary value problems in physical geodesy, Man. Geod., **10**, 195-207.

Sansò, F., 1988. The Wiener Integral and the Overdetermined Boundary Value Problems of Physical Geodesy, Manuscripta Geodaetica, **13**, 75-98.

Sansò, F. and G.Sona, 1995. The theory of optimal linear estimation for continuous fields of measurements, Man. Geod., **20**, 204-230.

Skorohod, A.V., 1974. Integration in Hilbert Space, Ergebnisse der Mathematik und ihrer Grenzgebiete, Band 79, Springer-Verlag.

Least Squares Adjustment of High Degree Spherical Harmonics

Wolf-Dieter Schuh

Technical University, Mathematical Geodesy and Geoinformatics, Austria

1 Introduction

In spite of new developments in computer design, the standard least squares procedure for the estimation of spherical harmonic coefficients for high degree gravitational fields exceeds present capabilities. Only gridded data can be computed, because of orthogonality relations a sparse normal equation system results. In general, observations can not be performed on a regular grid and especially in connection with orbit dependent measurements, where an area between two parallels is well covered, no gridded data set is established. This is the reason why in this case the least squares harmonic analysis results in a dense normal equation system. The knowledge about the sparse structure of gridded data can be used for tailored preconditioning strategies. Therefore an optimal reordering of the sparse normal equation system brings benefit for gridded data in a direct approach as well as for well covered, but not gridded data in an iterative approach. This paper reports on an optimal reordering scheme for high degree spherical harmonic analysis. Two different scenarios, only gridded data and combined systems with not gridded low order (dense) and gridded high order (sparse) measurements are investigated. In both cases an optimal reordering strategy brings about the possibility to decompose (e.g. Cholesky-factorization) the normal equations without any fill-in.

The mathematical representation of earth's gravity potential is conveniently done with normalized spherical harmonics,

$$V(r,\theta,\lambda) = \frac{GM}{r}\left\{1 + \sum_{\ell=1}^{\ell_{max}}\sum_{m=0}^{\ell}\left(\frac{a}{r}\right)^{\ell}\bar{P}_{\ell m}(\cos\theta)\left(\bar{C}_{\ell m}\cos m\lambda + \bar{S}_{\ell m}\sin m\lambda\right)\right\}.$$

(1)

GM ...	geocentric gravitational constant	ℓ ...	degree
a ...	semimajor axis	m ...	order
r ...	radius vector	ℓ_{max} ...	maximum degree
θ ...	polar distance (colatitude)	$\bar{P}_{\ell m}$ ...	fully normalized associated Legendre functions of the first kind
λ ...	longitude	$\bar{C}_{\ell m}, \bar{S}_{\ell m}$ ...	harmonic coefficients

All groups of observations can be expressed with the help of the zero, first or second order derivative of the potential. Since the early 1980s a growing number of spherical harmonic models of the earth's gravity field has become available up to very high degrees and orders. Expansions up to degree and order 180 or 360, the latter with more than 130 000 individual coefficients, are widely used for many purposes. All the applications of spherical harmonics can be divided into two tasks:

- determination of the harmonic coefficients $\bar{C}_{\ell m}$ and $\bar{S}_{\ell m}$ using observed quantities of the gravity potential (*'global spherical harmonic analysis'*) and
- computation of the gravity potential and/or derivatives thereof at a special point P $(r_P, \theta_P, \lambda_P)$ with known harmonic coefficients $\bar{C}_{\ell m}$ and $\bar{S}_{\ell m}$ (*'global spherical harmonic synthesis'*).

Due to types and accuracies of the observations, but also due to the point location and distribution of the measurements a more or less high degree expansion can be determined using least squares techniques or analytic approaches.

In the continuous case the global spherical harmonics have the nice property of orthogonality. In reality, in most of the cases only discrete function values are available and therefore, only orthogonalities in the discrete case are of interest. If we discretize the Legendre function the orthogonalities are lost, but this does not hold for the cosine and sine series under special assumptions. Therefore, if the data set meets these assumptions, orthogonalities between different orders m will occur and sparse normal equations are preserved. But these assumptions are very restrictive because all the data (along a parallel) should be

- of the same type,
- with homogeneous weights,
- on an equiangular grid,
- without any gap.

We denote such a data set as *'grid'* data. In addition the (anti-)symmetry

$$\bar{P}_{\ell m}(-\cos\theta) = (-1)^{(\ell-m)}\bar{P}_{\ell m}(\cos\theta) \tag{2}$$

of the Legendre function can be used if the data set is symmetric with respect to the equator. Orthogonalities appear between even and odd degree elements within the same order.

Ungridded data of satellite missions essentially cause new investigations. Due to high sampling rates quite a number of measurements with a good coverage can be collected. But this data sets are not gridded, because the measurements can only be performed along the orbit. On the other side, low orbit satellites bring about a lot of information to the higher order coefficients of the gravity potential also. In future missions it will be intended to use the satellite gravity gradiometry (SGG) technique to gain better information of the high frequency part of the gravity field. This technique brings about only poor information on the low frequency part and therefore it will be completed by a satellite-to-satellite tracking (SST) system. This allows to determine a high precision orbit as well as the low frequency part of the gravity field. The standard least squares procedure for the estimation of high degree spherical harmonic coefficients with these data sets exceeds present computer capabilities. Therefore, special techniques have to be developed to accomplish this task. Tailored procedures must be developed with the help of orthogonality considerations, special numbering schemes and optimized preconditioning strategies. Also the use of parallel resources and robust statistical methods have to be investigated.

This paper gives a report about orthogonality considerations in connection with numbering schemes. Especially the combination of gridded data for the high frequency part with ungridded informations within the low frequency part is of interest. Due to the orthogonality relations of gridded data sets a sparse normal equation system can be expected. Therefore an optimal numbering scheme allows an efficient storage and computation of these sparse systems. This knowledge can be used in two directions: First, if gridded data are available high degree spherical harmonic analysis can be performed very efficient (Colombo 1981) and secondly if the data sets are not gridded but cover the whole region between two parallels very regularly, then the information of the idealized sparse systems allows to find tailored preconditioning strategies to speed up iterative procedures very well (Schuh 1995).

2 Gridded Data: Orthogonalities and Numbering Schemes

The commutativity law of addition in Eq. (1) allows different choices for the ordering of the summation sequences. Fig. 1 illustrates the pattern of all the

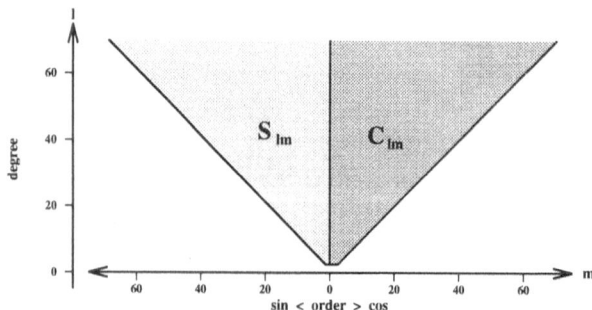

Fig. 1. Triangular scheme with information about the coefficients of the spherical harmonics.

spherical harmonic coefficients. This triangular scheme can be divided into two sub-triangles. The left triangle contains the sine coefficients and the right triangle contains the cosine coefficients. The degree ℓ increases from bottom to top, and the order increases from the center to the left (right) for the sine (cosine) coefficients.

The elements within a horizontal line contain coefficients of the same degree, for example $\bar{C}_{\ell,*}$ or $\bar{S}_{\ell,*}$, whereas vertical lines contain coefficients of the same order, $\bar{C}_{*,m}$ or $\bar{S}_{*,m}$.

To find short and terse key words for the numbering schemes we identify the sums only by their indices (ℓ... sum over all degrees, m... sum over all orders). The sequence of the sums from the left to the right is reflected by the sequence of the indices (first (outer) sum ... first character, second (nested) sum ... second character, and so on). The key words carry no information about the limits, the direction and the spacing step of the sum. Only increasing loops with step '1' are used. It is only necessary to restrict the indices to even and odd numbers. This is done with the subscript 'e' or 'o'. For example ℓ_e means only even degrees. Upper letters C and S symbolically stand for the $\bar{C}_{\ell m}$ and $\bar{S}_{\ell m}$ coefficients. Brackets are used to emphasize the processing sequence.

The numbering schemes can be divided into two groups. The first group, **organized per degree**, is characterized by an outer loop over all degrees ℓ and an inner loop with increasing order m. The cosine and sine coefficients may alternate or not. Two members of this group are:

$$\mathcal{NUM}\left\{\ell m(C+S)\right\} \quad \Longrightarrow \quad \sum_{\ell=2}^{\ell_{max}}\left(\bar{C}_{\ell 0}+\sum_{m=1}^{\ell}\left(\bar{C}_{\ell m}+\bar{S}_{\ell m}\right)\right)$$

$$\mathcal{NUM}\left\{\ell m(C)+\ell m(S)\right\} \quad \Longrightarrow \quad \sum_{\ell=2}^{\ell_{max}}\sum_{m=0}^{\ell}\bar{C}_{\ell m}+\sum_{\ell=2}^{\ell_{max}}\sum_{m=1}^{\ell}\bar{S}_{\ell m} \quad .$$

These two schemes are widely used because they allow a simple management of the coefficients, an easy assignment of the proper storage place for each element and an easy truncation of the spherical harmonic series at a certain degree and order.

The second group of numbering schemes, **organized per order**, is characterized by an outer loop over all orders m and an inner loop with increasing degree ℓ. Within each order first all cosine coefficients and then all sine coefficients are arranged. These numbering schemes reflect the sparsity of the normals due to the orthogonalities in a block-diagonal structure. Depending on the utilization of the symmetry with respect to the equator two schemes are used:

$$\mathcal{NUM}\left\{m(\ell C+\ell S)\right\} \quad \Longrightarrow \quad \sum_{\ell=2}^{\ell_{max}}\bar{C}_{\ell 0}+\sum_{m=1}^{\ell_{max}}\left(\sum_{\ell=\max(2,m)}^{\ell_{max}}\bar{C}_{\ell m}+\sum_{\ell=\max(2,m)}^{\ell_{max}}\bar{S}_{\ell m}\right)$$

$$\mathcal{NUM}\left\{m(\ell_e C+\ell_o C+\ell_e S+\ell_o S)\right\} \quad \Longrightarrow \quad \sum_{\ell=1}^{\text{Int}(\ell_{max}/2)}\bar{C}_{2\ell,0}+\sum_{\ell=1}^{\text{Int}(\ell_{max}/2)}\bar{C}_{2\ell+1,0}+$$

$$+\sum_{m=1}^{\ell_{max}}\left(\sum_{\ell=m}^{\text{Int}(\ell_{max}/2)}\bar{C}_{2\ell,m}+\sum_{\ell=m}^{\text{Int}(\ell_{max}/2)}\bar{C}_{2\ell+1,m}+\sum_{\ell=m}^{\text{Int}(\ell_{max}/2)}\bar{S}_{2\ell,m}+\sum_{\ell=m}^{\text{Int}(\ell_{max}/2)}\bar{S}_{2\ell+1,m}\right) \quad .$$

The resulting structure of the normal equations using gridded data distribution is a block-diagonal structure. In the first case the size of the block starts form $(\ell_{max}-1)\times(\ell_{max}-1)$ and ends by 1×1. In the second case each large block is represented by two blocks with half the size.

If we use gridded data we can use the last two numbering schemes to assemble block-diagonal structured normal equations and perform a high degree spherical analysis. This technique is based on a work by Colombo (1981).

3 Numbering Schemes for Combined Data Sets

Because of the poor performance of satellite gravity gradiometry (SGG) with respect to long wavelength, it is important to combine these data with other informations. Especially satellite-to-satellite tracking (SST) data build a nice supplementation, because they provide strong information at exactly those frequencies where SGG is poor. The low degree/order system coming from SST depends on the configuration of the tracking satellites. The configuration changes from situation to situation. This irregularity results in a dense normal equation system. Therefore the combination of a full low degree/order system with a block-diagonal high degree/order system is required. If the usual numbering scheme $\mathcal{NUM}\{m(\ell_e C + \ell_o C + \ell_e S + \ell_o S)\}$ is used, the full low degree/order system produces a *chess-board pattern* outside the block-diagonal structure. Fig. 2a illustrates this behaviour and in addition, Fig. 2b shows the fill-in elements which are generated during the reduction step (e.g. Cholesky-factorization).

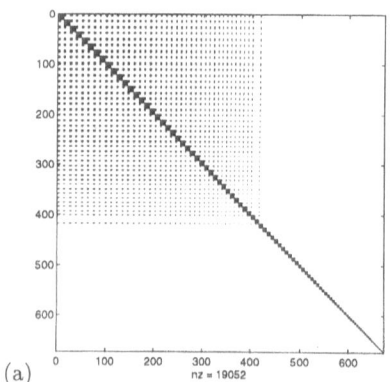
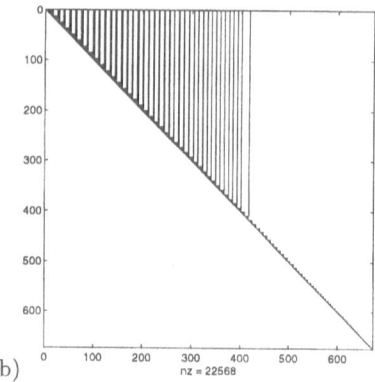

Fig. 2. Sparsity of the (reduced) normal equations, numbering scheme $\mathcal{NUM}\{m(\ell_e C + \ell_o C + \ell_e S + \ell_o S)\}$. Block-diagonal structure up to degree 25, full system up to degree 10. (a) normal equations (b) reduced normal equations.

An appropriate organization scheme to store the normals and to manage the fill-in is a profile structure (variable bandwidth). Unfortunately the storage requirements increase quadratically with respect to lower degrees as well as with respect to the higher degree field. Balmino (1993) and Bosch (1993) introduce a numbering scheme which works with different regions within the pattern of spherical harmonic coefficients. The first region is defined by the full

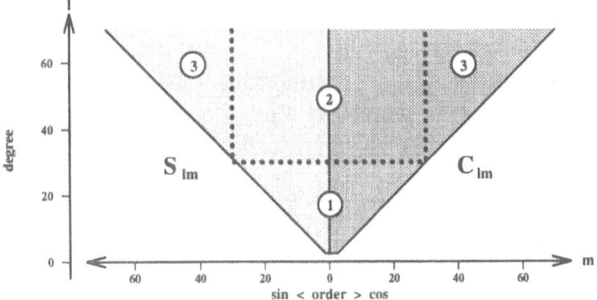

Fig. 3. Numbering scheme *block* $\mathcal{NUM}\{m(\ell_e C + \ell_o C + \ell_e S + \ell_o S)\}$.

low degree/order model (cf. Fig. 3). After these coefficients all coefficients with higher order are arranged. At last the coefficients with higher degree and order are numbered. Fig. 4a illustrates the normal equations based on this particular

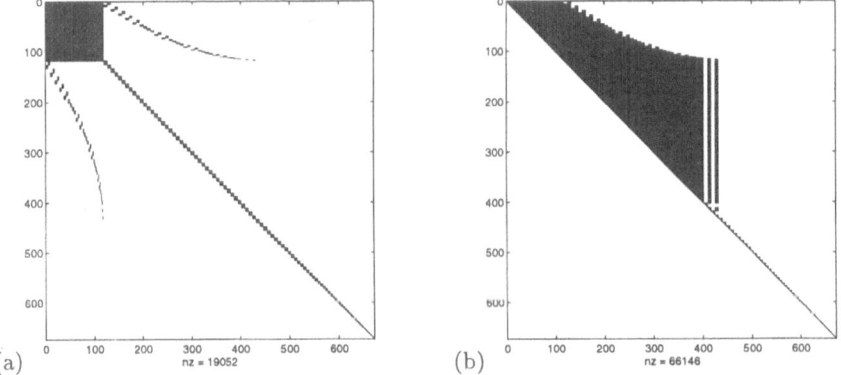

Fig. 4. Sparsity of the (reduced) normal equations, numbering scheme *block* $\mathcal{NUM}\{m(\ell_e C + \ell_o C + \ell_e S + \ell_o S)\}$. Block-diagonal structure up to degree 25, full system up to degree 10. (a) normal equations (b) reduced normal equations.

blocked numbering scheme. The non-zero elements are packed in the first part of the normals. Two thin lines, like the tails of a kite, represent the correlations between low and high degree coefficients within the same order. Unfortunately, during a factorization step the region below this tail is filled up with non-zero elements. Fig. 4b shows the result of this factorization step. This numbering scheme doesn't preserve the symmetry with respect to the equator and produces a lot of fill-in elements. The advantage of this method is the concentration of the fill-in elements to the first part of the normal equations.

It's not necessary to compute and compare the number of fill-in elements of the above numbering schemes, just because another numbering scheme can be employed which produces **no fill-in** elements during the whole factorization process.

The main idea behind the new numbering scheme is the reversal of the ordering. Not the figure of an ascending but of a descending kite should appear. Therefore first the high degree and high order elements are numbered. Next the high degree and low order elements are arranged and the full block of low degree and order comes at the end (cf. Fig. 5). After a first look to this structure we

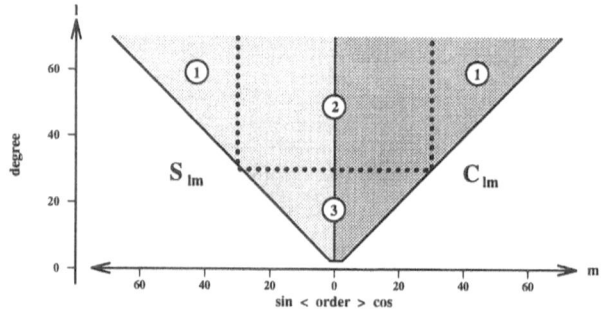

Fig. 5. Numbering scheme *reverse block* $\mathcal{NUM}\left\{m(\ell_e C + \ell_o C + \ell_e S + \ell_o S)\right\}$.

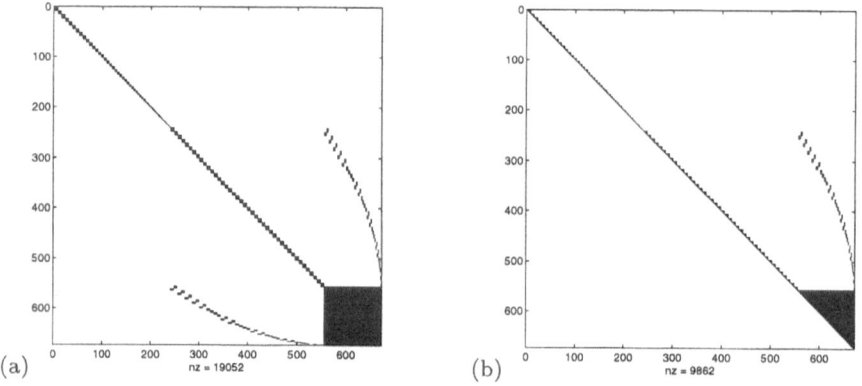

(a)

(b)

Fig. 6. Sparsity of the (reduced) normal equations employing *reverse block* numbering scheme $\mathcal{NUM}\left\{m(\ell_e C + \ell_o C + \ell_e S + \ell_o S)\right\}$. Block-diagonal structure up to degree 25, full system up to degree 10. (a) normal equations (b) reduced normal equations.

would expect only few fill-in elements in the area below the lateral tails of the kite. But an exact analysis shows, that in each column and row of this lateral tail region only one nonzero block exists. Therefore the typical scalar product of two columns during the factorization step vanishes and no fill-in elements are produced.

Just by reversing the order of block numbering we achieve an arrangement of the normals such that during the factorization step no fill-in elements are generated (cf. Fig. 6). The normals can therefore be stored in a fixed structure. Parallelism within the factorization process can be easily exploited and the amount of process intercommunication is reduced dramatically.

4 Concluding Remarks

This newly developed numbering scheme, especially for the combination of dense low degree fields and block-structured high degree fields produce no fill-in during the factorization step. Therefore very efficient solutions can be performed:
- with a combined data set (ungridded low frequency data and gridded high frequency data) a very efficient direct solution can be determined and
- in the case of ungridded low frequency data and irregular distributed high frequency data, which densely cover the area between two parallels, the knowledge about the sparse structure can be used to find optimal preconditioning techniques.

The fast iterative solution mainly bases on the possibility to find a representative matrix, which can be easily computed, quickly solved and on the other side represents as close as possible all informations of the whole system. If the block structured high degree model combined with a dense low degree model is used as a mask for the representative matrix both facts are fulfilled, easiest computation and a close relation between the behaviour of the rigorous dense system and the sparse preconditioner.

Numerical simulations demonstrate that these iterative techniques are well tailored and ideally suited for the least squares solution of harmonic coefficients of high degree models based on satellite-to-satellite tracking data and satellite gravity gradiometry data. The very favorable convergence rate with a spectral radius up to 0.04 for the preconditioned conjugate gradient algorithm permits the solution of a system with 2,597 unknowns within 3 to 5 iteration steps. The fast solution is based on the possibility to find a fit and proper preconditioner. This is possible because of the newly developed numbering scheme.

References

Balmino George (1993): The Spectra of the Topography of the Earth, Venus and Mars. Geophysical Research Letters, Vol. 20, No. 11, pp. 1063-1066.

Bosch Wolfgang (1993): A Rigorous Least Squares Combination of Low and High Degree Spherical Harmonics. Presented Paper on IAG General Meeting, Beijing 1993.

Colombo Oscar L. (1981): Numerical Methods for Harmonic Analysis on the Sphere. Department of Geodetic Science, Report No. 310, Ohio State University, Columbus, Ohio.

SCHUH Wolf-Dieter (1995): SST/SGG Tailored Numerical Solution Strategies. ESA-Project CIGAR III / Phase 2, WP 221, Final Report: Part 2.

Piecewise Polynomial Solutions to Linear Inverse Problems

Per Christian Hansen[1] and Klaus Mosegaard[2]

[1] UNI•C, Technical University of Denmark, Lyngby, Denmark
[2] Niels Bohr Institute for Astronomy, Physics and Geophysics, Copenhagen University, Copenhagen, Denmark

1 Introduction

Many of the systems studied in earth science, astronomy and physics display a high degree of order. Long term equilibration of a heterogeneous system in, e.g., a gravitational field results in a fractionation of material and the formation of *layers* of approximately uniform density. Examples of this is the existence of geological layers (sedimentary layers or layers formed by fractional crystallization in a magma chamber), and the internal density layering of the deep Earth, formed early in the Earth's history.

The detection of layered structures in a physical object often enables a reconstruction of its composition and its history of formation. For instance, the existence of a major boundary in the Earth at a depth of approximately $3000km$ (conjectured by Oldham 1906 and confirmed by Jeffreys 1939) was, together with the existence of the Earth's magnetic field, and astronomical information about the Earth's moment of inertia and the average composition of cosmical material, interpreted as the presence of a liquid core, consisting predominantly of iron.

Our knowledge of the internal structure of physical systems comes from experimental data and the inverse theory that connects data with a (parametrized) model of the system. In cases where the theory is linear, a number of well established inverse methods are available for the analysis of the properties of model parameters. One of the most widely used techniques is the least squares technique that minimizes the 2-norm of the data residual, i.e., the difference between observed and modeled data.

Write the linear inverse problem as $Kf = g$, where f represents the model (either a function or a parameter vector), g is the observed data, and K is an operator (e.g., a matrix) that connects the data with the model. Linear inverse problems are usually "almost underdetermined" in the sense that the system $Kf = g$ is close to an exactly underdetermined one (if K is a matrix, then K is "almost rank deficient"). This is an intrinsic property of K, and it is the heart of the difficulties associated with the mathematical and numerical treatment of inverse problems. In order to compute a stable solution by inverse methods, one must impose additional information about the desired solution, often in the form of prior information. See, e.g., Hanke & Hansen 1993 and Hansen 1994.

An important property of any inverse method is the way it (explicitly or implicitly) introduces prior information into the solution. Minimization of the 2-norm of the model f often corresponds to an assumption that the solution has *minimum energy*. One can also assume other a priori model properties as, for instance, *minimum flatness* and *minimum roughness*, which leads to minimization of the 2-norm of the first and the second derivative, respectively, of the model, cf. [9, §3.9]. The main reason for these a priori assumptions is often rather mathematical convenience than a physical knowledge about the considered system. This is especially true when systems that are likely to have an internal layering or shell structure are studied. None of the pure 2-norm methods will in this case give a physically (e.g., geologically) meaningful result, because these methods are unable to produce models containing sharp boundaries.

Other properties of the model f can be imposed by replacing the 2-norm of f, or one of its derivatives, by a different norm or norm-like function. For example, the entropy function of f is popular as an a priori assumption in certain applications (see Smith & Grandy 1985). The 1-norm has also been used successfully in some applications, see Levy & Fullager 1981 and Santosa & Symes 1986.

In this paper we present a new algorithm, called *PP-TSVD*, that incorporates the 1-norm in a different fashion. This enables us to compute models f that are piecewise polynomials, without specifying a priori the positions of the break points between the polynomial pieces. In particular, we can compute piecewise constant functions. Such functions can often be considered as the simplest models that describe certain phenomena, and—as discussed above—they are indeed useful in geophysics as well as other areas.

Our algorithm can also be used as a "preprocessor" for other algorithms, such as the one in Simcik & Linz 1994, in which the discontinuities must be explicitly incorporated. In this case, our algorithm can be used to detect and locate the discontinuities.

The *PP-TSVD* algorithm, plus some necessary background, is presented in §2, and in §3 we illustrate the use of our algorithm by several numerical examples. Throughout the paper, if x is a vector of length n then $\|x\|_1 = |x_1| + \cdots + |x_n|$ and $\|x\|_2 = (x_1^2 + \cdots + x_n^2)^{1/2}$ denote the 1- and 2-norm of x, respectively.

2 The Algorithm

Our algorithm is designed for treating linear systems of equations with an $m \times n$ coefficient matrix A and right-hand side b, and we assume that the elents of the solution vector x are "samples" of the function taht we wish to reconstruct. There are no restrictions on m and n. We are particularly interested in situations where A has a very large condition number—which is the mathematical formulation of saying that A is "almost rank deficient." Such linear systems typically arise from the discretization of ill-posed problems.

The basis for our algorithm is the *singular value decomposition (SVD)* of the coefficient matrix A. If $m \geq n$, then the SVD takes the form

$$A = U\Sigma V^T = \sum_{i=1}^{n} u_i \sigma_i v_i^T, \tag{1}$$

where $U = (u_1, \ldots, u_n)$ and $V = (v_1, \ldots, v_n)$ both have n orthonormal columns, i.e., $U^T U = V^T V = I_n$, and $\Sigma = \mathrm{diag}(\sigma_1, \ldots, \sigma_n)$ is a diagonal matrix whose diagonal elements σ_i, the singular values, satisfy

$$\sigma_1 \geq \sigma_2 \geq \cdots \geq \sigma_n \geq 0.$$

Ill-conditioned matrices are characterized by A having one or more very small singular values, relative to σ_1, and hence the condition number of A, given by σ_1/σ_n, is very large. For more details about the SVD see, e.g., Golub & Van Loan 1989 or §6.7 in Menke 1989.

2.1 Background: The $TSVD$ and $MTSVD$ Methods

A popular method for treating linear equations with a highly ill-conditioned coefficient matrix is the *truncated SVD* (*TSVD*) method. In this method, one simply ignores all the small singular values, and approximates A by a rank-k matrix A_k in which only the largest k singular values are retained:

$$A_k = \sum_{i=1}^{k} u_i \sigma_i v_i^T \quad \text{with} \quad k < n. \tag{2}$$

In this way, the "almost rank deficient" matrix A is replaced by an exactly rank-deficient one given by (2) that has a well defined null space of dimension $n - k$, spanned by the right singular vectors $v_{k+1}, \ldots, v_n$.

Then the original linear system of equations is replaced by the problem

$$\min \|x\|_2 \quad \text{subject to} \quad \|A_k x - b\|_2 = \text{minimum}, \tag{3}$$

and the solution to (3), called the $TSVD$ solution x_k, is given by

$$x_k = \sum_{i=1}^{k} \frac{u_i^T b}{\sigma_i} v_i. \tag{4}$$

The 2-norm of x_k satisfies $\|x_k\|_2^2 = \sum_{i=1}^{k} \sigma_i^{-2}(u_i^T b)^2$, and thus $\|x\|_2$ increases monotonically with k. The *truncation parameter* k controls the amount of stabilization imposed on x_k: the smaller the k, the smaller the 2-norm of x_k, and the less information from the right-hand side b is actually used.

As mentioned in the Introduction, it is often more appropriate to minimize the solution's seminorm $\|Lx\|_2$ where L is, e.g., a discrete approximation to a derivative operator, such as the banded matrices

$$L_1 = \begin{pmatrix} 1 & -1 & & \\ & \ddots & \ddots & \\ & & 1 & -1 \end{pmatrix} \quad \text{or} \quad L_2 = \begin{pmatrix} 1 & -2 & 1 & \\ & \ddots & \ddots & \ddots \\ & & 1 & -2 & 1 \end{pmatrix}. \tag{5}$$

Without loss of generality, we can assume that L has full row rank (see §4.2 in Hanke & Hansen 1993). For such problems, Hansen, Sekii & Shibahashi 1992 proposed the *modified TSVD (MTSVD)* method, in which the seminorm $\|Lx\|_2$ is introduced into (3), i.e.,

$$\min \|Lx\|_2 \qquad \text{subject to} \qquad \|A_k x - b\|_2 = \text{minimum}. \tag{6}$$

To express the *MTSVD* solution $x_{L,k}$ to (6) in closed form, it is convenient to introduce the matrix V_k consisting of A_k's null vectors,

$$V_k = (v_{k+1}, \ldots, v_n). \tag{7}$$

Then it is proved in Hansen, Sekii & Shibahashi 1992 that the *MTSVD* solution $x_{L,k}$ consists of the *TSVD* solution x_k plus a modification, lying entirely in the null space of A_k and ensuring that the seminorm $\|Lx\|_2$ is minimum:

$$x_{L,k} = x_k - V_k(LV_k)^\dagger Lx_k, \tag{8}$$

where $(LV_k)^\dagger$ is the pseudoinverse of LV_k. From a computational point of view, the vector $z_k = (LV_k)^\dagger Lx_k$ is simply the least squares solution to the problem $\min \|(LV_k)z - Lx_k\|_2$, and z_k can therefore be computed by standard least-squares software. See Hansen, Sekii & Shibahashi 1992 for details.

2.2 The *PP-TSVD* Method

Our new algorithm for computing piecewise polynomial solutions, represented as "samples" in the solution vector x, is derived from the *MTSVD* method by replacing the 2-norm of Lx with the 1-norm. Thus, we arive at the problem

$$\min \|Lx\|_1 \qquad \text{subject to} \qquad \|A_k x - b\|_2 = \text{minimum}. \tag{9}$$

We refer to this as the *PP-TSVD* method, and we emphasize that *TSVD* plays a central role in (9) as a means for treating the ill-conditioning of A in a numerically stable way. As in the *TSVD* and *MTSVD* methods, the truncation parameter k in the *PP-TSVD* method acts as a means for controlling the stabilization imposed on the solution.

The *PP-TSVD* solution $\bar{x}_{L,k}$, like the *MTSVD* solution $x_{L,k}$, consists of the *TSVD* solution x_k plus a modification $-V_k w_k$, lying in A_k's null space, that ensures the minimization of $\|Lx\|_1$:

$$\bar{x}_{L,k} = x_k - V_k w_k. \tag{10}$$

But now—with the 2-norm replaced by the 1-norm—there is no explicit expression for the vector w_k that solves the linear ℓ_1-problem

$$\min \|(LV_k)w - Lx_k\|_1. \tag{11}$$

Efficient software is available for solving (11) numerically. Traditional algorithms are based on the simplex algorithm (see Chapter 8 in Menke 1989). In our implementation we use a new algorithm based on continuation methods (Madsen

& Nielsen 1993); this algorithm has the same complexity bound as the simplex-based algorithm, but it was found to be faster in most cases (the Fortran program was kindly provided by Assoc. Prof. Hans B. Nielsen from the Technical University of Denmark).

Below we summarize our algorithm, and we remark that if k is only changed slightly in Step 8, then the "warm start" feature of the ℓ_1-solver in Madsen & Nielsen 1993, using the old w_k as start vector, is useful.

ALGORITHM *PP-TSVD*
1. Compute the SVD: $A = U \Sigma V^T$.
2. Choose an initial value of k.
3. Compute the $TSVD$ solution $x_k = \sum_{i=1}^{k} \sigma_i^{-1} u_i^T b \, v_i$.
4. Form the matrix $V_k = (v_{k+1}, \ldots, v_n)$.
5. Solve the linear ℓ_1-problem min $\|(LV_k)w - Lx_k\|_1$ for w_k.
6. Compute the $PP-TSVD$ solution $\bar{x}_{L,k} = x_k - V_k w_k$.
7. Inspect the solution and residual (semi)norms.
8. If necessary, adjust k and go to 3.

The key to understanding why the $PP-TSVD$ solution $\bar{x}_{L,k}$ represents a piecewise polynomial is to identify $L\bar{x}_{L,k}$ as the residual vector of the ℓ_1-problem (11). Moreover, we make use of the fact that the solution to an $m \times n$ linear ℓ_1-problem always satisfies at least n equations exactly, such that the corresponding residual vector has at least n zero elements (see, e.g., [15, Thm. 6.2]).

Assume that L is an $(n-p) \times n$ matrix that approximates the pth derivative operator, and that $k \geq p$. Since LV_k has dimensions $(n-p) \times (n-k)$, the residual vector $L\bar{x}_{L,k}$, of length $n-p$, has at least $n-k$ zero elements, because the 1-norm forces $n-k$ equations to be satisfied exactly. Hence, we compute a solution vector $\bar{x}_{L,k}$ such that the vector $L\bar{x}_{L,k}$ is a "sparse spike train" with at most $(n-p) - (n-k) = k - p$ nonzero elements. Since this vector represents a sequence of delta functions, we conclude that the solution itself, represented by $\bar{x}_{L,k}$ and being the pth integral of the sequence of delta functions, is a piecewise polynomial of degree $p-1$ with at most $k-p$ break points. Moreover, the derivative of order $p-1$ is discontinuous across the break points. E.g., if $p = 1$ such that L approximates the first derivative operator, then $\bar{x}_{L,k}$ represents a piecewise constant function with at most k discontinuities, and if $p = 2$ such that L approximates the second derivative operator, then $\bar{x}_{L,k}$ represents a continuous function consisting of at most $k-1$ straight lines.

Thus, we see precisely how the parameter k controls the stabilization of the solution $\bar{x}_{L,k}$: the smaller the k, the "simpler" the model—at the expense of neglecting information from the right-hand side b.

The choice of the appropriate value of the parameter k is outside the scope of this paper, but we note that standard parameter-choice methods for linear regularization methods (see §5.3 & §5.4 in Hanke & Hansen 1993) also apply to our $PP-TSVD$ method. In this connection, it is often useful to monitor the norm or seminorm of the solution and the residual norm as k varies. For our $PP-TSVD$ algorithm, one should plot either $\|\bar{x}_{L,k}\|_2$, $\|L\bar{x}_{L,k}\|_2$ or $\|L\bar{x}_{L,k}\|_1$ versus the residual norm $\|A\bar{x}_{L,k} - b\|_2$. All these norms are easy to compute. The seminorm $\|L\bar{x}_{L,k}\|_\ell$ is the ℓ-norm of the residual vector in (11) which is available as a by-product when solving (11), and using the fact that x_k and w_k are orthogonal we obtain

$$\|\bar{x}_{L,k}\|_2 = \left(\|x_k\|_2^2 + \|w_k\|_2^2\right)^{1/2} \tag{12}$$

$$\|A\bar{x}_{L,k} - b\|_2 = \left(\sum_{i=k+1}^{n}\left(u_i^T b - \sigma_i(w_k)_i\right)^2 + \|b^\perp\|_2^2\right)^{1/2}. \tag{13}$$

Here, $(w_k)_i$ is the ith component of w_k, and $b^\perp$ is the component of b that lies outside the column space of A. This component can readily be ignored when plotting the residual norm since it is independent of k.

3 Numerical Examples

In this section we illustrate the use of our *PP-TSVD* algorithm by means of some simple examples. All tests were carried out in MATLAB using double precision.

3.1 Detecting a Discontinuity

In our first example we illustrate that our *PP-TSVD* algorithm is indeed capable of detecting a discontinuity in the solution, while the *TSVD* and *MTSVD* methods do not share this capability.

The test problem is a 32×32 discretization of an integral equation

$$\int_{-\pi/2}^{\pi/2} K(s,t)\, f(t)\, dt = g(s), \qquad -\pi/2 \le s \le \pi/2$$

with kernel K given by

$$K(s,t) = (\cos(s) + \cos(t))\left(\frac{\sin(u)}{u}\right)^2, \qquad u = \pi\,(\sin(s) + \sin(t)).$$

The problem is identical to the model problem shaw from the MATLAB package REGULARIZATION TOOLS (Hansen 1994), except that we introduced a large discontinuity in the solution x. Figure 1 shows the matrix A together with the solution x and the right-hand side b. We computed the *TSVD*, *MTSVD* and *PP-TSVD* solutions with $L = L_1$, cf. (5), approximating the first derivative operator, and we used $k = 2, \ldots, 7$.

The results are shown in Fig. 2; notice the discontinuity between $i = 12$ and $i = 13$. Neither the *TSVD* method nor the *MTSVD* method is able to recover this discontinuity very well, and we notice that these methods also have difficulties in reconstructing the smooth parts of the solution—due to the discontinuity. The *PP-TSVD* method, on the other hand, captures the discontinuity when k is chosen large enough, in this example when $k \ge 4$, while the location is incorrect for $k < 4$. For $k \ge 4$ we see that one of the intervals degenerates into a single element of the solution, located at $i = 12$ or $i = 13$. The *PP-TSVD* method also does its best to try to recover the smooth parts of the solution, and does quite well for $k = 7$.

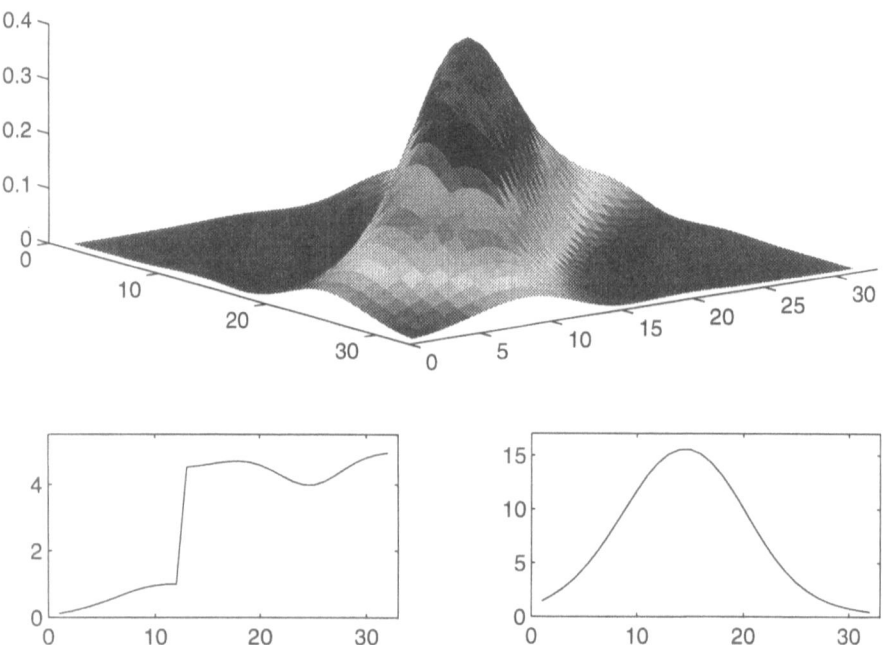

Fig. 1. The integral equation test problem: coefficient matrix A (top), solution x (bottom left), and right-hand side b (bottom right).

3.2 The Earth's Radial Density Function

The Earth's total mass M and moment of inertia I_2 (the second mass moment) are known from astronomical measurements. These two pieces of information carry information about the Earth's internal (radial) density distribution $\rho(r)$ through the linear equations:

$$M = 4\pi \int_0^R \rho(r)\, r^2\, dr$$

$$I_2 = \frac{8\pi}{3} \int_0^R \rho(r)\, r^4\, dr$$

where R is the radius of the Earth (we do not take the Earth's flattening into account). Measurements show that $M = 6.0 \cdot 10^{24} kg$ and $I_2 = 8.1 \cdot 10^{37} kg\, m^2$. The question is now: What is the simplest possible shell model of the Earth's density that is consistent with these observations? In order to answer this question, we discretize the above problem:

$$M = 4\pi \sum_{j=1}^n \rho\left((j - 0.5)R/n\right)\left((j - 0.5)R/n\right)^2$$

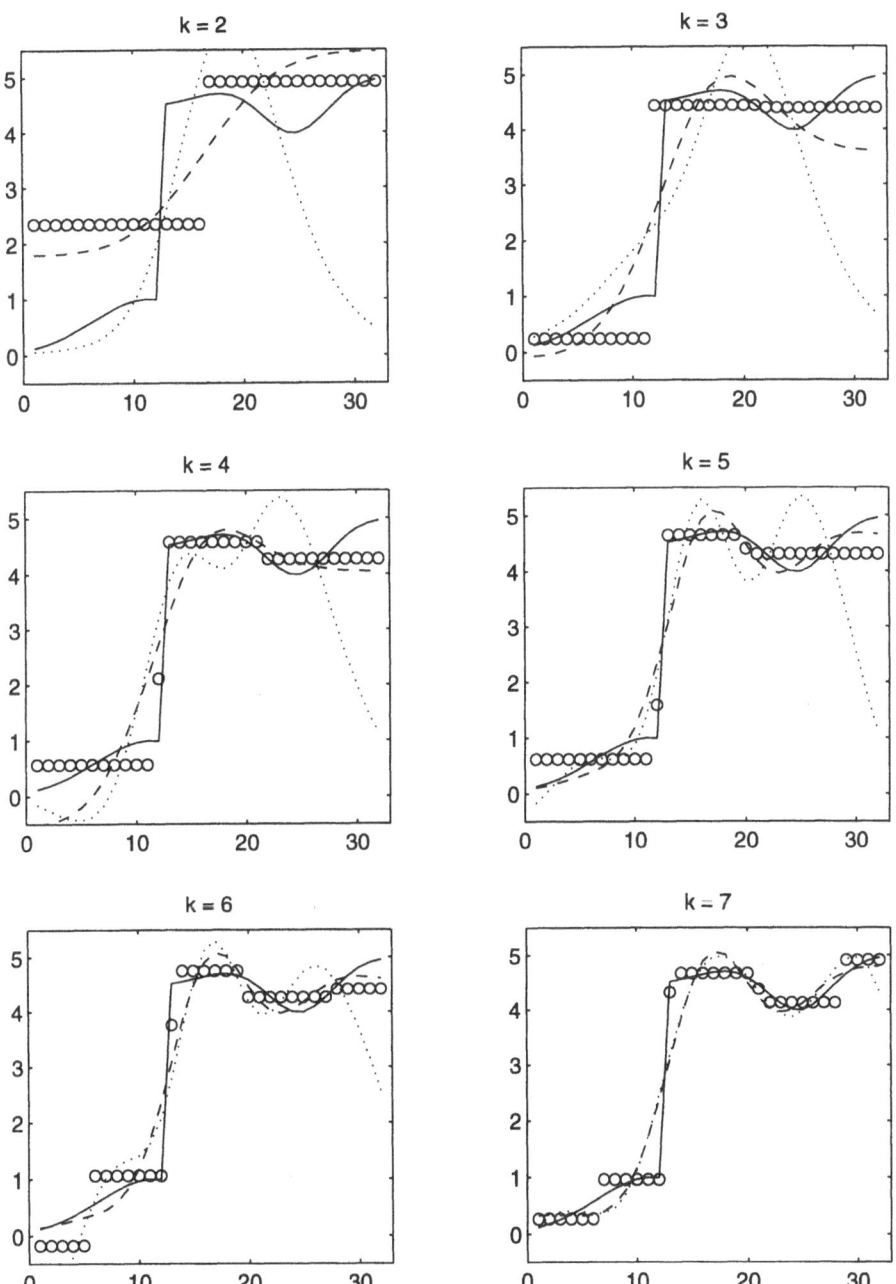

Fig. 2. Exact solution (solid line) and reconstructed solutions for the integral equation test problem using *TSVD* (dotted line), *MTSVD* (dashed line), and *PP-TSVD* (circles).

$$I_2 = \frac{8\pi}{3} \sum_{j=1}^{n} \rho\left((j-0.5)R/n\right)\left((j-0.5)R/n\right)^4$$

and apply our *PP-TSVD* method.

Using a model consisting of 32 radially sampled values of the Earth's density function from Jeffreys & Bullen 1967, we computed the solutions to this highly underdetermined problem with $m=2$ and $n=32$. In addition to the *PP-TSVD* solution with $k=m=2$, we computed the *TSVD* and *MTSVD* solutions, and we used $L=L_1$, see (5), equal to the first derivative operator. The computed solutions are shown in Fig. 3, and we see that given only M and I_2 we are not able to predict the density function by any of the methods. However, the *PP-TSVD* does provide the simplest possible density structure of the Earth.

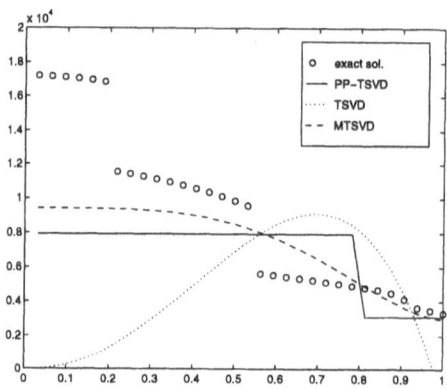

Fig. 3. Exact model (circles) and reconstructed solutions for the shell model of the Earth's density, given the mass M and the moment of inertia I_2. The abscissa axis is normalized with R.

Imagine that measurements of higher order moments $I_4, I_6, I_8, \ldots$ were also available. Would it then be possible to locate the Earth's core/mantle boundary and the inner core? We define the general ith-order moment by:

$$I_i = \frac{8\pi}{3} \int_0^R \rho(r)\, r^{n+2}\, dr, \tag{14}$$

and after discretization our problem becomes that of determining the internal radial density function $\rho(r)$ from the following set of equations:

$$M = 4\pi \sum_{j=1}^{n} \rho\left((j-0.5)R/n\right)\left((j-0.5)R/n\right)^2$$

$$I_{2i} = \frac{8\pi}{3} \sum_{j=1}^{n} \rho\left((j-0.5)R/n\right)\left((j-0.5)R/n\right)^{2i+2}, \quad i=1,\ldots,\ell.$$

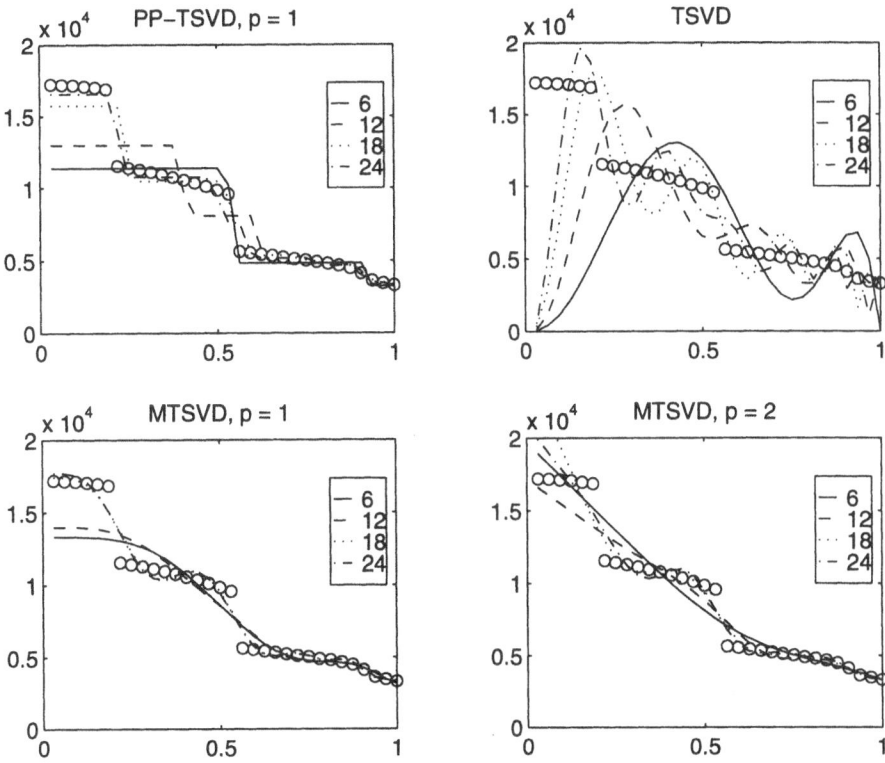

Fig. 4. Exact model (circles) and reconstructed solutions for the shell model of the Earth's density, given the mass M and moments of inertia $I_2, I_4, \ldots, I_\ell$. The solutions are computed for $\ell = 6, 12, 18, 24$. The abscissa axis is normalized with R.

We repeated our numerical tests with $\ell = 6, 12, 18$ and 24, and the results are shown in Fig. 4. In contrast to the *TSVD* and *MTSVD* solutions, the *PP-TSVD* solution does locate the core/mantel boundary (for $k \approx 24$) as well as the boundary between the inner and the outer core.

4 Conclusion

We have presented a new algorithm *PP-TSVD* that computes piecewise polynomial solutions to ill-posed problems, without a priori knowledge about the positions of the break points. In particular, we can compute piecewise constant functions that describe layered models. Such solutions are useful, e.g., in seismological problems, and the algorithm can also be used as a preprocessor for other methods where break points/discontinuities must be incorporated explicitly.

References

1. G. H. Golub & C. F. Van Loan, 1989, *Matrix Computations*, 2. Ed., Johns Hopkins University Press, Baltimore.

2. M. Hanke & P. C. Hansen, 1993, *Regularization methods for large-scale problems*, Surv. Math. Ind. **3**, 253–315.

3. P. C. Hansen, 1994, *Regularization Tools: a Matlab package for analysis and solution of discrete ill-posed problems*, Numer. Algo. **6**, 1–35.

4. P. C. Hansen, 1992, T. Sekii & H. Shibahashi, *The modified truncated SVD method for regularization in general form*, SIAM J. Sci. Stat. Comput. **13**, 1142–1150.

5. H. Jeffreys, 1939, *The times of P_cP and S_cS*, Mon. Not. R. A. S., Geophys. Suppl. **4**, 537.

6. H. Jeffreys & K. E. Bullen, 1967, *Seismological Tables*.

7. S. Levy & P. K. Fullager, 1981, *Reconstruction of a sparse spike train from a portion of its spectrum and application to high-resolution deconvolution*, Geophysics **46**, 1235–1243.

8. K. Madsen & H. B. Nielsen, 1993, *A finite smoothing algorithm for linear ℓ_1 estimation*, SIAM J. Optim. **3**, 223–235.

9. W. Menke, 1989, *Geophysical Data Analysis: Discrete Inverse Theory*, Academic Press, San Diego.

10. K. Mosegaard & P. D. Vestergaard, 1991, *A simulated annealing approach to seismic model optimization with sparse prior information*, Geophysical Prospecting **39**, 599–611.

11. R. D. Oldham, 1906, *Constitution of the interior of the Earth as revealed by earthquakes*, Quart. J. Geol. Soc. **62**, 456.

12. C. R. Smith & W. T. Grandy (Eds.), 1985, *Maximum-Entropy and Bayesian Methods in Inverse Problems*, D. Reidel Pub., Boston.

13. F. Santosa & W. W. Symes, 1986, *Linear inversion of band-limited reflection seismograms*, SIAM J. Sci. Stat. Comput. **7**, 1307–1330.

14. L. Simcik & P. Linz, 1994, *Qualitative regularization: resolving non-smooth solutions*, Report CSE-94-12, Dept. of Computer Science, Univ. of California, Davis.

15. G. A. Watson, *Approximation Theory and Numerical Methods*, Wiley, Chichester, 1980.

Regularization Methods for Almost Rank-deficient Nonlinear Problems

Jerry Eriksson and Per-Åke Wedin

Department of Computing Science, Umeå University, Umeå, Sweden

1 Tikhonov and truncated methods for nonlinear problems

Let us consider the nonlinear least squares problems

$$\min_{x \in \Re^n} F(x) = \frac{1}{2} \sum_{i=1}^{m} f_i^2(x), \tag{1}$$

where $f(x) \in \Re^m, m \geq n$ is a real–valued function twice continuously differentiable with the Taylor expansion

$$f(x + p) = f(x) + J(x)p + \mathcal{O}(\|p\|^2),$$

where J is the Jacobian matrix of f. In the Gauss–Newton method the given nonlinear problem is locally approximated with the linear least squares problem

$$\min_{p \in \Re^n} \frac{1}{2}\|J(x_k)p + f(x_k)\|^2, \tag{2}$$

where $\|\cdot\|$ as in the sequel denotes the Euclidean norm. If p_k solves this problem and α_k is the step–length along the search direction p_k then $x_{k+1} = x_k + p_k\alpha_k$ becomes the new iterate. When solving problem (1) regularization methods are often applied during the iterations to stabilize the approximating linear least squares problem (2). The Levenberg–Marquardt method and the trust–region method are based on Tikhonov regularization of the linear problem (2) (More 1978). Subspace minimization methods can be based on the truncated QR–method for the linear problem (2) (Lindström & Wedin 1988). These methods require a well defined solution point.

In this paper we discuss methods for inverse nonlinear problems (1) with almost rank–deficient[1] Jacobians $J(x)$. For a linear least squares problem

$$\min_{p \in \Re^n} \|Jp - b\|$$

---

[1] A matrix J that is either exactly rank–deficient or close to a rank–deficient matrix.

there are two direct regularization approaches; Tikhonov and truncated methods. We will generalize these techniques to the nonlinear inverse problem (1) by applying both approaches to a sequence of linear least squares problems derived from an approximation similar to (2) for the regularized nonlinear problem. In a way we combine the direct approaches to regularization for linear least squares with iterative methods for optimization. This approach is different from the ones taken in (Scherzer et. al. 1993) and (Hanke et. al. 1995) for regularization of (1) where iterative methods for linear regularization problems are generalized to nonlinear problems. Our approach is closer to the one used by Kaufman (Kaufman 1994) to solve large–scale practical problems, but differs in its intentions.

We want to classify the methods and find the crucial curvatures that both govern the local convergence of the essential methods and the conditioning of the solution. For *exactly rank–deficient* problems there are just two crucial condition numbers; one, $\mathcal{K}_r$, that depends on the normal curvature of $f(x)$ in the function space $\Re^m$ and one, $\mathcal{K}_0$, that depends on how the nonlinearity of f affects the regularization in the parameter space $\Re^n$. For an *almost rank–deficient* problem whose Jacobian $J(x)$ with singular values $\sigma_1 \geq \sigma_2 \geq \ldots \geq \sigma_n \geq 0$ is close to a matrix of rank r, the gap σ_r/σ_{r+1} has to be large to retain the nice properties of the methods for exactly rank–deficient problems. This is a statement with important qualifications similar to those for linear problems.

The regularized problem

$$\min_x \tfrac{1}{2} \|f(x)\|^2$$

$$\text{s.t.} \quad \tfrac{1}{2} \|x - c\|^2 \leq \frac{1}{2}\Delta^2 \tag{3}$$

has the Lagrange function

$$\mathcal{L}(x,\mu) = \frac{1}{2}\|f(x)\|^2 - \frac{1}{2}\mu^2(\Delta^2 - \|x - c\|^2).$$

Here c is a natural, chosen center. [2].

To every Δ there is a corresponding Tikhonov parameter $\mu \geq 0$ such that the augmented unconstrained nonlinear least squares problem

$$\min_x \frac{1}{2}\left\| \begin{pmatrix} f(x) \\ \mu(x - c) \end{pmatrix} \right\|^2 \tag{4}$$

has the same solution as (3). At x_k Problem (4) is approximated by the linear least squares problem

$$\min_{p \in \Re^n} \frac{1}{2}\left\| \begin{pmatrix} J(x_k)p + f(x_k) \\ \mu(x_k + p - c) \end{pmatrix} \right\|. \tag{5}$$

Let p_k solve Problem (5) and take $x_{k+1} = x_k + p_k\alpha_k$ as the new iterate. The choice of step–length $\alpha_k > 0$ will be discussed later.

---

[2] For the nonlinear problem where we linearize at a sequence of iterates $\{x_k\}$ it is most practical to introduce a center c already in the problem formulation.

If a nonlinear truncated SVD method is used, compute the search direction p_k as the solution of the local minimum norm problem

$$\min_{p} \|x_k + p - c\|$$

$$\text{s.t. } \|J_1 p + f(x_k)\| \text{ is minimal.} \tag{6}$$

Here $J_1 = U_1 \Sigma_1 V_1^T$ where $J(x_k)$ has the singular value decomposition

$$J = U\Sigma V^T = (U_1, U_2) \begin{pmatrix} \Sigma_1 \\ & \Sigma_2 \end{pmatrix} (V_1, V_2)^T = U_1 \Sigma_1 V_1^T + U_2 \Sigma_2 V_2^T,$$

with $\Sigma_1 = \text{diag}(\sigma_1, \ldots, \sigma_r)$ and $\Sigma_2 = \text{diag}(\sigma_{r+1}, \ldots, \sigma_n)$.

Hence, for both Tikhonov and truncation methods the given problem is approximated by a sequence of linear least squares problems (5) and (6), respectively.

Let us here only compare the search directions p_k computed at a certain point x_k for the two methods. A few calculations show that the search direction computed from the Tikhonov method (5) equals

$$p_{Tikh} = -((J^T J + \mu^2 I)^{-1} J^T, \mu^2 (J^T J + \mu^2 I)^{-1}) \begin{pmatrix} f(x_k) \\ (x_k - c) \end{pmatrix} \tag{7}$$

and the search direction computed with the truncation method equals

$$p_{Trunc} = -(J_1^+, P_{N(J_1)}) \begin{pmatrix} f(x_k) \\ (x_k - c) \end{pmatrix}, \tag{8}$$

where J_1^+ denotes the pseudo-inverse of J_1 and $P_{N(J_1)}$ is the ortogonal projection on the null-space of J_1. If $J(x_k)$ is a matrix of exact rank equal to r then

$$p_{Tikh} \rightarrow p_{Trunc} \quad \text{when } \mu \rightarrow 0.$$

It can further be shown that if the gap σ_r/σ_{r+1} is large then it is possible to choose μ such that the directions p_{Tikh} and p_{Trunc} are close.

2 Condition numbers to characterize the nonlinearity of the regularization problem locally

To understand how the nonlinearity of the function f affects the regularization problem we are going to start from the problem to minimize $\|x - c\|$ when $\|f(x)\|$ is minimal and the Jacobian $J(x)$ has *constant rank* $r < n$. The condition number $\|J^+\|\|J\|$ is the key concept for analyzing the stability of the *linear minimum norm* problem

$$\min_{x} \|x - c\| \quad \text{when } \|Jx - b\| \text{ is minimal}$$

that has the unique solution

$$x = (J^+, P_{N(J)}) \begin{pmatrix} b \\ c \end{pmatrix}.$$

For linear algebra problems c is usually the null–vector. The truncated method with $J_1 = J$ and $x_{k+1} = x_k + \alpha_k p_k$ in each step solves an approximating minimum norm problem (6) to get a search direction p_{Tikh} defined by (7). If the rank equals r everywhere this is the method of choice.

For problems that only take the exact lower rank r in the neighbourhood of the solution it may be more natural to use the method for a sequence of Tikhonov regularized problems (5) and let the regularization parameter μ decrease to zero.

With the step–length $\alpha_k = 1$ close to the solution and $\mu \searrow 0$ the truncated (6) and the Tikhonov regularized Gauss–Newton method (5) get the same asymptotic convergence rate $\max(\kappa_r, \kappa_0)$. Here

$$\mathcal{K}_r = \max_{v \in \mathcal{N}(J)^\perp} \left| \frac{v^T (\sum_{i=1}^m f_i f_i'') v}{v^T J^T J v} \right|$$

and

$$\mathcal{K}_0 = \max_{v \in \mathcal{N}(J)} \left| \frac{v^T (\sum_{i=1}^m (J^{+T}(\hat{x} - c))_i f_i'') v}{v^T v} \right|$$

both have the geometrical meaning that contains the essential difference between the linear and nonlinear minimum norm problem. [3]

For $f(x)$ and x in $\Re^2$ and the Jacobian $J(x)$ of constant rank 1, $\{y : y = f(x)\}$ and the set $\{x : J(x)^T f(x) = 0\}$ become curves in $\Re^2$. Let $\|u - \hat{f}\|$ and $\|v - \hat{x}\|$ be the radii of the osculating circles at the solution of these two curves in function space and parameter space, respectively. Then it can be proved that $\mathcal{K}_r = \|f(\hat{x})\| / \|f(\hat{x}) - u\|$ and $\mathcal{K}_0 = \|\hat{x} - c\| / \|\hat{x} - v\|$. These geometrical quantities are illustrated in Fig. 1 for f and x in $\Re^2$ and the Jacobian $J(x)$ of constant rank 1. The two methods converge if $\max(\mathcal{K}_r, \mathcal{K}_0) < 1$ and $\hat{x}$ is also a strong local minimum of the nonlinear minimum norm problem. An analysis of that problem based on the constant rank theorem [4] has been made by the authors (Eriksson et al. 1995:1).

We now turn to almost rank–deficient problems where there is a large gap σ_r / σ_{r+1}. Assume that μ is chosen suitably. Then the two condition numbers

$$\mathcal{K}_r = \max_{v \in \mathcal{N}(J_1)^\perp} \left| \frac{v^T (\sum_{i=1}^m f_i f_i'') v}{v^T J^T J v} \right|.$$

and

$$\mathcal{K}_0 = \max_{v \in \mathcal{N}(J_1)} \left| \frac{v^T (\sum_{i=1}^m (J_1^{+T}(\hat{x} - c))_i f_i'') v}{v^T v} \right|$$

play about the same role as in the exactly rank deficient case discussed above. Just as in the linear case the two kinds of regularization problems now have close but different solutions.

---

[3] An ordinary optimization method like the Levenberg–Marquardt method cannot be used to solve the minimum norm problem. Such a method terminates as soon as the gradient of $\frac{1}{2}\|f\|^2$ is zero. (see Eriksson et al. 1995:1).

[4] If $J(x)$ has constant rank r then $f(x)$ can be written as a composite function $f(x) = h(z(x))$ with $z \in \Re^r$ in a neighbourhood of the solution $\hat{x}$.

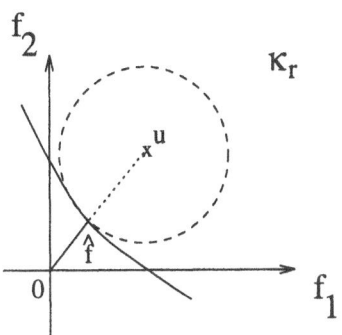
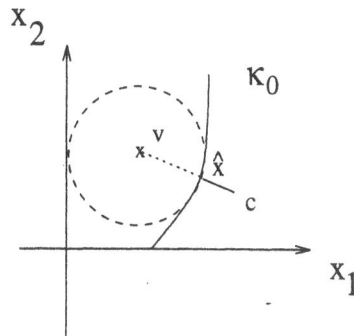

Fig. 1. Both the curvatures in $\Re^m$ and $\Re^n$ are crucial for the regularized problem. The radii of curvatures for Problem (1) are central and are shown in the figures.

3 Tools for nonlinear regularization

The existence of several local minima is from a computational point of view the most difficult property introduced when moving from linear to nonlinear least squares. Especially problems that need to be regularized are prone to have many local minima.

By Tikhonov regularization we can reduce this difficulty substantially. The reason for this is that Tikhonov regularization applied to a problem that need to be regularized reduces the curvature substantially. For a given direction $w \in \mathcal{N}(J^T)$ and direction $v \in \Re^n$ at a given point x the Tikhonov regularized problem (4) has the curvature

$$\mathcal{K} = \left| \frac{v^T \sum_{i=1}^m w_i f_i''}{v^T (J^T J + \mu^2 I_n) v} \right| \tag{9}$$

while the original nonlinear least squares problem has its curvature defined by the the same formula but with μ equal to zero. Take $v = v_n$ as the singular vector in $\Re^n$ corresponding to the smallest singular value σ_n. For a problem that needs regularization there is a substantial difference in size between the curvatures

$$\frac{|v_n^T \sum_{i=1}^m w_i f_i'' v_n|}{\mu^2 (1 + (\sigma_n/\mu)^2)}$$

for the regularized problem and

$$\frac{|v_n^T \sum_{i=1}^m w_i f_i'' v_n|}{\sigma_n^2}$$

for the ordinary nonlinear least squares problem.

Especially, note that this smoothing property of Tikhonov regularization is highly relevant also for very nonlinear problems like the weight computation for neural networks where the approximate rank may change during the computations. We are confident that a Tikhonov regularized method should be the method of choice for weight computations in neural networks. Tikhonov regularization (4) makes a great difference in efficiency and size of the computed weights compared to the Levenberg–Marquardt method where only a sequence of linear least squares problems are regularized around a new center x_k each time.

In the reports (Eriksson et al. 1995:1-4), we propose various implementations of the Tikhonov method. The idea is to work with a sequence of decreasing μ–values and in each step choose the step–length α_k to get sufficient decrease of the current regularized function

$$\frac{1}{2}\|f(x)\|^2 + \frac{1}{2}\mu^2\|x - c\|^2$$

to guarantee fairly fast global convergence [5].

As mentioned earlier the truncated method should only be used in those cases where the truncated QR and the SVD are used for the corresponding linear problem as in many applications in signal processing. However, in those cases truncated method should be the method of choice. To ensure convergence of the sequence $\{x_k\}$ a merit function is needed until x_k is so close to the solution that we can rely on the generally fast convergence with step–length $\alpha_k = 1$. We have used the merit function

$$\frac{1}{2}\|f(x_k + \alpha_k p_k) - J_2(x_k)\alpha_k p_k\|^2 + \frac{1}{2}\nu_k\|x_k + \alpha_k p_k - c\|^2$$

for a sequence of ν_k–values. Here $J_2 = U_2 \Sigma_2 V_2^T$

Our intention is to develop a set of lucid utility routines for nonlinear regularization. It should be possible to use these routines both as building blocks in regularization algorithms and as tools to analyze regularization assumptions. These tools will involve algorithms for line search, merit functions, termination criteria etc. A natural model for our work is Hansen's Matlab Toolbox (Hansen 1993). A useful function is an algorithm for computing the L–curve corresponding to Tikhonov regularization of the nonlinear problem (4). This L–curve need not look like an L–curve for a linear problem (even a linearization of the nonlinear problem for a suitable μ) as is evident from Fig. 2 but is just as useful for the determination of the suitable size of the regularization parameter μ.

---

[5] Global convergence means that it should not be possible for the algorithm to jam up at a point where the condition $J^T f + \mu^2(x - c) = 0$ is not satisfied or the μ–value is unacceptable large.

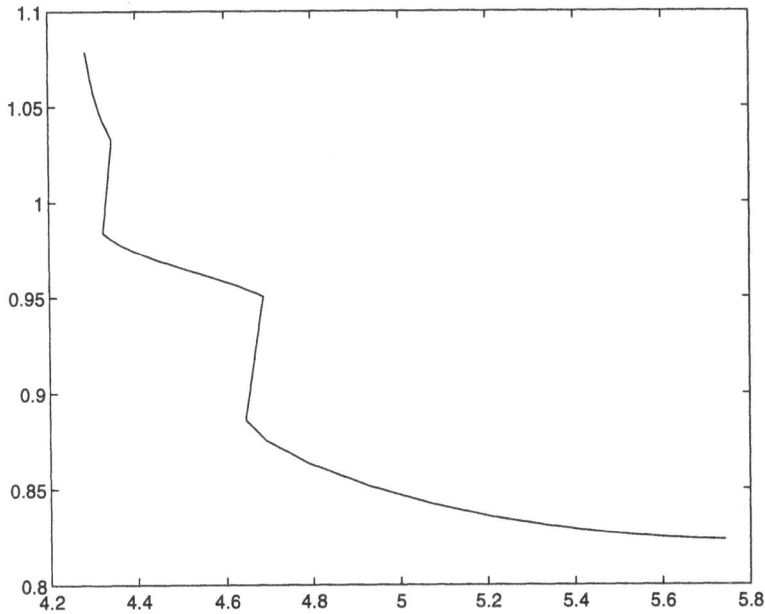

Fig. 2. This is an example of a nonlinear L–curve for a small almost rank–deficient problem. We compute the regularized solution $x(\mu)$ for 50 values of μ in the interval $\mu = (0.1, \ldots, 0.001)$. Then we plot $\|f(x(\mu))\|$ with values on the vertical axis against $\|x(\mu) - c\|$.

References

1. Eriksson J., Lindström P., and Wedin P–Å., 1995:1. A new regularization method for rank–deficient nonlinear least squares problems. Technical Report UMINF–95.01. Department of computing Science, University of Umeå, Sweden.
2. Eriksson J., Lindström P., and Wedin P–Å., 1995:2. Algorithms and benchmarks for training feed–forward neural networks Technical Report UMINF–95.02. Department of computing Science, University of Umeå, Sweden.
3. Eriksson J. and Wedin P–Å., 1995:3. Regularization methods for nonlinear least squares. Part I. rank–deficient problems. Technical Report UMINF. Department of computing Science, University of Umeå, Sweden.
4. Eriksson J. and Wedin P–Å., 1995:4. Regularization methods for nonlinear least squares. Part II. Almost rank–deficient problems. Technical Report UMINF. Department of computing Science, University of Umeå, Sweden.
5. Hansen P.C., 1993. Regularization Tools – A Matlab package for analysis and solution of discrete ill–posed problems. UNI•C, Danish Computing center for research

and education, Building 305, Technical University of Denmark, DK–2800 Lyngby, Denmark.

6. Kaufman L. and Neumaier A., 1994. Image reconstruction through regularization by envelope guided conjugate gradients., report 940819–14, Dept. of Computer science, AT&T Bell Laboratories.

7. Lindström P. and Wedin P–Å., 1988. Methods and software for nonlinear least squares problems. Technical Report UMINF–133.87. Inst. of Information Processing, University of Umeå, Sweden.

8. More J.J., 1978. The Levenberg–Marquardt algorithm: implementation and theory. Lecture notes in mathematics 630. Springer Verlag.

9. Hanke M., Neubauer A., and Scherzer O., 1995. A convergence analysis of the Landweber iteration for nonlinear ill–posed problems. Institut Für Mathematik, Johannes–Kepler–Universität, A–4040 Linz, Austria.

10. Scherzer O., Engl H. W., and Kunisch K., 1993. Optimal a posteriori parameter choice for Tikhonov regularization for solving nonlinear ill–posed problems. SIAM J. Numer. Anal., Vol. 30:1796–1838.

Constant Thermodynamic Speed Simulated Annealing

Bjarne Andresen[1] and J.M. Gordon[2]

[1] Ørsted Laboratory, Niels Bohr Institute, University of Copenhagen, Denmark
[2] Center for Energy and Environmental Physics, Jacob Blaustein Institute for Desert Research, Ben Gurion University of the Negev, Israel

Simulated Annealing

Simulated annealing is a global optimization procedure (Kirkpatrick et al. 1983) which exploits an analogy between combinatorial optimization problems and the statistical mechanics of physical systems. The analogy gives rise to an algorithm for finding near-optimal solutions to the given problem by simulating the cooling of the corresponding physical system. Just as Nature, under most conditions, manages to cool a macroscopic system into or very close to its ground state in a short period of time even though its number of degrees of freedom is of the order of Avogadro's number, so does simulated annealing rapidly find a good guess of the solution of the posed problem.

The simulated annealing algorithm is based on the Monte Carlo simulation of physical systems. It requires the definition of a state space $\Omega = \{\omega\}$ with an associated cost function (physical analog: energy) $E: \Omega \rightarrow R$ which is to be minimized in the optimization. At each point of the Monte Carlo random walk in the state space the system may make a jump to a neighboring state; this set of neighbors, known as the move class $N(\omega)$, must of course also be specified. The only control parameter of the algorithm is the temperature T of the heat bath in which the corresponding physical system is immersed.

The random walk inherent in the Monte Carlo simulation is accomplished by the Metropolis algorithm (Metropolis et al. 1953) which states that:

(i) At each step t of the algorithm a neighbor ω' of the current state ω_t is selected at random from the move class $N(\omega_t)$ to become the candidate for the next state.

(ii) It actually becomes the next state only with probability

$$P_{accept} = \begin{cases} 1 & \text{if } \Delta E \leq 0 \\ e^{-\Delta E/T_t} & \text{if } \Delta E > 0 \end{cases} , \tag{1}$$

where $\Delta E = E(\omega') - E(\omega_t)$ is the increase in cost for the move. If this candidate is accepted, then $\omega_{t+1} = \omega'$, otherwise $\omega_{t+1} = \omega_t$.

The only thing left to specify is the sequence of temperatures T_t appearing in the Boltzmann factor in P_{accept}, the so-called annealing schedule. Like in metallurgy, this cooling rate has a major influence on the final result. A quench is quick and dirty, often leaving the system stranded in metastable states high above the ground state/optimal solution. Slow annealing produces the best result but is computationally expensive.

This completes the formal definition of the simulated annealing algorithm, which in principle simply is repeated numerous times until a satisfactory result is obtained.

The Optimal Annealing Schedule

So far all suggested simulated annealing temperature paths (annealing schedules) have been of the a priori type and thus have not adjusted to the actual behavior of the system as the annealing progresses. Examples of such schedules are

$$T(t) = a\, e^{-t/b} \tag{2}$$

$$T(t) = \frac{a}{b+t} \tag{3}$$

$$T(t) = \frac{a}{\ln(b+t)} . \tag{4}$$

The real annealing of physical systems often has rough parts where the surrounding temperature must be decreased slowly due to phase transitions or regions of large heat capacity or slow internal relaxation. The same behavior is seen in the abstract systems, so annealing schedules which take such variations into account are preferable in order to keep computation time at a minimum for a given accuracy of the final result (Ruppeiner 1988).

Since asking a question (= one comparison of the energy function with its previous value in eq. (1)) in information theoretic terms is equivalent to producing one bit of entropy, the computationally most efficient procedure will be that temperature schedule $T(t)$ which overall produces minimum entropy. In the past we have derived various bounds and optimal paths for real thermodynamic systems using finite-time thermodynamics (Salamon and Berry 1983; Feldmann et al. 1985; Salamon et al. 1980b; Andresen 1983). The minimum-entropy production path which most readily generalizes to become the optimal simulated annealing schedule is the one calculated with thermodynamic length. In its simplest form it is defined through

$$\frac{dT}{dt} = -\frac{vT}{\varepsilon\sqrt{C}} \tag{5}$$

or equivalently

$$\frac{\langle E \rangle - E_{eq}(T)}{\sigma} = v. \tag{6}$$

In these expressions v is the (constant) thermodynamic speed (see next section for further explanation), C and ε are the heat capacity and internal relaxation time of the system, respectively, $\langle E \rangle$ and σ the corresponding mean energy and standard deviation of its natural fluctuations, and finally $E_{eq}(T)$ is the internal energy the system would have if it were in equilibrium with its surroundings at temperature T.

The physical interpretation of eq. (6) is that the environment should at all times be kept v standard deviations ahead of the system. Similarly eq. (5) indicates that the annealing should slow down where internal relaxation is slow and where large amounts of 'heat' have to be transferred out of the system (Salamon et al. 1988). (Transfer of 'heat' in this abstract case means populating many states with roughly the same energy, i. e. large degeneracy.) In case C and ε do not vary with temperature, eq. (5) integrates to the standard schedule eq. (2). The more realistic assumption of an Arrhenius-type relaxation time, $\varepsilon \sim \exp(a/T)$ for some constant a, and a

heat capacity $C \sim T^{-2}$ implies the vastly slower schedule eq. (4).
Reality is usually in between these extremes. Figure 1 shows the
successive decrease in the lowest energy seen for annealings on a
graph partitioning problem following different annealing schedules.

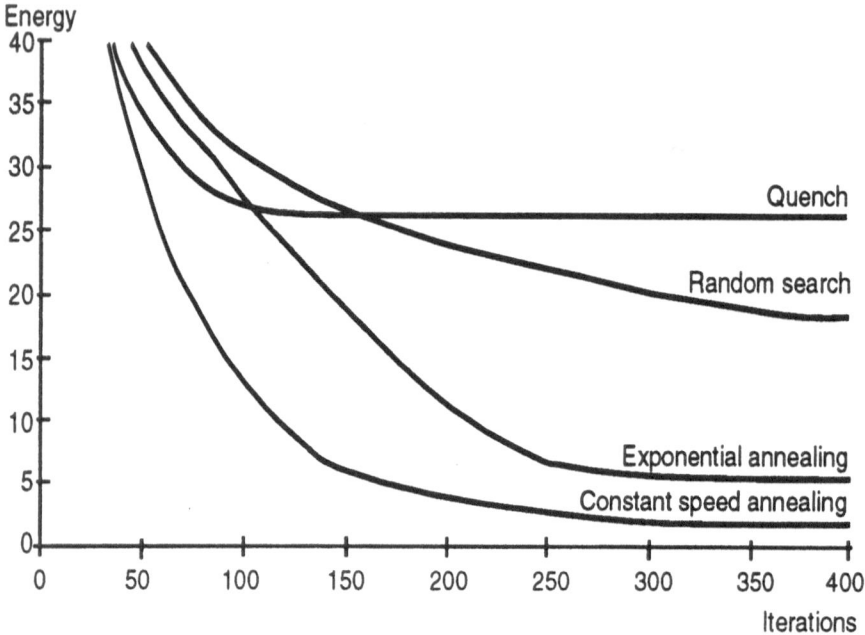

Figure 1. The lowest energy seen for a graph partitioning problem as the
annealing progresses, using different annealing schedules: quench (T=0), random search (T=∞), exponential (eq. (2)), constant thermodynamic speed (eq. (5)
or (6)). (From Pedersen 1989).

The extra temperature dependent variables of the constant thermodynamic speed schedule of course require additional computational effort. Since systems often change considerably in a few steps,
ergodicity is not fulfilled, so the use of time averages to obtain $\langle E \rangle$, σ,
C, and ε is usually not satisfactory. Instead we (Andresen et al. 1988)
suggest to run an ensemble of systems in parallel, i. e. with the same
annealing schedule, in the true spirit of the analogy to statistical
mechanics. Then these variables can be obtained anytime as ensemble averages in the usual fashion of statistical mechanics. This use
of ensemble annealing is particularly well suited for implementation on present day parallel computers.

Thermodynamic Speed

The constant thermodynamic speed strategy shown above in eq. (5) has proven superior to other proposed paths in a large number of instances (Ruppeiner 1988; Salomon et al. 1988; Andresen et al. 1988), but not in all cases (Hoffmann and Salomon 1990; Sibani et al. 1990). We have therefore taken a closer look at its original derivation (Salamon and Berry 1983) in an attempt to find the origin of the discrepancies.

The instantaneous entropy production dS^u can be expressed in terms of changes in the conjugate extensive (i. e. proportional to system size) variables X and intensive (i. e. independent of system size) variables Y by

$$dS^u = \Delta Y \cdot dX, \tag{7}$$

where

$$X = (U, V, N, ...)$$

$$Y = \left(\frac{\partial S}{\partial U}, \frac{\partial S}{\partial V}, \frac{\partial S}{\partial N}, \cdots\right) = \left(\frac{1}{T}, \frac{p}{T}, -\frac{\mu}{T}, \cdots\right); \tag{8}$$

μ denotes the chemical potential and S the entropy of the system; $\Delta Y = Y - Y^e$ is the difference in intensity between the system's value Y and the environment's value Y^e that gives rise to the flow of extensity dX from the environment to the system. The goal is to derive that operating strategy, or environment path, $Y^e(t)$ (t = time) which will take the system from X_i to X_f in time τ while producing as little entropy as possible along the way.

Limiting ourselves for simplicity to one pair of variables $(X, Y) = (U, 1/T)$, the next step in the derivation is a power expansion of ΔY around equilibrium in terms of ΔU:

$$\Delta Y = (1/T) - (1/T^e)$$

$$= \frac{\partial}{\partial U}\left(\frac{\partial S}{\partial U}\right)\Delta U + \frac{1}{2}\frac{\partial^2}{\partial U^2}\left(\frac{\partial S}{\partial U}\right)\Delta U^2 + \dots$$

$$= \frac{\partial^2 S}{\partial U^2}\Delta U + \frac{1}{2}\frac{\partial^3 S}{\partial U^3}\Delta U^2 + \dots, \tag{9}$$

where $\Delta U = U - U^e$ is the deviation of the system energy from its hypothetical value U^e if it were in equilibrium with the environment. The first term on the right-hand-side of eq. (9) includes the thermodynamic metric

$$D^2 S = \left\{\frac{\partial^2 S}{\partial X_i \partial X_j}\right\} \tag{10}$$

first introduced by Weinhold (1975) at a point and later extensively studied by Salamon and coworkers (Salamon et al. 1980a; Salamon et al. 1984; Nulton and Salamon 1985; Salamon et al. 1985). In statistical mechanics, where the entropy of a probability distribution $\{p_i\}$ takes the form

$$S(\{p_i\}) = -\sum_i p_i \ln p_i, \tag{11}$$

the metric $D^2 S$ is particularly simple, being the diagonal matrix (Feldmann et al. 1985)

$$D^2 S = \left\{\frac{1}{p_i}\right\}. \tag{12}$$

Next we need the phenomenological connection between the degree of deviation from equilibrium ΔU and the resulting extensity flow dU/dt. This is where the non-equilibrium nature of the process enters. However, since this will in general be a complicated relation, we use the functional form of its asymptotic behavior for small ΔU,

$$\frac{dU}{dt} = \frac{-1}{\varepsilon(T,U)}\Delta U. \tag{13}$$

As long as we use the instantaneous apparent relaxation time $\varepsilon(T,U)$, this is not an approximation. Rather, $\varepsilon(T,U)$ is the local inherent unit of time of the system and is defined by eq. (13). The natural

(dimensionless) time scale ξ for the process is therefore given by
(Nulton et al. 1985)

$$d\xi = dt/\varepsilon(T,U) \tag{14}$$

so that eq. (13) can be rewritten as

$$\frac{dU}{d\xi} = -\Delta U. \tag{15}$$

Some rearrangement leads to the following expression for the rate
of entropy production:

$$\frac{dS^u}{d\xi} = -\frac{\partial^2 S}{\partial U^2}\left(\frac{dU}{d\xi}\right)^2 + \frac{1}{2}\frac{\partial^3 S}{\partial U^3}\left(\frac{dU}{d\xi}\right)^3 + \dots$$

$$= v^2\left(1 + \frac{\theta(T)}{C\,T}\frac{dU}{d\xi} + \dots\right), \tag{16}$$

where we have defined

$$\theta(T) = 1 + \frac{T}{2\,C}\frac{\partial C}{\partial T}. \tag{17}$$

The first term on the right-hand-side of eq. (16),

$$v^2 = -\frac{dU}{d\xi}D^2S\frac{dU}{d\xi} = \left(\frac{dL_s}{d\xi}\right)^2 \tag{18}$$

is the square of the thermodynamic speed v, namely, it is the square
of the natural-time derivative of the thermodynamic length L_s
(Nulton and Salamon 1988) calculated with the entropy metric $-D^2S$.
Further arguments (Andresen and Gordon 1994) show that the
optimal temperature schedule is obtained when $dS^u/d\xi$ is constant.
The corresponding differential equation for the environment temperature is

$$\frac{dT}{dt}\sqrt{1 + \frac{\theta(T)\,\varepsilon\,(dT/dt)}{T} + \dots} = \text{constant} \times \frac{T}{\varepsilon\,\sqrt{C}}. \tag{19}$$

Two main results emerge from the present analysis:

• Along the optimal path the rate of entropy production, *measured in local time*, is constant.
• The original result, eq. (5), is the leading term in a power series away from equilibrium only.

The same procedure of calculating metric bounds for dynamic systems has been applied to coding of messages (Flick et al. 1987) and to economics (Salamon et al. 1987). As long as the system in question is accurately described by statistical mechanics, these bounds and the paths that achieve them will provide the most efficient solution within the allotted time, i. e. arrive at the lowest expected energy.

Acknowledgments

We would like to thank Drs. Jim Nulton and Peter Salamon for numerous inspiring discussions on thermodynamic speed. BA gratefully acknowledges a travel grant from the Danish Ministry of Education, and the generous hospitality of the Blaustein International Center and the Center for Energy and Environmental Physics, Sede Boqer, Israel.

References

Andresen, B. 1983. *Finite-time thermodynamics* (Physics Laboratory II, University of Copenhagen).

Andresen, B.; Gordon, J. M. 1994. "Constant thermodynamic speed for minimizing entropy production in thermodynamic processes and simulated annealing". *Phys. Rev. E* **50**, 4346–4351.

Andresen, B.; Hoffmann, K. H.; Mosegaard, K.; Nulton, J.; Pedersen, J. M.; Salamon, P. 1988. "On lumped models for thermodynamic properties of simulated annealing problems". *J. Phys. (France)* **49**, 1485–1492.

Feldmann, T.; Andresen, B.; Qi, A.; Salamon, P. 1985. "Thermodynamic lengths and intrinsic time scales in molecular relaxation". *J. Chem. Phys.* **83**, 5849–5853.

Flick, J. D.; Salamon, P.; Andresen, B. 1987. "Metric bounds on losses in adaptive coding". *Info. Sci.* **42**, 239–253.

Hoffmann, K. H.; Salamon, P. 1990. "The optimal simulated annealing schedule for a simple model". *J. Phys. A: Math. Gen.* **23**, 3511–3523.

Kirkpatrick, S.; Gelatt, C. D. Jr.; Vecchi, M. P. 1983. "Optimization by simulated annealing". *Science* **220**, 671–680.

Metropolis, N.; Rosenbluth, A. W.; Rosenbluth, M. N.; Teller, A. H.; Teller, E. 1953. "Equation of state calculations by fast computing machines". *J. Chem. Phys.* **21**, 1087–1092.

Nulton, J.; Salamon, P. 1985. "Geometry of the ideal gas". *Phys. Rev. A* **31**, 2520–2524.

Nulton, J.; Salamon, P. 1988. "Statistical mechanics of combinatorial optimization". *Phys. Rev. A* **37**, 1351–1356.

Nulton, J.; Salamon, P.; Andresen, B. 1985 (private discussions).

Pedersen, J. M. 1989. "Simulated annealing and finite-time thermodynamics". Ph. D. thesis. (Physics Laboratory, University of Copenhagen).

Ruppeiner, G. 1988. "Implementation of an adaptive, constant thermodynamic speed simulated annealing schedule". *Nucl. Phys. B (Proc. Suppl.)* **5A**, 116–121.

Salamon, P.; Andresen, B.; Gait, P. D.; Berry, R. S. 1980. "The significance of Weinhold's length". *J. Chem. Phys.* **73** 1001–1002; erratum *ibid.* **73**, 5407E.

Salamon, P.; Berry, R. S. 1983. "Thermodynamic length and dissipated availability". *Phys. Rev. Lett.* **51**, 1127–1130.

Salamon, P.; Komlos, J.; Andresen, B.; Nulton J. 1987. "A geometric view of welfare gains with non-instantaneous adjustment". *Math. Soc. Sci.* **13**, 153–163.

Salamon, P.; Nitzan, A.; Andresen, B.; Berry, R. S. 1980. "Minimum entropy production and the optimization of heat engines". *Phys. Rev. A* **21**, 2115–2129.

Salamon, P.; Nulton, J.; Berry, R. S. 1985. "Length in statistical mechanics". *J. Chem. Phys.* **82**, 2433–2436.

Salamon, P.; Nulton, J.; Ihrig, E. 1984. "On the relation between entropy and energy versions of thermodynamic length". *J. Chem. Phys.* **80**, 436–437.

Salamon, P.; Nulton, J.; Robinson, J.; Pedersen, J. M.; Ruppeiner, G.; Liao, L. 1988. "Simulated annealing with constant thermodynamic speed". *Comp. Phys. Comm.* **49**, 423–248.

Sibani, P.; Pedersen, J. M.; Hoffmann, K. H.; Salamon, P. "Monte Carlo dynamics of optimization problems: A scaling description". 1990. *Phys. Rev. A* **42**, 7080–7086.

Weinhold, F. 1975. "Metric geometry of equilibrium thermodynamics". *J. Chem. Phys.* **63**, 2479–2483.

Mean Field Reconstruction with Snaky Edge Hints

Peter Alshede Philipsen, Lars Kai Hansen and Peter Toft

CONNECT, Electronics Institute, Technical University of Denmark, Denmark

1 Introduction

Reconstruction of imagery of a non-ideal imaging system is a fundamental aim of computer vision. Geman and Geman [Geman(1984)] introduced Metropolis sampling from Gibbs distributions as a simulation tool for visual reconstruction and showed that a *Simulated Annealing* strategy could improve the efficiency of the sampling process. The sampling process is implemented as a stochastic neural network with symmetric connections. Peterson and Anderson [Peterson(1987)] applied the *Mean Field* approximation, and observed substantial improvements in speed and performance.

In the next section the Bayesian approach to reconstruction is outlined and we study the so-called *Weak Membrane* model, as a model example. The Weak Membrane is a popular vehicle for piece wise smooth reconstruction and involves edge units (called line processes in [Geman(1984)]). Edge unit control has shown to be a major challenge in applications of the Weak Membrane. However, recently there has been substantial progress in use of deformable models for contour (2D) and surface (3D) modeling, see, e.g., [Cohen(1993), Chen(1994)]. In this contribution we suggest to combine the two approaches, in particular we show how a "Snake" contour model may be used to produce efficient edge hints to the Weak Membrane. Finally, section three contains experiments and concluding remarks.

2 Bayesian Visual Reconstruction

The basic idea is to consider both the source (un-degraded) signal and the processes of the imaging system as stochastic processes. The Bayes formula can then be used to obtain the distribution $P(V|d)$ of the reconstructed signal

V, conditioned on the observed degraded signal d:

$$P(V|d) = \frac{P(d|V)P(V)}{P(d)} \tag{1}$$

This conditional distribution is the product of the distribution of the imaging system process: $P(d|V) \equiv P(V \rightarrow d)$, and the *prior* distribution of the reconstructed signal $P(V)$. $P(V|d)$ of Equation (1) is referred to as the *posterior* distribution. A useful estimate of the reconstructed signal is given by the location of the mode of the posterior distribution, the so-called *Maximum A Posteriori* estimate.

2.1 The Weak Membrane Model

The *prior* distribution reflects our general insight into the objects being imaged, for example expressing that the image represents extended 3D structures etc. The Weak Membrane is a simple model for reconstruction of piece-wise continuous intensity surfaces (2D signals) from noisy observations [Geman(1984)]. In the lattice $(m, n) \in Z^2$ version the reconstructed surface is described by the intensity values $V_{m,n}$. The prior information formalizes the expectation that neighbor intensity values should be close, except when they are disconnected by the rare occasion of an active edge-unit. We introduce horizontal edges $h_{m,n}$ and vertical edges $v_{m,n}$. The *prior* probability distribution for a membrane reads:

$$P(V, h, v) = Z_1^{-1} \exp\left(-E_{\text{prior}}(y, h, v)\right) \tag{2}$$

where $E_{\text{prior}}(y, h, v)$ is the energy or cost function:

$$E_{\text{prior}}(V, h, v) =$$
$$\frac{\nu}{2}\left[\sum_{m,n}(1 - h_{m,n})(V_{m,n} - V_{m+1,n})^2 + (1 - v_{m,n})(V_{m,n} - V_{m,n+1})^2\right]$$
$$+ \sum_{m,n}\mu^h_{m,n}h_{m,n} + \sum_{m,n}\mu^v_{m,n}v_{m,n} - \sum_{m,n}\gamma_{m,n}h_{m,n}h_{m,n+1}v_{m,n}v_{m+1,n} \tag{3}$$

and Z_1^{-1} is a normalization constant. This is a *Gibbs* distribution[1]. Note that the last terms in the cost function act as local *chemical potentials* for control of the number of active edge-units (i.e., $h_{m,n} = +1$).

---

[1] A distribution of the form $P(x) = Z^{-1} \exp\left(-E(x)/T\right)$, where $E(x)$ is a cost-function, bounded from below, and T is a parameter.

The degradation process produces the measurements $(V_{m,n} \to d_{m,n})$. In the Weak Membrane example we will for simplicity study addition of zero-mean, white Gaussian noise [2] with variance σ^2 :

$$P[d|V] = Z_2^{-1} \exp\left(-\frac{1}{2\sigma^2} \sum_{m,n} (V_{m,n} - d_{m,n})^2\right) \qquad (4)$$

Using Bayes formula (1), we can combine Equations (2) and (4), to obtain the parameterized posterior distribution:

$$P[V, h, v|d] = Z^{-1} \exp(-E(V, h, v, d)) \qquad (5)$$

where Z is a normalization constant and the energy function is given by

$$E(V, h, v, d) = \frac{1}{2\sigma^2} \sum_{m,n} (V_{m,n} - d_{m,n})^2 + E_{\text{prior}}(V, h, v). \qquad (6)$$

2.2   Network design

Inspecting the posterior distribution we note that the three-component process (V, h, v) realizes a *Compound Random Markov Field*. This property ensures that the neural network implementation only involves neighbor connectivity. To enhance the sampling efficiency we use a simulated annealing strategy as recommended by Geman and Geman [Geman(1984)]. The scheme is implemented by introducing a *temperature T* in the Gibbs distribution and designing the sampler with a decreasing sequence of temperatures endinged at $T = 1$. The temperature dependent distribution reads:

$$P[V, h, v|d] = Z_T^{-1} \exp\left(-\frac{E[V, h, v, d]}{T}\right) \qquad (7)$$

As investigated by Geman and Geman, sampling of this distribution is too slow for most applications. It is therefore recommended to invoke a deterministic approximation scheme to obtain the necessary averages [Blake(1989), Peterson(1987)]. The self-consistent Mean Field equations for the Weak Membrane model in the k'th iteration read (see also [Hertz(1991)] for an introduction):

$$
\begin{aligned}
V_{m,n}^{k+1} &= \kappa^{-1}\left[\sigma^{-2} d_{m,n} + (1 - h_{m,n}^k)V_{m+1,n}^k + (1 - h_{m-1,n}^k)V_{m-1,n}^k\right.\\
&\left. \quad + (1 - v_{m,n}^k)V_{m,n+1}^k + (1 - v_{m,n-1}^k)V_{m,n-1}^k\right]\\
h_{m,n}^{k+1} &= \tanh\left[\beta\left(\frac{1}{2}(V_{m,n}^k - V_{m+1,n}^k)^2 - \mu_{m,n}^h - \gamma_{m,n}\Gamma_{m,n}^h(k)\right)\right] \qquad (8)\\
v_{m,n}^{k+1} &= \tanh\left[\beta\left(\frac{1}{2}(V_{m,n}^k - V_{m,n+1}^k)^2 - \mu_{m,n}^v - \gamma_{m,n}\Gamma_{m,n}^v(k)\right)\right]
\end{aligned}
$$

---

[2]It is possible to incorporate any other model of the degeneration process.

where $\beta = 1/T$ and

$$\Gamma_{m,n}^h(k) = h_{m,n+1}^k v_{m,n}^k v_{m+1,n}^k + h_{m,n-1}^k v_{m,n-1}^k v_{m+1,n-1}^k \tag{9}$$

$$\Gamma_{m,n}^v(k) = v_{m+1,n}^k h_{m,n}^k h_{m,n+1}^k + v_{m-1,n}^k h_{m-1,n}^k h_{m-1,n+1}^k \tag{10}$$

$$\kappa = \sigma^{-2} + 4 - h_{m,n}^k - h_{m-1,n}^k - v_{m,n}^k - v_{m,n-1}^k \tag{11}$$

We have used the same symbols for the averaged quantities as for the stochastic ones, but that should not lead to confusion since the two never occur in the same expression. The coupled equations are solved by straightforward iteration as shown in Equation (8), defining a recursive nearest neighbor connected *cellular neural network*.

2.3 Snake hints

There has been much recent progress in the use of deformable models for edge and surface identification, see, e.g., [Cohen(1993)]. In this presentation we suggest to produce strong edge hints for the Weak Membrane through a Snake deformable model. The Snake is defined to be a periodic set of N points in the visual field of an image: $\mathbf{r}_j = (x_j, y_j)$. Periodicity meaning $\mathbf{r}_{j+pN} = \mathbf{r}_j, p \in Z$. The Snake energy function consists of a form control part E_{form} and a match part E_{match}.

$$E = E_{\text{form}} + E_{\text{match}} \tag{12}$$

In this simple version the form control preserves total length,

$$E_{\text{form}} = \frac{a}{2}(d - d_0)^2 \tag{13}$$

$$d = \sum_j |\mathbf{r}_j - \mathbf{r}_{j+1}| \tag{14}$$

where d_0 is the initial Snake length and a is a parameter to control the balance between the two parts in the energy function. More complex form controls can be designed that preserve shape, corners, etc. The Snake match energy is designed to ensure that the Snake points track edge contours in the image field. This effect may be obtained by letting the Snake points seek local maxima in the gradient energy map of the image, $G(\mathbf{r})$, evaluated from the observed image. The total Snake match energy is then given by,

$$E_{\text{match}} = -\sum_j G(\mathbf{r}_j). \tag{15}$$

Snake dynamics is established through gradient descent,

$$\frac{1}{\eta}\frac{\partial \mathbf{r}_j}{\partial t} = -\frac{\partial E}{\partial \mathbf{r}_j} = a(d - d_0)\frac{\partial d}{\partial \mathbf{r}_j} + \frac{\partial G(\mathbf{r}_j)}{\partial \mathbf{r}_j} \tag{16}$$

where η is a parameter to control the step size in the gradient direction.

3   Experiments and concluding remarks

In numerous image processing applications quite detailed prior information can be devised. In, e.g., brain scan reconstruction detailed atlases are known describing the generic brain topography under various scanning modes. To illustrate the potential of using strong edge hints in brain scans, we want to reconstruct a head phantom shown in Figure 1. This is done from a noise corrupted image shown in Figure 2. During the reconstruction process we also want to estimate edges, due to the interpretation of the image.

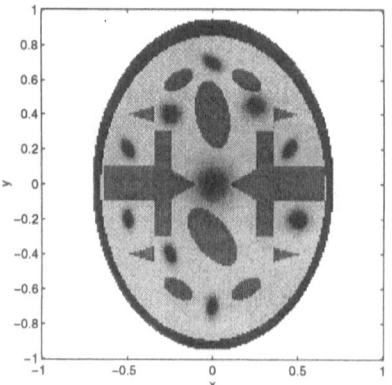

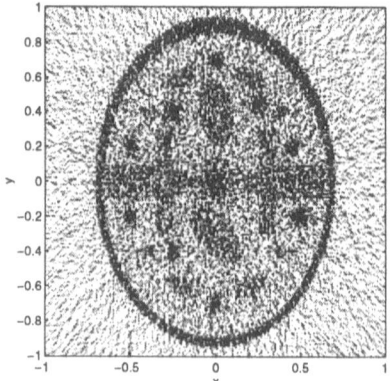

Figure 1: Original head phantom without noise.

Figure 2: Noise corrupted head phantom shown in same color scale.

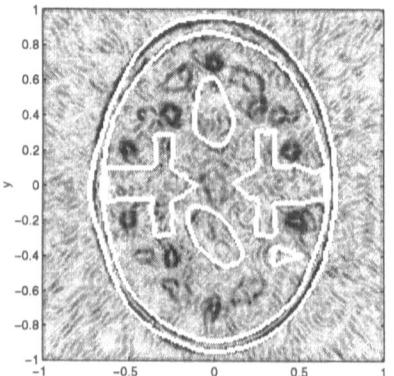

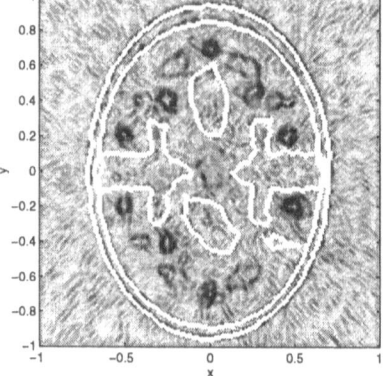

Figure 3: Edge energy map with initial snake positions shown in white.

Figure 4: Edge energy map with equilibrated Snake positions shown in white.

Snakes are initialized in seven generic positions, as shown in Figure 3. For this particular instance the Snake equilibrates in about 100 iterations as illustrated in Figure 4. The approach to equilibrium is quite sensitive to the topography of the image and careful control of η is necessary, Equation (16). This problem may be relieved by use of a pseudo-second order search direction instead of the gradient of the edge energy map.

Subsequently, hints are created by modulating the chemical potentials $\mu^h_{m,n}$, $\mu^v_{m,n}$ and $\gamma_{m,n}$. This modulation implies that edge units are strongly suppressed outside of the region suggested by the Snakes. Finally, we relax the cellular network as defined in Equation (8). The result of the cellular net is presented in Figures 5 (without Snakes) and 6 (using Snakes).

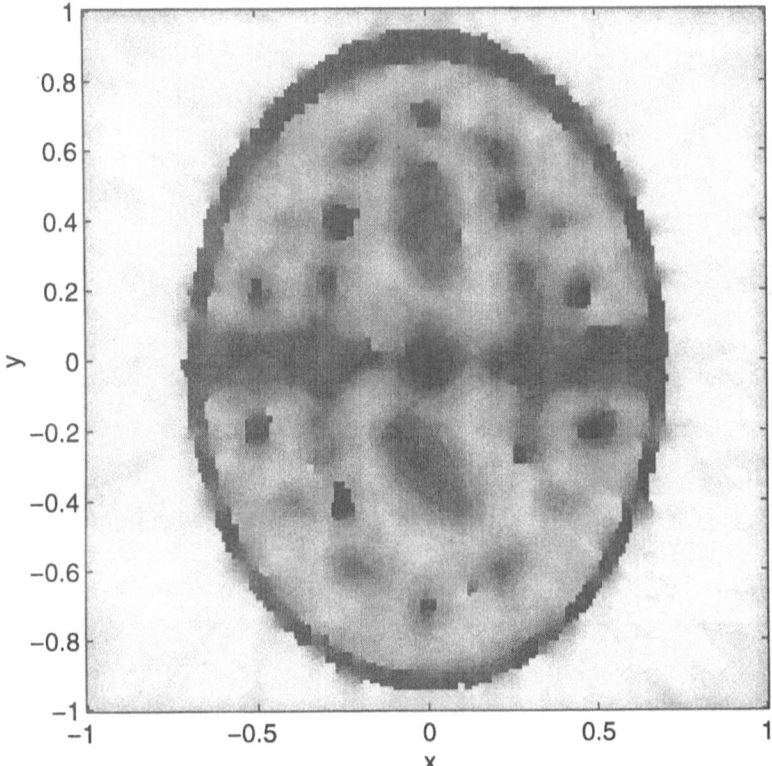

Figure 5: The restored head phantom without using Snakes edge priors to the Mean Field Annealing.

Note that the intensity levels inside the regions of the Snakes have been estimated more closely using Snakes. This might be of significant importance in, e.g., brain activation studies, in which the minute differences in brain activity between activated and resting states are investigated. Using Snakes and Mean Field Annealing enhances the Signal to Noise ratio (SNR) with 12.5 dB. Without Snakes the improvement is 11.3 dB. The improvement is due to a better estimation of edges.

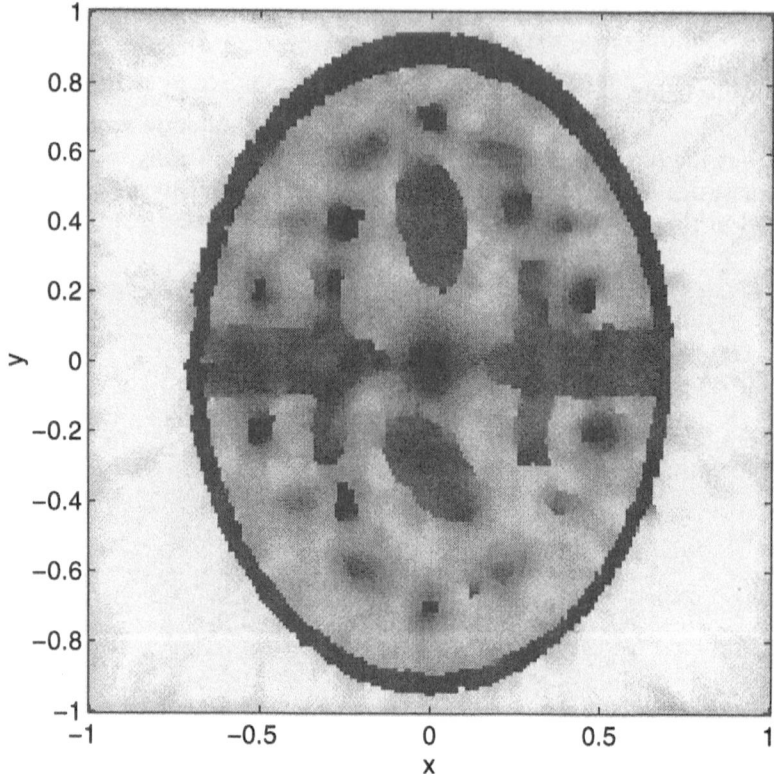

Figure 6: The restored head phantom using Snakes as edge priors to the Mean Field Annealing.

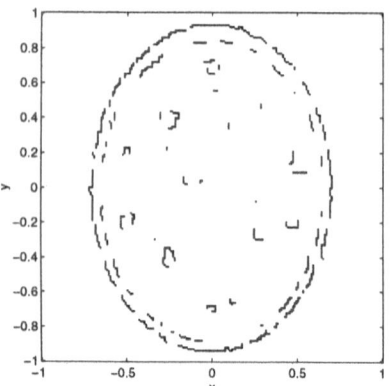

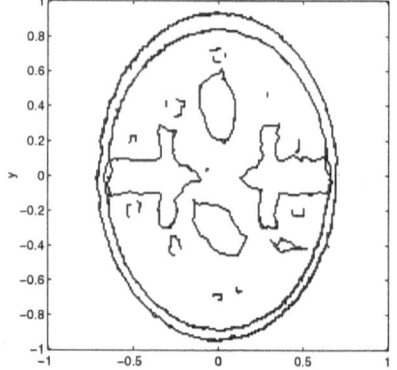

Figure 7: The estimated edges without using Snakes. Active edges are marked black.

Figure 8: Estimated edges using Snakes as priors. Active edges are marked black.

In Figures 7 and 8 are shown the active edges (i.e., $h_{m,n} = +1$ or $v_{m,n} = +1$) without using Snakes respectively with Snakes. As seen from Figure 8 it is possible to incorporate strong priors in order to obtain closed edge contours. This can be used to segment the image.

In conclusion we have shown that strong structural priors may be introduced in the Weak Membrane model by invoking deformable models like Snakes. The present study was based on a head phantom; we are currently pursuing the viability of the approach in the context of Positron Emission Tomography.

Acknowledgment

This research is supported by the Danish Research Councils for the Natural and Technical Sciences through the Danish Computational Neural Network Center.

References

[Blake(1989)] A. Blake: *Comparison of the Efficiency of Deterministic and Stochastic Algorithms for Visual Reconstruction*. IEEE Transactions on Pattern Analysis and Machine Intelligence **PAMI-11** 2-12 (1989).

[Cohen(1993)] L. Cohen and I. Cohen: *Finite Element Methods for Active Contour Models and Balloons for 2-D and 3-D Images*. IEEE Transactions on Pattern Analysis and Machine Intelligence **PAMI-15** 1133-1147 (1993).

[Chen(1994)] X. Ouyang, W.H. Wong, V.E. Johnson, X. Hu, C.-T. Chen: *Incorporation of Correlated Structural Images in PET Image Reconstruction*. IEEE Transactions on Medical Imaging **13** 627-640 (1994).

[Geman(1984)] D. Geman and S. Geman: *Stochastic Relaxation, Gibbs distributions and the Bayesian Restoration of Images*. IEEE Transactions on Pattern Analysis and Machine Intelligence **PAMI-6** 721-741 (1984).

[Hertz(1991)] J. Hertz, A. Krogh and R.G. Palmer: *Introduction to the Theory of Neural Computation*. Addison Wesley, New York (1991).

[Peterson(1987)] C. Peterson and J.R. Anderson: *A Mean Field Learning Algorithm for Neural Networks*. Complex Systems **1** 995-1019, (1987).

[Roth(1990)] M.W. Roth: *Survey of Neural Network Technology for Automatic Target Recognition*. IEEE Transactions on Neural Networks **1**, 28-43, (1990).

An Object-Oriented Toolbox for Studying Optimization Problems

H. Lydia Deng, Wences Gouveia and John Scales

Center for Wave Phenomena, Colorado School of Mines, Golden, Colorado, USA

Introduction

We have developed the CWP object-oriented optimization library (COOOL) as a tool for studying optimization problems and to aid in the development of new optimization software. COOOL consists of a collection of C++ *classes* (encapsulations of abstract ideas), a variety of optimization algorithms implemented using these classes, and a collection of test functions, which can be used to evaluate new algorithms. To use one of the optimization algorithms, a user codes an objective function (and its derivatives if available) according to a simple input/output model — in any language. Moreover, the object classes themselves can facilitate the development of new optimization and numerical linear algebra software since the routine aspects of such coding will benefit from the reusability of code inherent in the object-oriented philosophy. After giving a brief description of the library, we illustrate its application on a problem from our current research.

Overview

A wide variety of problems in science can be thought of as solutions to optimization problems: the maximizing or minimizing of a function, possibly subject to constraints or penalties. In geophysics, for example, because we can never truly "know" the earth's subsurface properties except by direct observation. Instead, we seek interpretations that are most likely in some sense. In real applications, the functions characterizing the likelihood of such interpretations are often complicated entities existing in high-dimensional spaces. Knowing just which optimization tool is appropriate to solve a given problem is half the battle. By having a varied set of tools available one might be able to "browse" through various choices to find the most desirable match between an algorithm and the problem at hand. Further, all optimization problems share certain basic features. They all must have some concept of a set of unknown parameters (**model**), and they all involve communication between a computational module whose job is to update

models (**optimization method**), and an **objective function**, whose job it is
to measure which models are good and which are bad.

Our motivation in putting together this library was threefold. First, we wanted
to have a varied collection of optimization routines and realistic test functions
so that we could study the computational complexity of problems of the sort en-
countered in geophysics. Second, we wanted to have a programming environment
that facilitated the rapid development of new algorithms so that there would be
maximum re-use of existing code and maximum flexibility in mating optimization codes with objective functions on a variety of hardware platforms. Finally,
we wanted to encourage the development of a test-bed of geophysical optimization problems in order to stimulate interest in geophysical optimization problems
among applied mathematicians.

Object-Oriented Programming (OOP) represents an approach to software
development in which the system is organized as a collection of *objects* that are
encapsulations of data structures and behavior. There are many languages sup-
porting the OO paradigm. In C++, which we used to build COOOL, *classes* are
user-defined types to represent objects in the software. COOOL is a collection of
C++ classes, each of which serves for some specific functions in an optimization.
Taking advantages of *encapsulation, sharing,* and *inheritance* (Pohl, 1993) features provided by the OO paradigm, COOOL are designed to facilitate maximal
flexibility and re-usability.

Optimizing with COOOL

Design of COOOL

The three criteria behind the development of COOOL are:

1. It should provide a consistent application programming interface (API) for
   solving optimization problems.
2. It should allow incremental development of the library.
3. It should allow for application packages to be easily built from the existing
   library.

By consistent, we mean that optimization algorithms as well as objective-
function formats should be relatively transparent to application users. This provides
the flexibility for users to choose optimization methods easily and concentrate on
the specific problem rather than struggling to fit the requirements of the various
optimization algorithms.

Figure 1 shows the structure that COOOL uses to achieve these goals. Level
0 contains generic classes, such as **Vector, Matrix, List, Astring,** etc. These
classes are basic element of COOOL for handling algebraic computations. Special classes, such as **SpaMatrix** and **DiagMatrix**, handle sparse and diagonal matrices efficiently. Level 1 is the main level of COOOL, containing classes
for three major components in any optimization problems: unknown parameters (**Model**), objective functions (**ObjFcn**), and optimization algorithms (**Optima**). All these classes can be easily accessed (either adding or modifying) with

minimum influence on others. Level 2 is the application level where problem-specific packages can be built with existing classes from lower levels. For example, we should be able to build a package for a certain travel-time inversion problem with the flexibility of choosing any of the mathematical optimization methods.

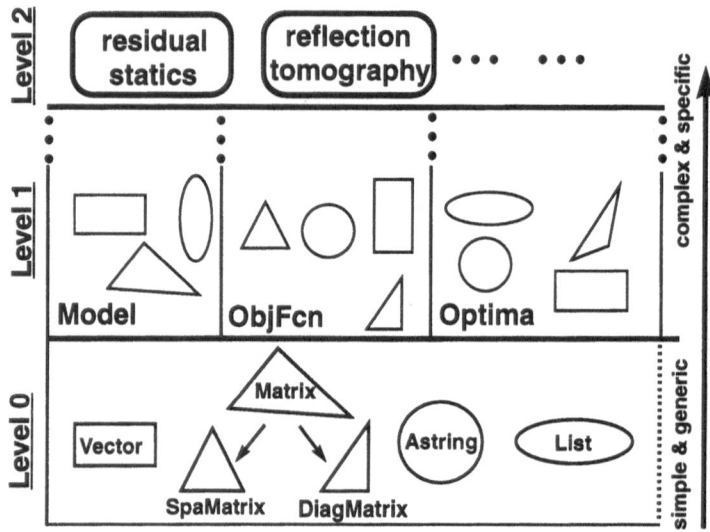

Fig. 1. Structure of COOOL. Level 0 contains generic algebraic data structures. Level 1 contains prototypes for three necessary components in any optimization problems: **Model, ObjFcn,** and **Optima**. Both levels are extensible with minimum influence on existing classes. Application packages at Level 2 can be easily built with lower-level classes.

Optimization Methods in COOOL

Figure 2 shows a classification of the optimization methods included in the preliminary release of the library. We divide the methods into two main classes: linear solvers and local optimization methods. Global optimization methods will be part of a future release.

Linear Solvers. All of the linear solvers in COOOL are iterative. Descriptions of most algorithms can be found in (Scales & Smith, 1994). Presently COOOL implements two flavors of row action methods: **ART** (Algebraic Reconstruction Technique) and **SIRT** (Simultaneous Iterative Reconstruction Technique). These methods are especially attractive when problems are too big to fit into memory. When the non-zero matrix elements fit into memory, **CGLS** (Conjugate Gradient Least Squares) is very attractive. Finally, we include a version of the **IRLS**

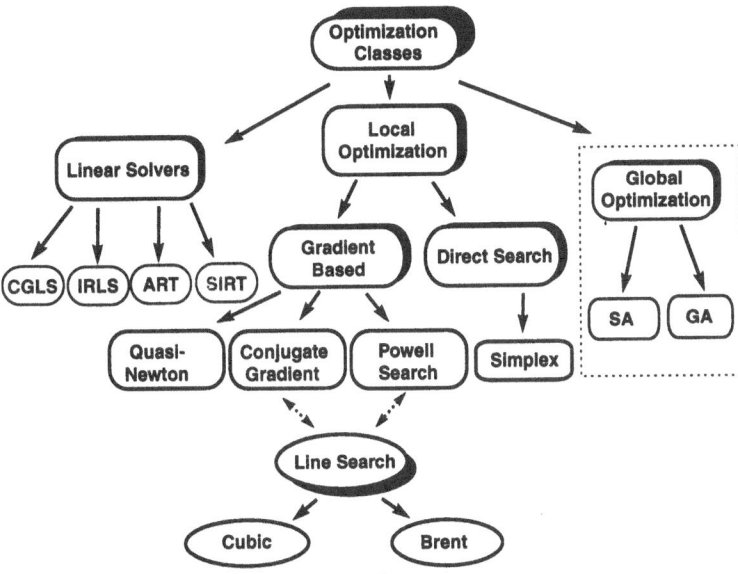

Fig. 2. An overview of the optimization algorithms contained in COOOL. Included are linear solvers for rectangular linear systems using both least squares and general ℓ_p norms and local methods based on direct search and quasi-Newton methods. Monte Carlo global optimization methods will be part of a future release of COOOL.

(Iteratively Re-weighted Least-Squares) algorithm to efficiently solve linear systems in the ℓ_p norm. This algorithm takes advantage of certain approximations to achieve nearly the speed of conventional least squares methods while being able to robustly handle long-tailed noise distributions.

Local Optimization Methods. The algorithms presented here can be applied to quadratic and non-quadratic objective functions alike. The term **local** refers both to the fact that only information about a function from the neighborhood of the current approximation is used in updating the approximation, and that these methods usually converge to extrema near the starting models. As a result, the global structure of an objective function is unknown to a local method. Some of these techniques, such as Downhill Simplex and Powell's method do not require explicit derivative information of the objective function. Others, such as the quasi-Newton methods require at least the gradient. In the latter case, if analytic expressions are not available for the derivatives, a module for finite-difference calculation of the gradient is provided. COOOL also includes non-quadratic generalizations of the conjugate gradient method incorporating two different kind of line search procedures.

Objective Functions

Objective functions are those functions to be minimized (or maximized). Several analytical test functions are implemented in the library. They include a generalized N-dimensional quadratic function, N-dimensional Rosenbrock function, and a two-dimensional multi-modal analytical function.

For realistic problems, the formulation of objective functions may vary tremendously. A user can write the objective function as a stand-alone Unix executable which reads models from its standard input and writes numbers to its standard output. COOOL initiates this executable (using ideas developed by Don Libes in his *Expect* package) in such a way as to take over the objective function's input and output. (This communication between COOOL and the objective function is called a pseudo-tty in Unix and is somewhat like a pipe.) The net result is that the objective function can be written in any language and may take advantage of hardware-specific features. This communication model is illustrated in Fig. 3. COOOL sends out a flag to the user-defined objective function file, indicating whether the function value or its derivatives is needed, followed by a set of model parameters. The objective function reads the flag and the model from standard input, and writes the result to standard output; COOOL will then capture these information and put them to an optimization algorithm for an updated model. If there is no analytical gradient information available, COOOL will approximate the gradient vector using a finite-difference technique. This simple communication model allows us to consider algorithms as diverse as downhill simplex, Newton's method, and simulated annealing within a unified framework.

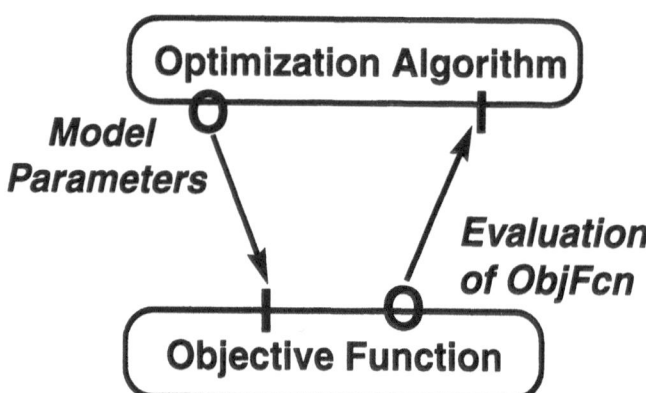

Fig. 3. A model of communication between an objective function and an optimization class. The user-defined stand-alone objective function are responsible for evaluating the value or gradient for a set of parameters according to the message sent by the optimization algorithm.

A key point in COOOL is that communication between optimization algorithms and objective functions is transparent to users. To use COOOL, a user only needs to construct several objects choosing from the library; namely, a **Model**, an **ObjFcn**, and an **Optima**. The interior computation and communication among these objects are handled by COOOL. Pseudo-code for solving a descretized optimization problem using COOOL is shown in Fig. 4. If the optimization problem is continuous or without boundary, **Model** can be constructed without those arguments, the implementations of **ObjFcn** and **Optima** are not affected. Although COOOL is written in C++, little knowledge of C++ is needed for using it. Evaluations of objective functions may be the most computationally intensive step for many optimization problems; COOOL allows users to code them in the most efficient language available.

```
main(argc, **argv)
{
        // Construct a constrained Model with n unknowns
        Model m(n, upper, lower, Δm);
        // Construct an ObjFcn: user-defined function, statics
        ObjFcn *f = new ObjFcn("statics", ...);
        // Construct an Optima: choose an optimization method
        Optima *opt = new ConjugateGradient(f, imax, tol, ...);

        // Initialize the model to m₀
        m = m₀;

        // COOOL returns an optimal model optm
        Model optm = opt → optimizer(m);
}
```

Fig. 4. Pseudo-code for using COOOL to solve an optimization problem. Constraints of model parameters are separated from optimization algorithms. The boundary [*lower*, *upper*]. If the model is descretized, the grid-size Δm is specified when constructing the **Model**. object.

An Application of COOOL to Residual Statics Correction

COOOL can be used to solve very general kinds of optimization problems. Here we briefly show examples of how we have used this library to analyze a hard optimization problem from exploration seismology.

Residual statics are time shifts observed in reflection seismic data and are attributed to localized heterogeneities in the near surface. These time shifts result in a decrease of the stacking quality and can therefore be estimated by optimizing some measure of stacking quality such as the *stacking power function*. The stacking power function is the sum of squares of the stacked sections over a time-window which contains most significant events (Ronen & Claerbout, 1985; Rothman, 1985),

$$SP(\mathbf{m}) = \sum_{cmp} \sum_{time} \left[\sum_{offset} Trace(\mathbf{m}) \right]^2 \tag{1}$$

where $\mathbf{m}$ is the parameter vector of time-shifts. The function $SP(\mathbf{m})$ in (1) is generally a high dimensional function with many local maxima. Figure 5 shows a simple 3-parameter residual statics objecive function.

Here, we show two examples of using COOOL to study such an optimization problem. First, we show the result of a multi-grid approach in solving this optimization problem. Next we use COOOL to study a multi-resolution-analysis (MRA) of a one-dimensional statics problem. Both examples use the non-linear conjugate gradient method in COOOL as the searching algorithm.

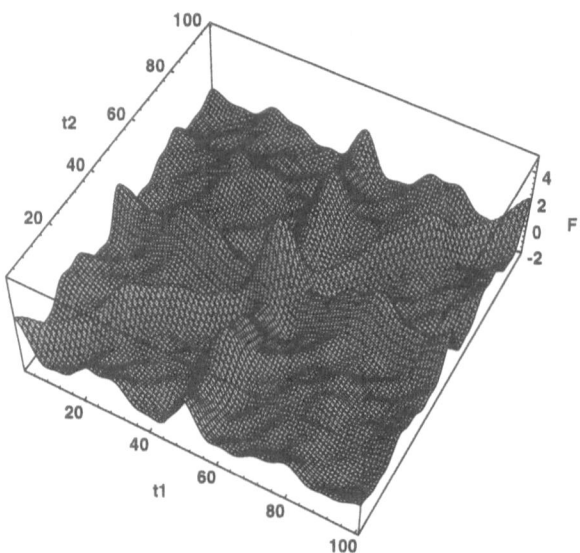

Fig. 5. A 2-dimensional stacking-power function.

Multi-grid Residual Statics Correction

The left figure of Fig. 6 shows a stacked section acquired over a permafrost zone without statics corrections. The input for COOOL is the filtered pre-stacked data within the time-window, indicated by the black bar in Fig. 6. The right figure of Fig. 6 shows the result of applying the COOOL non-quadratic version of **ConjugateGradient** to optimize the stacking power in successively wider frequency bands. It is clear that the continuity of the reflections was restored, and the resolution and signal-to-noise ratio improved. This multi-grid approach was suggested by Bunks, et al. (1995).

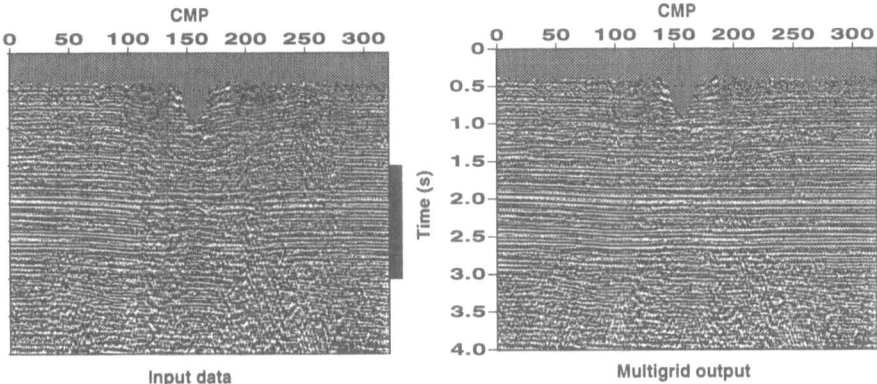

Fig. 6. The left figure shows a stacked section acquired over a permafrost zone without statics corrections. The right figure shows the stacked section after surface-consistent statics are corrected. Statics are estimated using COOOL by optimizing the stacking power in successively wider frequency bands. The black bar on the right shows the time window of data used to compute the statics corrections.

Multi-Resolution Analysis for Simplifying an Objective Function

Multiresolution analysis (MRA) projects the input signal to a set of nesting subspaces which represent a series of successive coarser resolution levels. Using MRA, we intend to study the behavior of complex objective functions at various resolution levels.

Consider a simple version of the statics problem: two identical traces with an unknown shift. For this problem, the objective function (1) is one-dimensional. Maximizing this stacking power is equivalent to minimizing the following mean-squared error function,

$$E(\delta) = \sum_{i=0}^{N-1} (P_0(i - \delta) - P_1(i))^2, \qquad (2)$$

where $P_0(t)$ and $P_1(t)$ are the data traces, N is the number of samples per trace, and δ is the unknown time-shift. For two traces containing a shifted Ricker-wavelet, Fig. 7 shows mean-squared error functions when input data are decomposed to successively coarser resolution levels. The bases for the decomposition is a symmetric, shift-invariant wavelet bases developed by Saito and Beylkin

(1993). A detailed discussion on this work can be found in (Deng, 1995). Randomly choosing 50 initial models between $[-0.2, 0.2]$ s, Figure 8 shows histograms of the converged time-shifts obtained by a non-quadratic **ConjugateGradient** tool provided in COOOL. These results show that the MRA increases the chance for local search methods to find the global extrema.

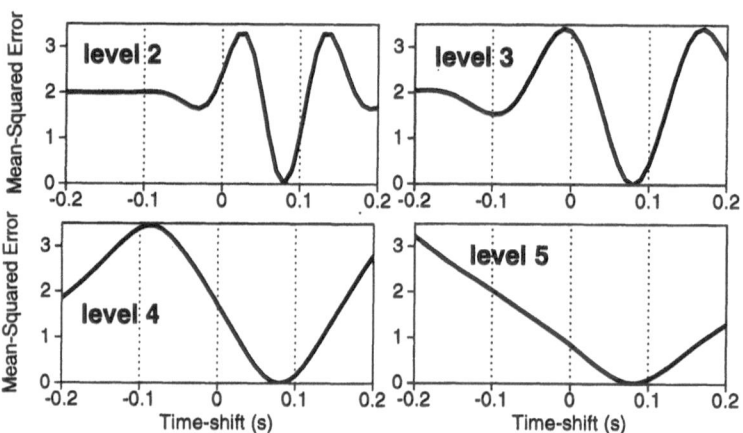

Fig. 7. The mean-squared error functions for two shifted traces containing Ricker wavelets at various resolution levels. Input is decomposed by an MRA using a symmetric, shift-invariant wavelets bases.

Conclusion

The CWP Object-Oriented Optimization Library (COOOL) is a collection of C++ classes for studying and solving optimization problems. It was developed using the freely available GNU compiler *gcc*. The library contains the basic building blocks for the efficient design of numerical linear algebra and optimization software; it also comes with a variety of unconstrained optimization algorithms and test objective functions drawn from our own research. The only requirement for using one of the optimization methods is that a simple model of communication be followed. This allows us to use exactly the same code to optimize functions tailored for a variety of hardware, no matter what programming language is used. Further, since we have provided class libraries containing building blocks for general purpose optimization and numerical linear algebra software, the development of new algorithms should be greatly aided.

COOOL is now freely available via anonymous ftp at

`ftp.cwp.mines.edu/pub/cwpcodes/coool`,

and on the WWW at

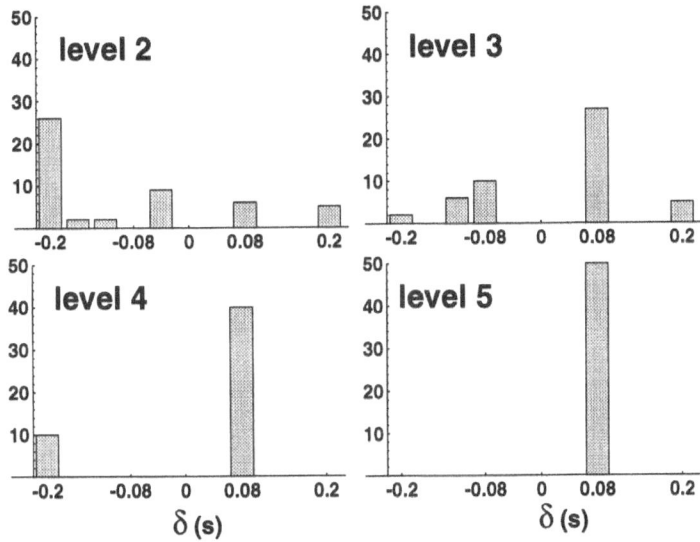

Fig. 8. Histograms of the obtained time-shifts of 50 conjugate-gradient optimization experiments for data at corresponding resolution levels as in Fig. 7. Initial models are chosen randomly between $[-0.2, 0.2]$ s. The horizontal axis is the number of shift-samples, where the sample interval is 0.01 s, and the grid size of the histograms is 4 samples. All 50 experiments found the true solution when the data are decomposed to resolution level 5.

http://www.cwp.mines.edu/cwpcodes/coool

Postscript and HTML versions of a technical report on COOOL by the authors are available by anonymous ftp or WWW from the same addresses. Any bug reports or suggestions should be sent to

optima@dix.mines.edu.

Acknowledgment

Our inspiration for implementing optimization software in an object oriented fashion is Martin Smith. We have been influenced by the classes Martin Smith and Terri Fischer developed for the Uni-Processor Genetic Algorithm (UGA). The communication between the objective function and the optimization algorithms is managed via *EXPECT*, written by Don Libes of the National Institute of Standards and Technology (NIST) and adapted to our purposes from UGA. We would like to thank Pechorageofika for the permission to use the seismic data set in this work, and Gregory Beylkin for his help on the MRA.

This work was partially supported by the sponsors of the Consortium Project on Seismic Inverse Methods for Complex Structures at the Center for Wave Phenomena, Colorado School of Mines, and the Shell Foundation.

References

Bunks, C., Saleck, F. M., Zaleski, S., & Chavent, G. 1995. Multiscale seismic waveform inversion. *Geophysics*, **60**(5), 1457–1473.

Deng, H. L. 1995. *Using Multi-Resolution Analysis to Study the Complexity of Inverse Calculations*. Tech. rept. CWP-183. Center for Wave Phenomena, Colorado School of Mines.

Pohl, I. 1993. *Objected-oriented programming using C++*. The Benjamin/Cummings Publishing Companhy, Inc.

Ronen, J., & Claerbout, J. 1985. Surface-consistent residual statics estimation by stack-power maximization. *Geophysics*, **50**, 2759–2767.

Rothman, D. H. 1985. Nonlinear inversion, statistical mechanics, and residual statics estimation. *Geophysics*, **50**, 2797–2807.

Saito, N., & Beylkin, G. 1993. Multiresolution representations using the autocorrelation functions of compactly supported wavelets. *IEEE Transactions on Signal Processing*, **41**, 3585–3590.

Scales, J. A., & Smith, M. 1994. *Introductory Inverse Theory*. Samizdat Press. http://landau.mines.edu/~samizdat.

Maximum Mixing Method

Jens Hjorth

Institute of Astronomy, Cambridge, UK

1 Motivation and Introduction

This investigation considers the general problem of inferring the 'true' intensity distribution from a blurred, noisy, and partially corrupted observed signal. In particular, a formalism is developed which supplements the popular Maximum Entropy Method (MEM) and at the same time provides new insight into the foundations of this powerful reconstruction technique.

There are several motivations behind this exploratory study. MEM is generally discussed from a Bayesian or axiomatic point of view, and is as such developed essentially within a *probabilistic* framework. Despite its strengths this is also the weakness of this approach. It is by no means clear why one should associate a distribution of, say, counts in a detector with a probability distribution, even though both share the common features of being positive and normalizable. Hence, any attempt to do so is fundamentally of a somewhat *ad hoc* nature. One would therefore like to establish a coupling between some physical (stochastic) process and the blurring mechanism to account for the interpretation of the signal as a probability distribution. Or, one should abandon the probabilistic (information theoretic) approach and seek to justify MEM on other grounds or develop entirely new techniques. Indeed, several authors consider MEM merely as a useful regularizing technique. Others have proposed new entropies (in very different contexts) which may be applied to the inversion problem in question.

The problem dealt with is of a very general nature and applies to a wide range of scientific inversion problems. The present preliminary report tries to address several of the relevant outstanding issues.

1.1 Formulation of the problem

Consider a distribution $\{x_i\}$, $i = 1, \cdots, N$, $x_i \geq 0$ normalized,

$$1 = \sum_{i=1}^{N} x_i, \tag{1}$$

and subject to the constraints

$$y_r = \sum_{i=1}^{N} A_{ri} x_i, \quad r = 1, \cdots, M; \; M < N - 1. \tag{2}$$

Here $\{x_i\}$ may be identified with the 'true' (but unknown) distribution that we wish to infer about, A_{ri} with a known broadening function (point spread function, response function, wavelet, kernel...), and the constraints, y_r, with the data, i.e., the 'observed' distribution that we wish to deconvolve. The broadening function, A_{ri}, is also normalized (i.e., it only redistributes flux),

$$1 = \sum_{i=1}^{N} A_{ri} = \sum_{r=1}^{M} A_{ri}. \tag{3}$$

A 'simple' inversion method cannot recover the underlying distribution, since there are many realizations of the 'true' distribution that can give rise to the observed distribution, especially when the latter is affected by noise and 'missing' information (this is the case when $M < N - 1$. In fact, one may have $M \ll N$). Thus, some regularization is needed. The general problem considered in this investigation is that of maximizing a functional, H, under the constraints listed above. In other words, we seek the distribution which maximizes the quantity

$$H - \lambda_o \sum_{i=1}^{N} x_i - \sum_{r=1}^{M} \lambda_r y_r, \tag{4}$$

where λ_0 and λ_r are Lagrange multipliers.

As described in the introduction, information-theoretic arguments strongly favour the Boltzmann–Gibbs–Jaynes–Shannon entropy as the only correct choice of H,

$$S[x] = -\sum_{i=1}^{N} x_i (\ln \left(\frac{x_i}{m_i} \right) - 1), \tag{5}$$

where m_i is the so-called *prior* distribution. Jaynes (1957) has put it this way:

> "...in making inferences on the basis of partial information we must use that probability distribution which has maximum entropy subject to whatever is known. This is the only unbiased assignment we can make; to use any other would amount to arbitrary assumption of information which by hypothesis we do not have."

On the other hand, instead of considering Eq. (4) as an entropy subject to constraints one may alternatively see H as a regularization functional or a penalty function. The variety of successful regularizing functions suggests that the detailed form of the entropy/penalty function may not be of much importance as long as it discourages ripples (rapid variations) and negative values.

The controversy, however, remains unresolved and is at times the subject of spirited debates (e.g., Titterington vs. Skilling 1984). It would therefore be useful to explore the situation systematically, and to test whether inversion or estimation schemes based on maximum-entropy smoothing are better that the rather *ad hoc* ones widely used. The formalism presented here is very well suited for this purpose. Reviews that are central to this issue are Narayan & Nityananda (1986) and Donoho et al. (1992).

2 H-functions

In this investigation we consider the class of functions,

$$H[x] = -\sum_{i=1}^{N} m_i C\left(\frac{x_i}{m_i}\right),\tag{6}$$

where C is a convex function, $C''(x) \equiv d^2C/dx^2 > 0$. An H-function plays a role of a generalized entropy, either as a so-called collisionless entropy (Tremaine, Hénon & Lynden-Bell 1986) or through an H-theorem (second law of thermodynamics, $dH/dt \geq 0$) provided x obeys a master equation (van Kampen 1981).

A dynamical process is said to be *mixing* (Wehrl 1978) if *any* H increases as a consequence of the dynamics. Let H_B be the value of the H-function for the blurred data, and H_O that of the original intensity distribution. Then

$$H_B = -\sum_{j=1}^{N} C\left(\sum_{i=1}^{N} A_{ji}x_i\right) \geq -\sum_{j=1}^{N}\sum_{i=1}^{N} A_{ji}C(x_i)$$

$$= -\sum_{i=1}^{N}\sum_{j=1}^{N} A_{ji}C(x_i) = -\sum_{i=1}^{N} C(x_i)\sum_{j=1}^{N} A_{ji}\tag{7}$$

$$= -\sum_{i=1}^{N} C(x_i) = H_O.\tag{8}$$

Here the inequality follows from the mathematical properties of convex functions (the so-called 'convex-function theorem', see e.g. Grandy 1987), and the third line follows from Eq. (3). Thus, *the H-function of a blurred signal is always larger than that of the original, i.e., the blurred signal is more mixed.*

This is a purely mathematical statement: Any H-function with the observed distribution as argument is larger than that of an H-function with the true distribution as an argument. The physics behind this process is considerably more complicated. At the conference, however, mixing was demonstrated to occur for a light ray from a distant star (point source) that passes through the turbulent atmosphere and hits a detector.

3 Maximum Mixing Method (MMM)

The appealing properties of H-functions encourage us to take the idea seriously and further develop an alternative (or supplement) to MEM, namely what may be dubbed a Maximum Mixing Method (MMM). In this theory, instead of searching for the most 'probable' state (the data representing the most 'disordered' state) we limit ourselves to looking for a distribution that leads to a more mixed observed state.

From equation (4) it follows that the solution is of the form

$$\frac{x_i}{m_i} = C'^{(-1)} \left(-\lambda_0 - \sum_{r=1}^{M} \lambda_r A_{ri} \right), \tag{9}$$

where $C'^{(-1)}(z)$ denotes the inverse function of the first derivative of C with respect to x. If $H = S$ then $C'^{(-1)}(z) = \exp(z)$. Hence, the problem has been recast to one of determining the Lagrange multipliers (of which there are only $M + 1$).

It can be shown that maximizing H under a set of linear constraints is equivalent to minimizing a convex potential function. Specifically, it can be proven that the problem of finding the distribution $\{x_i\}$ (N points) which maximizes Eq. (6) (y_r, A_{ri}, C known) can be recast into the problem of finding a *unique minimum* of

$$V(\tilde{x}) \equiv H[\tilde{x}] + \sum_{i=1}^{N} (\tilde{x}_i - x_i) C' \left(\frac{\tilde{x}_i}{m_i} \right) \tag{10}$$

(here $\tilde{x}$ is a trial distribution) if and only if

$$C''(x) > 0. \tag{11}$$

The procedure outlined thus seeks to find a least mixed state (by minimizing V), i.e., the deconvolution is a 'de-mixing' procedure.

4 Superresolution, smoothing, and constraints on H

One aspect of 'statistical' inference is that of superresolution, i.e., that peaks in the restored array become sharper than those in the original one. The present formalism makes experiments with various choices of C relatively straightforward. The important function during the iterative deconvolution is $C'^{(-1)}(z)$ cf. Eq. (9). If $C'^{(-1)}(z) = z$ then all parts of the peak is treated equally, i.e., no improvement in resolution is expected. For superresolution to occur high intensities must be favoured at the expense of lower intensities. Thus, $C'^{(-1)}(z)$ must be a convex function, i.e.,

$$(C'^{(-1)})''(z) = -\frac{C'''(C'^{(-1)}(z))}{C''(C'^{(-1)}(z))^3} > 0 \tag{12}$$

and because of convexity of C ($C''(x) > 0$) a requirement for superresolution is $C'''(x) < 0$. Conversely, smoothing is achieved for $C'''(x) > 0$. This is especially relevant when $M \ll N$. To enforce positivity of the solution (9) one must also require that $C'^{(-1)}$ be a non-negative function of its arguments. We thus have

$$C'' > 0 ; \qquad C''' < 0 ; \qquad C'^{(-1)} \geq 0 \qquad (13)$$

for superresolution. Such conditions can be used to put limits on possible entropies or to construct new 'entropies'.

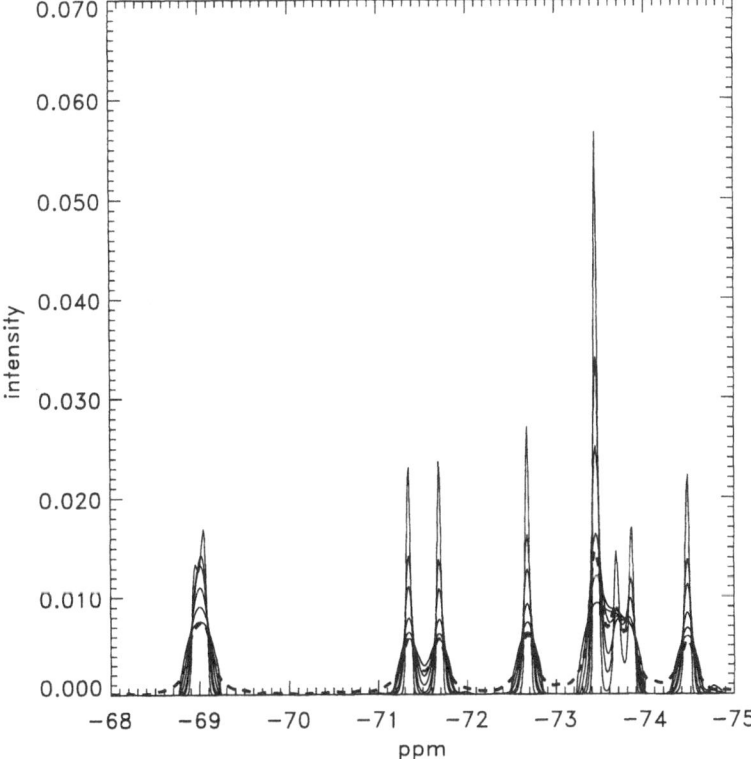

Fig. 1. Deconvolutions of solid-state ^{29}Si MAS NMR spectrum of 3CaO·SiO$_2$ spectrum using various choices of $C'^{(-1)}$. $C'^{(-1)}(z) = (z/n)^n$ with $n = 0.5, 1, 2, 5, 10, 100$ (for increasing resolution) For the highest resolution cases the left peak at -69 ppm splits up in two (see also Skibsted et al. 1990). The thick dashed line is the original spectrum.

As a test example we have considered the solid-state ^{29}Si MAS NMR spectrum of 3CaO·SiO$_2$ (Skibsted, Hjorth & Jakobsen 1990). The broadening function was derived from one of the peaks itself. The deconvolution was run for 100 iterations for various choices of $C'^{(-1)}$. The results shown in Figure 1 confirm that (i) any convex C is suitable in this approach and that (ii) the above considerations regarding superresolution versus smoothing are correct.

5 Discussion and Conclusion

The unique feature of MEM is that $C'^{(-1)}(z) = \exp(z)$ amplifies all scales equally. Narayan & Nityananda (1986) have shown that this leads to Gaussian deconvolved peaks. In MMM different scales are treated differently, depending on the choice of C. This gives different peak shapes, but also allows one to experiment with the degree of peak sharpening as a function of peak height. In fact, despite its strong information-theoretic background, MEM is known to redistribute flux incorrectly during deconvolution, thus making the method problematic if the goal is to get correct intensities out. MMM could remedy this problem by using an alternative to the entropy.

In conclusion, some ideas connecting the physics of blurring with a proposed reconstruction scheme, dubbed Maximum Mixing Method, have been presented. It has been shown that this physically motivated, non-information theoretic, non-probabilistic, non-Bayesian approach can be turned into a powerful deconvolution technique, competitive with, and having as a special case, the Maximum Entropy Method. Further work within the proposed framework is required to fully explore the consequences of the theory. A paper including proofs and examples is in preparation.

6 Acknowledgement

This work is supported by the Danish Natural Science Research Council.

References

Donoho, D. L., Johnstone, I. M., Hoch, J. C., Stern, A. S., 1992. Maximum entropy and the nearly black object. *Journal of the Royal Statistical Society* **B 54**, 41–81.

Grandy, W. T., 1987. *Foundations of Statistical Mechanics*, Vol. I, Reidel, Dordrecht.

Jaynes, E. T., 1957. Information theory and statistical mechanics. *Physical Review* **106**, 620–630; **108**, 171–190.

Narayan, R. & Nityananda, R., 1986. Maximum entropy image restoration in astronomy. *Annual Reviews of Astronomy and Astrophysics* **24**, 127–170.

Skibsted, J., Hjorth, J. & Jakobsen, H. J., 1990. Correlation between ^{29}Si NMR chemical shifts and mean Si–O bond lengths for calcium silicates. *Chemical Physics Letters* **172**, 279–283.

Skilling, J., 1984. The maximum entropy method. *Nature* **309**, 748–749; **312**, 382.

Titterington, D. M., 1984. The maximum entropy method for data analysis. *Nature* **312**, 381–382.

Tremaine, S., Hénon, M. & Lynden-Bell, D., 1986. H-functions and mixing in violent relaxation. *Monthly Notices of the Royal Astronomical Society* **219**, 285–297.

van Kampen, N. G., 1981. *Stochastic Processes in Physics and Chemistry*, North-Holland, Amsterdam.

Wehrl, A., 1978. General properties of entropy. *Reviews of Modern Physics* **50**, 221–260.

Subject Index

Lecture Notes in Earth Sciences